Travel Demand Forecasting 2018

Transportation Research Record: Journal of the Transportation Research Board Volume 2672, Issue 49

ISBN: 978-1-5108-8255-3

Printed from e-media with permission by:

Curran Associates, Inc.
57 Morehouse Lane
Red Hook, NY 12571

proceedings.com

Some format issues inherent in the e-media version may also appear in this print version.

Published by Sage Publications USA

ISBN (Print) 978-1-5108-8255-3
ISBN 2018 Printed Set (All Issues) 978-1-5108-7735-1

Printed by Curran Associates, Inc. (2019)

Additional copies of this publication are available from:

Curran Associates, Inc.
57 Morehouse Lane
Red Hook, NY 12571 USA
Phone: 845-758-0400
Fax: 845-758-2633
Email: curran@proceedings.com
Web: www.proceedings.com

Volume 2672 Number 49 December 2018

TRR
JOURNAL OF THE TRANSPORTATION RESEARCH BOARD

Contents

Travel Demand Forecasting

Articles

TRB
TRANSPORTATION RESEARCH BOARD

CORE SPONSORS

Transportation Departments of the 50 States and the District of Columbia

Federal Government

U.S. Department of Transportation
- Federal Aviation Administration (FAA)
- Federal Highway Administration
- Federal Motor Carrier Safety Administration (FMCSA)
- Federal Railroad Administration (FRA)
- Federal Transit Administration
- National Highway Traffic Safety Administration (NHTSA)
- Office of the Assistant Secretary of Transportation for Research and Technology (OST-R)

U.S. Air Force Civil Engineer Center (AFCEC)
U.S. Coast Guard (USCG)
U.S. Department of Energy
U.S. Department of the Interior (DOI)
U.S. Environmental Protection Agency (EPA)

Nongovernmental Organizations

American Public Transportation Association (APTA)
Association of American Railroads (AAR)
California Air Resources Board

The National Academies of
SCIENCES • ENGINEERING • MEDICINE

The **National Academy of Sciences** was established in 1863 by an Act of Congress, signed by President Lincoln, as a private, nongovernmental institution to advise the nation on issues related to science and technology. Members are elected by their peers for outstanding contributions to research. Dr. Marcia McNutt is president.

The **National Academy of Engineering** was established in 1964 under the charter of the National Academy of Sciences to bring the practices of engineering to advising the nation. Members are elected by their peers for extraordinary contributions to engineering. Dr. C. D. Mote, Jr., is president.

The **National Academy of Medicine** (formerly the Institute of Medicine) was established in 1970 under the charter of the National Academy of Sciences to advise the nation on medical and health issues. Members are elected by their peers for distinguished contributions to medicine and health. Dr. Victor J. Dzau is president.

The three Academies work together as the National Academies of Sciences, Engineering, and Medicine to provide independent, objective analysis and advice to the nation and conduct other activities to solve complex problems and inform public policy decisions. The Academies also encourage education and research, recognize outstanding contributions to knowledge, and increase public understanding in matters of science, engineering, and medicine.

Learn more about the National Academies of Sciences, Engineering, and Medicine at **www.national-academies.org.**

The **Transportation Research Board** is one of seven major programs of the National Academies of Sciences, Engineering, and Medicine. The mission of the Transportation Research Board is to increase the benefits that transportation contributes to society by providing leadership in transportation innovation and progress through research and information exchange, conducted within a setting that is objective, interdisciplinary, and multimodal. The Board's varied activities annually engage about 7,000 engineers, scientists, and other transportation researchers and practitioners from the public and private sectors and academia, all of whom contribute their expertise in the public interest. The program is supported by state transportation departments, federal agencies including the component administrations of the U.S. Department of Transportation, and other organizations and individuals interested in the development of transportation.

Learn more about the Transportation Research Board at **www.TRB.org**.

TRR
JOURNAL OF THE TRANSPORTATION RESEARCH BOARD

Article

Transportation Research Record
2018, Vol. 2672(49) 1–11

DOI: 10.1177/0361198118778926
journals.sagepub.com/home/trr
SAGE

Tour-Based Path Analysis of Long-Distance Non-Commute Travel Behavior in California

Adam W. Davis[1], Elizabeth C. McBride[1], Krzysztof Janowicz[2], Rui Zhu[2], and Konstadinos G. Goulias[1]

Abstract

Data of long-distance tours by each household from an 8-week California Household Travel Survey travel log are analyzed in this paper. Each tour record contains summary data from a single-day diary, household sociodemographic information, and place of residence characteristics. Each tour contains a main trip, selected tours with a main trip that is not a commute trip, and added destination descriptors from Foursquare. Path analysis is used on three censored variables (tour miles by air, miles driving, and miles by public transportation) and two categorical variables (main trip tour purpose) and number of overnight stays. Annual household income is a major determinant of air travel and multiple overnight stays. Moreover, travel by air is more likely to be in large dense destinations with high attractiveness ratings. In contrast, car ownership is a major determinant of more mileage accrued driving a car to urban environments. In terms of main trip purpose, work and shopping are not associated with overnight stays and vacation/sightseeing is the purpose associated with longer overnight stays. Moreover, California destinations are more likely to be in tours completed within a day. It was also found that vacation, sightseeing, and leisure trips are more likely to be combined with other vacation and sightseeing trips. In contrast, business and combined business and leisure trips are less likely to be combined with vacation trips. There were also substantial and significant differences in long-distance tour behavior among residents of urban versus rural environments.

Long-distance travel is increasingly recognized as a major contributor to greenhouse gas emissions (*1*) and a possible biodiversity threat (*2*). Congestion, maintenance costs, and security are also some of the long-distance travel impacts (*3*). California, in support of its initiatives aiming at curbing greenhouse emissions with a plethora of policy actions, also considers long-distance travel important for policy analysis at the state level (*4*). For this reason the California Household Travel Survey (CHTS) included a long distance added travel log to capture sufficient number of trips for the long-distance travel component in the statewide simulation. One of the objectives of CHTS is to also derive prototypical traveler profiles that encompass heterogeneity in long-distance travel behavior to understand long-distance travel. An extensive analysis of CHTS data, with focus on long commute trips, revealed a strong relationship between residential location characteristics and distance of the commute trip (*5, 6*). Using CHTS these studies found that long-distance travel and cost of housing are strongly related, and higher-income and employed persons travel more (*6*).

There is lack of consensus in the literature undermining comparisons across spatial and social contexts. First, the literature contains a variety of definitions for "long-distance" travel, including ones based on distance (e.g., longer than 50 miles, 100 miles, or longer than 100 kilometers) and travel time (e.g., 40 minutes). Second, long-distance travel researchers have also considered a variety of indicators including number of long-distance trips, activity before and/or after travel to workplace, mode used, time of day of trip, and destination (*6–16*). Most studies, however, did not address trip chaining (e.g., people going to a business location, then to a leisure destination, and then back home). In addition, very little analysis is found in differentiating trips with an overnight

[1]Department of Geography and GeoTrans Lab, University of California, Santa Barbara, CA
[2]Department of Geography and STKO Lab, University of California, Santa Barbara, CA

Corresponding Author:
Address correspondence to Konstadinos G. Goulias:
goulias@geog.ucsb.edu

stay, despite the important differences between these trips and daily commuting. The choice of analysis in past studies was presumably driven by: (a) an emphasis on trips to and from work; and (b) a focus on a single trip by an individual person as the unit of analysis, instead of multiple trips from household members. This focus on commute trips is also reflected in the multitude of person factors used to explain variation in travel behavior in past research. A few studies included household structure and location characteristics but not the totality of activities at destinations away from home and the mutual influence among these activities. Long-distance trips are often chains of trips with different purposes such as business and leisure, visiting friends and sightseeing, and education and vacation. This is another neglected aspect in the analysis of long-distance travel behavior.

Taking advantage of the CHTS source of information, this study attempts to fill some of these gaps by analyzing tours of households that include a main long-distance trip with a non-routine commute purpose at the end of this main trip. This paper explores behavioral facets of non-commute long-distance travel using a type of structural equation model (SEM) called path analysis to identify determinants of travel and relationships among modes used, distance traveled, number of overnight stays, and destination types chosen by Californians. The authors also experiment with the use of social media data (Foursquare) to describe destination characteristics and their significance in explaining long-distance tour behavior. Relationships are explored among activities at the end of the main long-distance trip jointly with all other trips connected to this main trip, the role of social and demographic characteristics of the travelers, and the relationships with attractiveness indicators derived from Foursquare. To the best of the authors' knowledge, this type of analysis has never been done before.

Data Used

The CHTS data used here are from a simplified travel log that asked respondents from households to report all the trips 50 miles or longer they made in the 8 weeks preceding the day they were assigned a full travel day diary. Also used were the household demographics survey component of CHTS and a selection of data from the single-day place-based diary [for a detailed description of CHTS see NUSTATS (*15*)]. The range of variables that can be analyzed depends heavily on the accuracy with which respondents reported their trips; they were generally richer in information and more accurate in location reporting in the daily diary. However, the long-distance travel log contains data that span periods longer than 24 hours and therefore a better record of trips with overnight stays away from home (*16*, *17*).

In order to consider as wide a range of tours and destinations as possible, the study starts by taking a generous approach to identifying tours. Long-distance tours are sequences of long-distance trips made by a household with end-to-end continuity (the destination location of each trip is the origin of the next trip), beginning with a trip from home and ending with a trip back to home. The pool of potential tours is expanded by running the tour identification process on households' 8-week long-distance logs sorted both by reported long-distance trip number and by date, keeping the results of whichever method produces a larger number of complete home-based tours for a given household (trip number ordering worked better for 91% of households). This analysis also includes tours with incomplete records, but steps are taken to account for uncertainty in duration and distance: in tours missing the from-home or to-home trip, the mode-specific mileage from whichever of those trips is present is double-counted in the tour total. This tour identification process leaves 23,511 full or partial tours to analyze (*16*).

The following steps were taken to identify the primary leg of each tour to extract the primary purpose and destination characteristics:

- Disqualify trips ending at an airport, with their purpose coded as "Return Home," or ending within 30 miles of home. Some tours have all their destinations eliminated by this process, and 21,584 tours remain.
- 2,128 tours have multiple destinations; the longest-duration destination is selected for each of these.
- 522 tours have multiple destinations with equally long durations. To break these ties, the destination farthest from home is selected.
- 76 ties remain; these are resolved by selecting the first tour in order.
- Last, for this analysis, all commute tours are eliminated (tours that include one or more trips to usual work location; tours with business trips that are distinct from work trips in the log are retained).

The aim is to investigate tour characteristics and destination choice at the level of single tours. To do this, each tour is collapsed to a single record that contains a mix of household-level variables (income, household size, number of household cars, etc.), tour-level variables (total distance by each mode, total duration, and presence/absence of multiple purposes), and variables that pertain to the tour's primary leg (single trip purpose and measures of destination attractiveness for the primary destination). For this analysis, the primary destination is considered to be the destination at which members of the

Table 1. List of Variables Used in the Analysis (Household, Tour, and Destination Levels)

Definition	Level	Mean
Household size	HH	2.55
Number of employees	HH	1.33
Number of students	HH	0.63
Number of licensed drivers	HH	2.03
Number of cars	HH	2.08
Number of bikes	HH	1.95
Income (category, treated as numeric)	HH	6.04
Homeowner	HH	0.87
Number of trips in daily diary	HH	10.35
Hispanic household	HH	14.7%
Total density (emps/km^2) around home	HH	1,123.54
Total driving distance (miles)	Tour	276.46
Total passenger distance (miles)	Tour	51.50
Total flying distance (miles)	Tour	485.84
Total ground transit distance (miles)	Tour	18.22
Total other mode distance (miles)	Tour	4.17
Total unknown mode distance (miles)	Tour	5.41
Any trips in tour for business (work-related meeting/convention/seminar)	Tour	14.2%
Any trips in tour for both business and pleasure	Tour	2.8%
Any trips in tour for school-related activity	Tour	2.4%
Any trips in tour for visiting friends/family/relatives	Tour	35.3%
Any trips in tour for medical purpose	Tour	5.0%
Any trips in tour for vacation/sightseeing	Tour	18.7%
Any trips in tour for outdoor recreation (sports, fishing, hunting, camping, boating, etc)	Tour	8.0%
Any trips in tour for entertainment (theater, concert, sports event, gambling, etc)	Tour	8.3%
Any trips in tour for personal business (e.g., shopping)	Tour	7.1%
Area of convex hull of 50 POIs around primary destination (square meters)	Dest	39,265,844.44
Log of convex hull area (ln m^2)	Dest	16.47
Entropy of POI types at 50 POIs around primary destination	Dest	4.68
Median checkins at 50 POIs around primary destination	Dest	1,052.33
Median rating at 50 POIs around primary destination	Dest	7.21
Median users at 50 POIs around primary destination	Dest	499.26
Destination in California	Dest	0.79

Note: POI = point of interest.

household stayed for the most time during the tour. Table 1 shows a list of the variables analyzed here and their averages. Table 2 shows the frequencies of the categorical variables.

In the method described here the authors experimented with the derivation and use of place attractiveness indicators from Foursquare. For each destination, the space covered by the fifty closest reporting locations (points of interest (POIs) with checkin numbers) was estimated. This space is different depending on the place characteristics (e.g., low-density environments have a wider space) and the willingness of Foursquare users to check in. The second variable is the median number of individuals checking in these locations (checkins in Table 1). It is assumed that the larger this indicator, the more attractive the target destination is. The diversity of the destination's neighborhood is another indicator to quantify attractiveness. In this work, the entropy of feature types of the fifty closest POIs at a destination is adopted to measure diversity. Generally speaking, the larger the diversity of a neighborhood, the more attractive it is, as it affords a variety of activities. Furthermore, in order to quantify the semantic of these places, the text of the Foursquare reviews was also analyzed to identify unique topics characterizing each destination; this not reported here because it produced a wide variety of groups of words reported by Foursquare users (these groups are called topics) and did not lend themselves to typical statistical modeling. Finally, also extracted were the median ratings and median number of users by the fifty closest POIs as indicators of destination attractiveness (also reported in Table 1). The larger these two indicators are, the more attractive a destination is. Each household's residential area is also classified by the business density surrounding each household's dwelling into center, suburb, exurb, and rural. Business establishment density for California and a database from Dunn&Bradstreet was used (see Goulias et al. (*16*)).

Table 2. List of Variables Used in the Analysis (Categories and Counts)

Variable	Level	Value	Count
Household block group center class	HH	Center	4,238
		Suburb	2,677
		Exurb	5,363
		Rural	5,451
Household block group category	HH	Center	4,238
		High-density suburb/exurb	4,879
		Low-density suburb/exurb or high-density rural	4,444
		Low-density rural	4,168
Home type	HH	Single home	15,756
		Apartment	1,313
		Other	660
Tour duration	Tour	One overnight	2,244
		Single-day	6,382
		Two-six overnights	6,070
		Seven + overnights	2,403
		Unknown	630
Day of the week of tour start	Tour	Sunday	2,331
		Monday	1,746
		Tuesday	1,752
		Wednesday	1,967
		Thursday	2,282
		Friday	3,473
		Saturday	3,548
		Unknown	630
Primary purpose of tour	Tour	Business (meeting/convention/seminar)	2,370
		Combined business and pleasure	466
		School-related activity	398
		Visit friends/family/relatives	5,939
		Medical	833
		Personal business	1,170
		Vacation/sightseeing	3,123
		Outdoor recreation	1,353
		Entertainment	1,364
		Drive someone else / DK / RF	713
Primary purpose (simplified)	Tour	Business	2,836
		Personal business	2,003
		Recreational	12,890
Season	Tour	Shoulder (other)	6,250
		Summer (May 15–Sep 15)	6,846
		Winter (Nov 15–Mar 15)	4,633
Region in USA	Dest	California	13,973
		Pacific	1,908
		Southwest	267
		Plains	154
		Midwest	227
		Northeast	368
		Southeast	404
		Not in the USA	418
		DK/RF	9
		NA	1

Path Analysis

Path analysis is, in essence, the simultaneous estimation of a set of regression equations in which some endogenous variables are entered as determinants of other endogenous variables. Path analysis SEM can be defined as:

$$y = By + \Gamma x + \zeta \quad (1)$$

where

$y = p$ *by* 1 vector of observed endogenous variables.
$x = q$ *by* 1 vector of observed exogenous variables.
$B = p$ *by* p matrix of coefficients of the y-variables.
$\Gamma = p$ *by* q matrix of coefficients of the x-variables.
$\zeta = p$ *by* 1 vector of equation errors.

This is a covariance-based model because the difference between the sample covariance and the covariance predicted by the model is minimized. The underlying theory of this estimation procedure is that the population covariance matrix of the observed variables (Σ) is a function of a set of parameters:

$$\Sigma = \Sigma(\vartheta) = \begin{bmatrix} \textit{covariance matrix of } y & \textit{covariance matrix of } y \textit{ and } x \\ \textit{covariance matrix of } x \textit{ and } y & \textit{covariance matrix of } x \end{bmatrix}$$

$$= \begin{bmatrix} (I-B)^{-1}(\Gamma\Phi\Gamma' + \Psi)\left[(I-B)^{-1}\right]' & (I-B)^{-1}\Gamma\Phi \\ \Phi\Gamma'\left[(I-B)^{-1}\right]' & \Phi \end{bmatrix} \quad (2)$$

where Φ = covariance matrix of x.

Ψ = covariance matrix of ζ.

The matrix $\sum(\theta)$ consists of four matrices. The unknown parameters $B, \Gamma, \Phi, \text{and} \Psi$ are estimated by finding the parameters such that the covariance matrix ($\hat{\sum}$) implied by the model is as close as possible to the sample covariance matrix (S). To know when the estimates are as close as possible, a fitting function that is to be minimized is defined. Mplus, the software used here, employs maximum likelihood estimation with robust computation of the standard errors of coefficient estimates (https://www.statmodel.com).

Models and Findings

Some key details and simplifications were introduced to make the analysis tractable. First, in the software used, the number of miles flown, driven, and by public transportation are declared as censored variables to account for the large number of tours that may have zero miles for any of these modes. Second, the categories of the main purpose of the tour are reduced to three (*3*). The first category is for relatively flexible, not mandatory, trips to visit relatives, vacation, outdoor recreation and related; the second category is for business and combined business and leisure trips; and the last category is for shopping and medical. A Logit model was used for this categorical variable and shopping/medical is the reference category. Therefore, the regression coefficients should be interpreted in a relative way as explained shortly.

The overnight stays were recorded in a way that tours with same day (no overnight stays) are used as the reference category in a second Logit model used in this path analysis. Figure 1 shows the path diagram for the path analysis model, developed using a variety of different sequences of relationships until a best-fitting model is found. The left-hand side variables are the exogenous variables (determinants of the travel indicators) and in this formulation are the variables that "motivate" households to behave in a certain way. In this case, in addition to the sociodemographic and place of residence variables, also included was the decision to stay in California for the long-distance trip as determinant. This model is also called recursive in which a set of variables are determined first and then another set is considered to be a function of exogenous variables and a function of the first column of dependent variables. This forms a cascade of relationships. The miles flown, driven, and by public transportation are first in the cascade, followed by the overnight stays and main trip purpose of the long-distance tour. The arrows in Figure 1 are the regression coefficients with blue the positive coefficients and red the negative. These are also shown in Tables 3–9 with their significance. A negative coefficient in the Tables indicates negative significant correlation between the dependent variable and the explanatory variables and a positive the opposite.

The amount of air miles flown is influenced by household characteristics, traits of the tour, and the Foursquare indicators. In summary, this model shows that households with higher income are more likely to travel by air for longer distances, but larger households with many cars, living in exurbs and rural environments, and Hispanic households are less likely to travel by air long distances. They are more likely to fly long distance when one or more trips in a tour are for work-related business and/or vacation. The Foursquare relationships show tours with more air miles are more likely to be in denser areas (e.g., big cities) that received higher Foursquare ratings. When the destination is in California the tours are more likely to have a smaller number of miles flown.

The number of miles driven by car shows that tours with a higher number of mileage are done by households with more employed persons, with higher car ownership levels, living in exurbs and rural environments. Larger households are less likely to make tours with many miles driven, and households that do not live in single homes are less likely to drive far. Trip purposes have similarity to the air miles for vacation but the opposite sign for work-related business. The Foursquare variables show that these are also tours in denser areas but not with the same higher ratings of attractiveness as the air miles.

The analysis for miles riding public transportation (presumably train) shows that households of lower income, with a lower number of cars, and living in central

Table 3. Regression of Miles Traveled by Air

Variable	Estimate	S.E.	*p*-value
Any trips in tour for business (meeting/convention/seminar)	0.267	0.016	0.000
Any trips in tour for vacation/sightseeing	0.073	0.016	0.000
Destination in California	–0.857	0.016	0.000
Household size	–0.018	0.006	0.003
Vehicles in household	–0.032	0.008	0.000
Household is Hispanic	–0.049	0.020	0.017
Household block group category	–0.078	0.006	0.000
Household annual income	0.052	0.003	0.000
Foursquare: convex hull area of 50 nearest POIs (log km^2)	–0.025	0.004	0.000
Foursquare: median number of ratings at 50 POIs around destination	0.014	0.003	0.000
Intercept	0.240	0.075	0.001

Table 4. Regression of Miles Traveled Driving

Variable	Estimate	S.E.	*p*-value
Any trips in tour for business (meeting/convention/seminar)	–0.010	0.001	0.000
Any trips in tour for vacation/sightseeing	0.010	0.001	0.000
Destination in California	–0.008	0.001	0.000
Season of tour	0.002	0.001	0.000
Household size	–0.001	0.000	0.033
Number employed in household	0.001	0.001	0.024
Household trips in daily diary	0.000	0.000	0.000
Vehicles in household	0.004	0.001	0.000
Household home type	–0.001	0.000	0.005
Household block group category	0.003	0.000	0.000
Household annual income	–0.001	0.000	0.033
Foursquare: convex hull area of 50 nearest POIs (log km^2)	0.001	0.000	0.007
Foursquare: median users at 50 POIs around destination	–0.023	0.008	0.004
Foursquare: median number of ratings at 50 POIs around destination	0.000	0.000	0.063
Intercept	0.003	0.006	0.682

Table 5. Regression of Miles Traveled by Transit

Variable	Estimate	S.E.	*p*-value
Any trips in tour for business (meeting/convention/seminar)	–0.021	0.012	0.075
Any trips in tour for vacation/sightseeing	0.020	0.009	0.026
Any trips in tour for outdoor recreation	–0.041	0.017	0.018
Destination in California	–0.036	0.008	0.000
Household trips in daily diary	0.001	0.000	0.077
Vehicles in household	–0.020	0.005	0.000
Household block group category	–0.010	0.003	0.003
Household annual income	–0.007	0.002	0.001
Foursquare: convex hull area of 50 nearest POIs (log km^2)	–0.016	0.002	0.000
Intercept	0.027	0.037	0.453

locations are more likely to make tours with more public transportation miles. Secondary trip purposes also play an important role similar to the miles driven, with the added effect of trips to outdoor/recreation trips that are less likely to be done by traveling long distances in public transportation. These tours are also more likely at destinations in central city environments as the Foursquare logarea variable shows.

Main trip purposes in this model are grouped together for the first category for school-related activity, visit

Table 6. Regression of Main Trip Purpose

Main trip purpose is vacation, sightseeing, leisure			
Variable	Estimate	S.E.	*p*-value
Any trips in tour for vacation/sightseeing	3.492	0.245	0.000
Household trips in daily diary	0.018	0.004	0.000
Household block group category	–0.482	0.027	0.000
Household annual income	0.119	0.014	0.000
Foursquare: convex hull area of 50 nearest POIs (log km^2)	0.157	0.016	0.000
Total miles by air in tour	3.596	0.549	0.000
Intercept	–0.454	0.285	0.111
Main trip purpose is both business and leisure			
Variable	Estimate	S.E.	*p*-value
Any trips in tour for vacation/sightseeing	–0.521	0.326	0.110
Household trips in daily diary	0.023	0.004	0.000
Household block group category	–0.395	0.031	0.000
Household annual income	0.259	0.016	0.000
Foursquare: convex hull area of 50 nearest POIs (log km^2)	0.006	0.019	0.768
Total miles by air in tour	5.339	0.554	0.000
Intercept	–0.468	0.334	0.162

Table 7. Regression of Seven or More Overnights

Variable	Estimate	S.E.	*p*-value
Any trips in tour for business (meeting/convention/seminar)	–0.547	0.090	0.000
Any trips in tour for both business and pleasure	0.715	0.159	0.000
Any trips in tour for vacation/sightseeing	1.714	0.071	0.000
Any trips in tour for outdoor recreation	0.484	0.104	0.000
Any trips in tour for personal business (e.g., shopping)	–0.636	0.114	0.000
Destination in California	–1.869	0.079	0.000
Month of tour	–0.018	0.009	0.036
Household size	–0.159	0.025	0.000
Number employed in household	–0.063	0.036	0.082
Household block group category	–0.111	0.024	0.000
Household annual income	0.062	0.015	0.000
Foursquare: convex hull area of 50 nearest POIs (log km^2)	0.065	0.017	0.000
Total miles by air in tour	9.189	0.668	0.000
Intercept	–0.550	0.300	0.070

Table 8. Regression of Two to Six Overnights

Variable	Estimate	S.E.	*p*-value
Any trips in tour for business (meeting/convention/seminar)	–0.264	0.060	0.000
Any trips in tour for both business and pleasure	0.733	0.121	0.000
Any trips in tour for vacation/sightseeing	1.283	0.059	0.000
Any trips in tour for outdoor recreation	0.608	0.069	0.000
Any trips in tour for personal business (e.g., shopping)	–1.073	0.085	0.000
Destination in California	–1.517	0.068	0.000
Month of tour	–0.002	0.006	0.702
Household size	–0.130	0.018	0.000
Number employed in household	0.073	0.026	0.005
Household block group category	–0.138	0.018	0.000
Household annual income	0.072	0.011	0.000
Foursquare: convex hull area of 50 nearest POIs (log km^2)	0.027	0.012	0.030
Total miles by air in tour	7.320	0.663	0.000
Intercept	0.737	0.227	0.001

Table 9. Regression of One Overnight

Variable	Estimate	S.E.	*p*-value
Any trips in tour for business (meeting/convention/seminar)	–0.095	0.072	0.187
Any trips in tour for both business and pleasure	0.394	0.157	0.012
Any trips in tour for vacation/sightseeing	0.689	0.077	0.000
Any trips in tour for outdoor recreation	0.119	0.096	0.212
Any trips in tour for personal business (e.g., shopping)	–0.885	0.108	0.000
Destination in California	–0.455	0.096	0.000
Month of tour	0.007	0.008	0.395
Household size	–0.114	0.023	0.000
Number employed in household	0.089	0.033	0.007
Household block group category	–0.077	0.023	0.001
Household annual income	0.058	0.014	0.000
Foursquare: convex hull area of 50 nearest POIs (log km^2)	–0.010	0.016	0.538
Total miles by air in tour	5.941	0.717	0.000
Intercept	–0.521	0.290	0.072

friends/family/relatives, vacation/sightseeing, outdoor recreation (sports, fishing, hunting, camping, boating, etc.), entertainment (theater, concert, sports event, gambling, etc.). The estimated model for this group shows the propensity of engaging in this type of activity in the main trip is positively correlated with secondary trips for vacation and sightseeing. In addition, tours with a higher number of miles also have a higher propensity to be done for this type of purpose. Central city dwellers and of higher incomes are more likely to engage in this type of tour purpose. Interestingly, households with more daily trips are also more likely to have this type of tour purpose. The positive coefficient of the variable logarea indicates this type of purpose/activity is associated with less dense destinations.

The main trip purposes in the second category are for business (work-related meeting/convention/seminar) and for combined business and pleasure. The propensity of tours having this purpose in the main trip is negatively correlated with vacation/sightseeing in the secondary trip. Households do not seem to combine these purposes. The rest of the variables show many similarities with the previous propensity. However, the logarea is not significantly different than zero, indicating tours of this type are at low-density and high-density destinations.

The last block is the analysis of the overnight stays. Recall the first category corresponds to tours that are longer than 7 nights, the second category with tours that are between 2 and 6 overnight stays, the third is for one overnight stay, and the reference category is same-day long-distance travel. The negative coefficients for the secondary trip purpose, work-related business, show that tours containing this type of purpose are more likely to be without overnight stays. The same happens for shopping. In contrast, secondary trips that combine business and pleasure, vacation/sightseeing, and outdoor recreation are more likely to have at least an overnight stay. Vacation/sightseeing is also the purpose with the largest coefficients (and therefore probability) of being in the many overnight stays. This is exactly as expected from other sources of information. California destinations are more likely to be in the same-day travel. Longer overnight stays are more likely to be in the earlier parts of the year (however, this needs further scrutiny).

The household characteristics show an interesting pattern. Only household income is positively associated with the propensity of tours of longer than a week. Household income and the number of employed persons are positively associated with the 2 to 6-day tours and single-night tours. Household size is negatively associated with overnight stays, indicating constraints in the ability of households to spend the night outside home. Rural households are less likely to have tours with overnight stays.

Summary

In this paper data are analyzed from the CHTS long-distance travel log, assembled into tours for each responding household to an 8-week long travel log combined with data from a single-day diary and household sociodemographic information. The study identified in each tour a main trip, selected tour with a non-commute main trip, and added destination descriptors from Foursquare. Path analysis was used on three censored variables (tour miles by air, miles driving, and miles by public transportation) and two categorical variables (main trip tour purpose) and number of overnight stays. Household income was found to be major determinant of air travel, with wealthier households combining work-related trips with leisure. Moreover, travel by air of longer distance is more likely to be in large, dense destinations. In contrast, household income is negatively correlated with mileage accrued driving a car, which is positively correlated with higher car

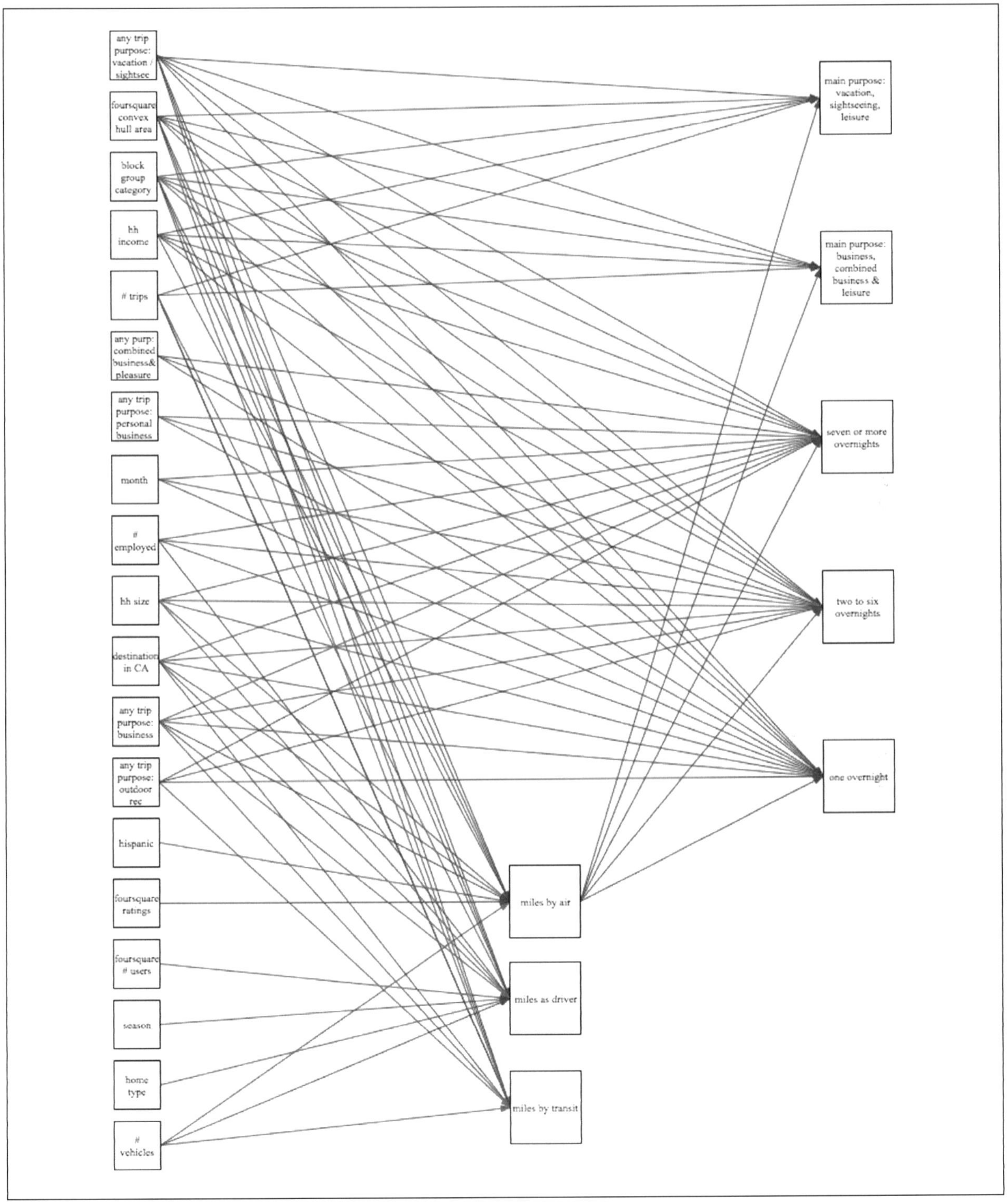

Figure 1. Path diagram of the path analysis model.

ownership. Urban dwellers with lower car ownership and income tend to travel long distances by public transportation. In terms of main trip purpose, work and shopping are not associated with overnight stays, and vacation/sightseeing is the purpose associated with longer overnight stays. In addition, higher household income is positively correlated with multiple overnight stays. Moreover, California destinations are more likely to be in same-day tours. Using Foursquare data to describe destinations and their attractiveness produced interesting results (it was possible to distinguish lower density from higher density destinations and their relationship to travel behavior facets). However, Foursquare user ratings were not informative enough. The tours analyzed here show that when the main trip is for vacation, sightseeing, and leisure, it is also more likely to be combined with other vacation and sightseeing trips. In contrast, when the main trip is business and combined business and leisure, this trip is less likely to be combined with vacation trips. Another key finding is the substantial and significant differences in long-distance tour behavior among residents of urban versus rural environments while controlling for differences in social and demographic characteristics. This is a reflection of different contexts and heterogeneous preferences that are probed in more depth in a companion paper (*17*).

The data used here did not allow a clear identification of the composition of the traveling party, and this would have been a key determinant of the choices households make. Household size and number of household workers functioned as proxies, but only indicated that larger households tend not to have overnight stays (presumably because of cost), but not enough to help in understanding decision making with respect to traveling party size and composition. In addition, the simplifications imposed for analytical tractability (e.g., grouping different trip purposes together) did not allow a more fine-grained analysis of relationships among trip purposes. Another key limitation is the absence of activity duration at different destinations and the lack of destination geographic precision to study routes and activity opportunities at destinations. Moreover, the Foursquare data used here employ the simplest possible indicators, leaving out topics and recommendations for each location in combination with other social media data (*18*). More statistical features, from thematic, temporal and/or spatial perspectives, could be adapted in order to better model the semantics of the destination neighborhood (*19–21*). Also, aside from the area of the convex hulls, the design of the attractiveness indicators did not take into account the explicit spatial patterns of POIs in the neighborhood. For example, the POIs that are far away from a destination could be discounted in determining attractiveness. Moreover, different arrangements of POIs might show different levels of attractiveness (*22*). All these factors are left for future work.

Acknowledgments

Funding for this research was provided by the California Department of Transportation (CALTRANS) through the University of California Center on Economic Competitiveness in Transportation (UCCONNECT). Four anonymous reviewers provided extremely valuable comments that helped us improve the paper.

Author Contributions

The authors confirm contribution to the paper as follows: study conception and design: Davis, McBride, Zhu, Janowicz, Goulias; data collection: Davis, McBride, Zhu; analysis and interpretation of results: Davis, McBride, Zhu, Janowicz, Goulias; draft manuscript preparation: Davis, McBride, Goulias. All authors reviewed the results and approved the final version of the manuscript.

References

1. Christensen, L. Environmental Impact of Long Distance Travel. *Transportation Research Procedia*, Vol. 14, 2016, pp. 850–859.
2. Gössling, S., C. B. Hansson, O. Hörstmeier, and S. Saggel Ecological Footprint Analysis as a Tool to Assess Tourism Sustainability. *Ecological Economics*, Vol. 43, No. 2, 2002, pp. 199–211.
3. Pourabdollahi, Z., R. Tillery, M. Gawade, T. Hill, G. Mathews, and J. Worrell Statewide Tourism Travel Demand Forecasting: A Behavior-Based Modeling Framework for the State of Florida. Presented at 96th Annual Meeting of the Transportation Research Board, Washington, D.C., 2017.
4. *Cambridge Systematics, Inc.* California Statewide Travel Demand Model, Version 2.0: Model Overview, 2014. http://www.dot.ca.gov/hq/tpp/offices/omsp/statewide_modeling/Files/Documentation/FR1_CSTDM_Model_Overview.pdf. Accessed February 23, 2018.
5. Mitra, S. K. *Land Use, Land Value, and Transportation: Essays on Accessibility, Carless Households, and Long-distance Travel.* PhD dissertation, UC Irvine (Accessed from ProQuest), 2016.
6. Mitra, S. K., and J. D. M. Saphores. *Determinants of Long-Distance Commuting: Evidence from the 2012 California Household Travel Survey*, No. 17-01286, 2017.
7. Georggi, N. L., and R. M. Pendyala. An Analysis of Long-distance Travel Behavior of the Elderly and the Low-income. Paper Presented at the Conference the Long and short. Transportation Research Circular E-C026—Personal Travel: The Long and Short of It, Transportation Research Board, Washington D.C., 2000.
8. Axhausen, K. W. (2001). Methodological research for a European survey of long-distance travel. *Proc., Conference Personal Travel: The Long and Short of It.* Washington, D.C., 2001, pp. 321–342.
9. Beckman, J. D., and K. G. Goulias. Immigration, residential location, car ownership, and commuting behavior: a multivariate latent class analysis from California. *Transportation*, Vol. 35, No. 5, 2008, pp. 655–671.
10. LaMondia, J. J., and C. R. Bhat. A Conceptual and Methodological Framework of Daily and Long Distance Leisure Activity-Travel Behavior. Presented at 90th Annual Meeting of the Transportation Research Board, Washington, D.C., 2011.

11. CALTRANS. Interregional Transportation Strategic Plan: Update 2015. http://www.dot.ca.gov/hq/tpp/offices/omsp/system_planning/itsp.html. Accessed November 10, 2017.
12. Shahrin, N., A. P. M. Som, and J. Jusoh. Long Journey Travel to Tourist Destination: A Review Paper. *SHS Web of Conferences, EDP Sciences*, Vol. 12, 2014, p. 01099.
13. Holz-Rau, C., J. Scheiner, and K. Sicks. Travel Distances in Daily Travel and Long-distance Travel: What Role is Played by Urban Form? *Environment and Planning A*, Vol. 46, No. 2, 2014, pp. 488–507.
14. Bernardin, V. L. Jr, N. Ferdous, H. Sadrsadat, S. Trevino, and C. C. Chen. *Integration of National Long Distance Passenger Travel Demand Model with Tennessee Statewide Model and Calibration to Big Data.* Presented at 96th Annual Meeting of the Transportation Research Board, Washington, D.C., 2017.
15. NUSTATS. *2010-2012 California Household Travel Survey*. Final Report. Austin, TX, 2013.
16. Goulias, K. G., A.W. Davis, E. C. McBride, K. Janowicz, and R. Zhu. *Long Distance Travel in CHTS and Social Media Augmentation.* Final Report. Contract Number: 65A0529, Task Order 047. University of California Santa Barbara. Department of Geography, GeoTrans and STKO Laboratories, Submitted to the California Department of Transportation, May 2017.
17. Davis, A. W., E. C. McBride, and K. G. Goulias. A Latent Class Patterns Recognition and Data Quality Assessment of Non-Commute Long Distance Travel in California. *Transportation Research Record: Journal of the Transportation Research Board*, 2018. (In press).
18. Gao, S., K. Janowicz, and H. Couclelis. Extracting Urban Functional Regions from Points of Interest and Human Activities on Location-based Social Networks. *Transactions in GIS*, Vol. 21, No. 3, 2017, pp. 446–467.
19. Adams, B., and K. Janowicz. Thematic Signatures for Cleansing and Enriching Place-related Linked Data. *International Journal of Geographical Information Science*, Vol. 29, No. 4, 2015, pp. 556–579.
20. McKenzie, G., K. Janowicz, S. Gao, and L. Gong. How Where is When? On the Regional Variability and Resolution of Geosocial Temporal Signatures for Points of Interest. *Computers, Environment and Urban Systems*, Vol. 54, 2015, pp. 336–346.
21. Zhu, R., Y. Hu, K. Janowicz, and G. McKenzie. Spatial Signatures for Geographic Feature Types: Examining Gazetteer Ontologies Using Spatial Statistics. *Transactions in GIS*, Vol. 20, No. 3, 2016, pp. 333–355.
22. Zhu, R., P. C. Kyriakidis, and K. Janowicz. Beyond Pairs: Generalizing the Geo-dipole for Quantifying Spatial Patterns in Geographic Fields. *The Annual International Conference on Geographic Information Science*, pp. 331–348. Springer, Cham, 2017.

The Standing Committee on Transportation Demand Forecasting (ADB40) peer-reviewed this paper (18-00160).

Article

TRR
JOURNAL OF THE TRANSPORTATION RESEARCH BOARD

Transportation Research Record
2018, Vol. 2672(49) 12–22

DOI: 10.1177/0361198118801339
journals.sagepub.com/home/trr
SAGE

Three Methods for Anticipating and Understanding Uncertainty of Outputs from Transportation and Land Use Models

Guangmin Wang[1] and Kara M. Kockelman[2]

Abstract

This study demonstrates three methods for uncertainty propagation in transportation and land-use models (LUMs): Local Sensitivity Analysis with Interaction (LSAI), Monte Carlo (MC), and Bayesian Melding (BM). Two case-study settings are used to illustrate how these methods work, allowing for inter-method comparisons. LSAI can provide the sign of change implied by changes in model inputs, the relative importance of changes in different inputs, and a decomposition of changes in outputs due to the impact of inputs' individual and interactive. LSAI is limited to relatively small-size problems because its computing time rises exponentially with the number of (groups of) inputs. Moreover, LSAI obtains only point estimates, while MC and BM methods can deliver entire distributions of each output through an understanding of the uncertainty in all model inputs and parameters. MC delivers each output's distribution and requires hundreds of samples, especially for more accurate results. Fortunately, MC methods are especially useful for high-dimensional problems because convergence rates are not a function of model dimensionality and errors depend only on sample size and input uncertainties. BM delivers posterior distributions for model outputs, using prior probability distributions and likelihoods of inputs and parameters, along with validation of/comparison to intermediate model outputs. A BM approach can be extremely expensive, in terms of computing time, since it requires several hundred model runs.

Scientific computing involves large-scale simulations to represent real-world phenomena, like the evolution of cities and their traffic patterns. With improvements in computing capabilities and more efficient algorithms, results can be more accurately simulated to represent real-world conditions. Models allow for accurate forecasts over longer prediction periods. Predictions are typically affected by uncertainties in input data and model parameters, and by incomplete knowledge of underlying behaviors (*1*). Models (and the systems they represent) are often explicitly stochastic (*2*), with random components being generated and used throughout the predictive process. It is very important for system design optimization and policy-making to capture and represent this uncertainty information appropriately in model results. Much work has been done in land-use forecasting (*3–5*).

Few studies have begun to examine uncertainty propagation in the context of the integrated land-use transportation modeling framework because of the required integration of metropolitan and statewide transportation plans with land-use plans by the Intermodal Surface Transportation Efficiency Act (ISTEA) of 1991 (*6*).

Obviously, one must first identify sources of uncertainty (*7, 8*), to carry out a probabilistic analysis of a system. Pradhan and Kockelman (*3*) reviewed the literature on the sources of uncertainty in land-use transportation models. Later, Sevcikova et al. (*9*) reviewed the key sources of uncertainty for UrbanSim land-use modeling outputs.

Uncertainty quantification is the process of representing imperfectly known or understood inputs and parameters, propagating this variability through the model system and then characterizing the uncertainty in the model's results. The outcome usually comes with attached "error bars" to indicate uncertainty ranges. Generally, the uncertainty is usually represented by interval mathematics (*10*), fuzzy theory (*11*), and probabilistic

[1]School of Economics and Management, China University of Geosciences, Wuhan, China
[2]Department of Civil, Architectural and Environmental Engineering, The University of Texas at Austin, Austin, TX

Corresponding Author:
Address correspondence to Kara M. Kockelman: kkockelm@mail.utexas.edu

analysis (*8*). In the probabilistic analysis approach, uncertainties described by probability distributions are associated with model inputs to estimate the outputs' probability distributions by using UrbanSim (*3*). However, the methods to research uncertainty propagation in the integrated land-use transportation modeling framework still attract attentions.

This work demonstrates and compares methods for uncertainty propagation in complex transportation and land-use models (LUMs). The following sections provide a literature review, introduction and two test examples for uncertainty propagation in transportation and LUMs, along with a summary of findings.

Literature Review

Many researchers have examined different methods for anticipating and understanding output uncertainties. This paper introduces three methods for forecasting uncertainty in land-use and transportation model outputs: Local Sensitivity Analysis with Interaction (LSAI), Monte Carlo (MC) method, and Bayesian Melding (BM) method. Two reasonably transparent transportation settings are then used to illustrate and review these methods.

LSAI

Sensitivity analysis is the study of how uncertainty in the output of a mathematical model or system (numerical or otherwise) can be apportioned to different sources of uncertainty in its inputs (*12*). There are many different ways of conducting sensitivity analyses; however, the various analyses may not produce identical results in answering these question. Thus, Hamby (*13*) summarized many available methods to conduct sensitivity analyses by presenting details of the types of sensitivity analyses utilized for various modeling situations. Hamby (*14*) compared the assessment of several methods and intended to demonstrate calculation rigor and parameter sensitivity rankings resulting from various sensitivity analysis techniques.

Local sensitivity analysis is the assessment of the local impact of inputs' variation on the model response by concentrating on the sensitivity in the vicinity of a set of input values. Such sensitivity is often evaluated through gradients or partial derivatives of the output functions at these inputs, so other inputs' values are held constant when studying the local sensitivity of a specific input. Kockelman (*15*) ran a gravity-based land-use model (G-LUM) over three alternative scenarios, with total employment counts (EMP), household counts (HH) and link impedances or travel times (TT) increasing by 50%, utilizing each set individually.

To overcome the limitations of local methods, Saltelli et al. (*12*) used the global sensitivity analysis in contrast to local sensitivity analysis to consider the entire range of input variations.

Besides the joint sensitivity of generic model outputs for small changes in all inputs (including parameters) addressed in (*16*, *17*), Saltelli and Tarantola (*18*) defined group sensitivity indices in the context of a global sensitivity analysis. As far as finite changes are concerned, Borgonovo (*19*) introduced sensitivity measures for individual exogenous variables. To find sensitivity measures for specific input factor sets (typically, sets of parameters), Borgonovo and Peccati (*20*) proved that a change in model output is decomposed as a function of factors with the same structure as the parameter decomposition and introduced factor finite-change sensitivity indices (FCSI) for model parameters to investigate the relationship between the factor FCSIs and parameter FCSIs. Then, Borgonovo et al. (*21*) used the G-LUM by Kockelman (*15*) to illustrate LSAI techniques and found that the outputs respond almost additively to variations in model inputs over the given scenarios. Wang and Kockelman (*22*) applied LSAI to evaluate random-utility-based multiregional input-output models by producing FCSI for the variation of inputs under different scenarios.

MC Method

MC techniques have been widely used for uncertainty propagation due to their conceptual simplicity and ease of implementation. In MC techniques, one samples random variables, runs an ensemble of simulations, and simply presents the distribution of outputs, for all uncertainty statistics. A reasonable result usually requires many ensemble runs, which is computationally expensive for large-scale systems. More efficient approaches need to be developed to represent and propagate uncertainties in large-scale simulations (*23*).

Zhao and Kockelman (*24*) conducted a study of the propagation of uncertainty in four-step travel demand models by using MC simulation to quantify variability in model outputs. Krishnamurthy and Kockelman (*25*) researched propagation of uncertainty in transportation LUMs through multivariate MC sampling of 200 scenarios. Harvey and Deakin (*26*) conducted a similar study to consider uncertainty in population growth, fuel price and household income levels in the Los Angeles region. Thompson et al. (*27*) examined the impact of higher-than-projected population estimates on emission trends for metropolitan regions in California. Saadi et al. (*28*) integrated Markov Chain MC simulation and profiling-based methods to capture the behavioral complexity and the great heterogeneity of agents of the true population for large-scale microsimulation scenarios of transportation and urban systems. Clay and Johnston (*29*) aimed to research which sources of uncertainty have the largest impact on outputs in a fully integrated land-use and transportation forecasting models. They ran all possible combinations of each variable at each level of uncertainty to analyze the impacts of uncertainty on the model's outputs. Clay et al. (*30*) used point-estimate inputs to research how uncertainty affects the Large Zone Economic Module outputs by tracking the effects of this uncertainty through the various submodels to the model outputs.

BM Method

Pradhan and Kockelman (*3*) examined the propagation of uncertainty in the context of UrbanSim. However, this study analyzed only the sensitivity to a small sample of selected input values and explored the effect of these changes on outputs, in addition to simple stochastic simulation error from variation in random seeds.

Another approach to extrapolate prediction accuracy for LUMs was taken in (*31*). Models calibrated at different time points were used to simulate the present land-cover change and estimate how accurately the model will predict the future through a measure derived by a validation with empirical data. Sevcikova et al. (*9*) developed and then applied BM by extending an earlier method to agent-based stochastic models for calibrating a stochastic model system with respect to uncertainty. This method encodes all available information about model inputs and outputs in terms of prior probability distributions and likelihoods, and used Bayes' theorem to obtain a posterior distribution for any quantity. However, BM can be very computationally intensive since it requires several hundred runs of the model. Thus, they provided various ways to reduce the required time (*9*). Based on work (*9, 32*), Sevcikova et al. (*2*) described the first incorporation of uncertainty assessment into the development of an official land-use forecast published by a metropolitan planning organization. They demonstrated how BM for assessing uncertainty could be used to support the application of an academically founded land-use model.

Introduction to Three Methods

Introduction to LSAI

Mathematical models are used to denote input-output mappings as follows:

$$y = f(\mathbf{x}), \quad f : \Omega_{\mathbf{X}} \to \mathbb{R} \tag{1}$$

where y is the output of interest, $\Omega_{\mathbf{X}} \subseteq \mathbb{R}^K$ and $\mathbf{x} = (x_1, x_2, \cdots, x_K)$, $\mathbf{x} \in \Omega_{\mathbf{X}}$ is the vector of inputs, and K is the number of inputs whose variations are of interest. Here, the components $x_i (i = 1, 2, \cdots, K)$ are supposed to be independent.

Therefore, the base-case output of the simulation $y^0 = f(\mathbf{x}^0)$ can be obtained by the simulation with inputs to a base-case scenario, $\mathbf{x}^0$. Similarly, different outputs depend on alternative values of inputs, $y^s = f(\mathbf{x}^s)$, where $s = 1, 2, \cdots, S$. The analyst knows the response of the output in each scenario, although they have no information about the sources of change (*21*). The change from scenario 0 to scenario 1 of the inputs induces the change $\Delta y = y^1 - y^0$ in the output. And the change can be decomposed by using by a multivariate Taylor expansion of Δy when supposing that $f(\mathbf{x})$ is r ($r \leqslant K$) times differentiable at $\mathbf{x}^0$ (*3, 14, 20*):

$$\Delta y = y^1 - y^0 = \sum_{k_1=1}^{K} f'_{k_1}(\mathbf{x}^0)\Delta \boldsymbol{x}_{k_1} + \sum_{k_1<k_2}^{K} f''_{k_1,k_2}(\mathbf{x}^0)\Delta \boldsymbol{x}_{k_1}\Delta \boldsymbol{x}_{k_2} + \ldots + \sum_{k_1<k_2<\cdots<k_r}^{K} f^r_{k_1,k_2,\cdots,k_r}(\mathbf{x}^0)\Delta \boldsymbol{x}_{k_1}\Delta \boldsymbol{x}_{k_2}\cdots\Delta \boldsymbol{x}_{k_r} + \cdots + o(||h||^r) \tag{2}$$

$h = max_{1 \leqslant i \leqslant r}\Delta \boldsymbol{x}_{k_i}$, and $o(||h||^r)$ denotes the infinitely small quantity of $||h||$.

Then, $f(\mathbf{x})$. is K times differentiable at $\mathbf{x}^0$, and one can obtain the following:

$$\Delta y = f(\mathbf{x}^1) - f(\mathbf{x}^0) = \sum_{k_1=1}^{K} \Delta_{k_1} f + \sum_{k_1<k_2}^{K} \Delta_{k_1,k_2} f + \sum_{k_1<k_2<k_3}^{K} \Delta_{k_1,k_2,k_3} f \cdots + \Delta_{1,2,\cdots,K} f \tag{3}$$

where

$$\begin{cases} \Delta_{k_1} f = f\left(x^1_{k_1}, \mathbf{x}^0_{\sim k_1}\right) - f(\mathbf{x}^0) \\ \Delta_{k_1,k_2} f = f\left(x^1_{k_1}, x^1_{k_2}, \mathbf{x}^0_{\sim(k_1,k_2)}\right) - \Delta_{k_1} f - \Delta_{k_2} f - f(\mathbf{x}^0) \\ \Delta_{k_1,k_2,k_3} f = f\left(x^1_{k_1}, x^1_{k_2}, x^1_{k_3}, \mathbf{x}^0_{\sim(k,k_2,k_3)}\right) - \Delta_{k_1,k_2} f \\ \quad -\Delta_{k_1,k_3} f - \Delta_{k_2,k_3} f - \Delta_{k_1} f - \Delta_{k_2} f - \Delta_{k_3} f - f(\mathbf{x}^0) \end{cases} \tag{4}$$

and where $\left(x^1_{k_1}, \mathbf{x}^0_{\sim k_1}\right)$denotes that the k_1th element of the $\mathbf{x}$ vector, $x^1_{k_1}$ is set at its Scenario 1, while all other variables are at their Scenario 0 values. Based on such a decomposition, FCSI can be computed as follows:

$$\varphi^r_{k_1,k_2,\cdots,k_r} = \Delta_{k_1,k_2,\cdots,k_r} f \tag{5}$$

where $k_1, k_2, \cdots, k_r$ denote a group of r indices and $\varphi^r_{k_1,k_2,\cdots,k_r}$ is the portion of Δy due to the interaction of inputs corresponding to the selected indices.

For the k_i th ($k_i = 1, 2, \ldots, K$) element of the **x** vector, x_{k_i}, the first-order FCSI are $\varphi^1_{k_i} = \Delta_{k_i} f$ and the total-order indices are $\varphi^T_{k_i} = \Delta_{k_i} f + \sum_{k_i < k_2}^{K} \Delta_{k_i, k_2} f + \cdots + \Delta_{1,2,\cdots,K} f$, where $\varphi^T_{k_i}$ is the total contribution of x_{k_i} to Δy, and is the sum of the individual contribution of x_{k_i}, plus all the contributions due to the interaction of x_{k_i} with the remaining inputs. Thus, the index $\varphi^I_{k_i} = \varphi^T_{k_i} - \varphi^1_{k_i}$ represents the effect of interactions associated with x_{k_i}.

According to $\varphi^T_{k_i}$'s definition, it can be computed as follows:

$$\varphi^T_{k_i} = f(\mathbf{x}^1) - f\left(x^0_{k_i}, \mathbf{x}^1_{\sim k_i}\right) \tag{6}$$

where $f(\mathbf{x}^1)$ is the value of the output in scenario 1 and $f\left(x^0_{k_i}, \mathbf{x}^1_{\sim k_i}\right)$ is the point obtained with x_{k_i} that remains at the base-case scenario but all other inputs at scenario 1.

As discussed in the literature (*18*, *33*), the sign of the first-order indices ($\varphi^1_{k_i}$) is the sign change in y due to the individual change in x_{k_i}. The sign of $\varphi^r_{k_1, k_2, \cdots, k_r}$ is the sign of the interaction between the inputs x_{k_1}, x_{k_2} and x_{k_r}. The total-order indices ($\varphi^T_{k_i}$) are the appropriate sensitivity measures, since they deliver not only the individual effects of inputs, but also account for interaction effects of inputs. The magnitudes of $\varphi^r_{k_1, k_2, \cdots, k_r}$ provide the natural sensitivity measures.

All FCSI can be computed by use of 2^K simulations if there are K (or group of) exogenous variables whose variations are of interest. The triplet $\left(\varphi^1_{k_i}, \varphi^I_{k_i}, \varphi^T_{k_i}\right)$ can be computed at the cost of *2K* simulations, instead of 2^K. This computational burden reduction result makes the sensitivity measures applicable also to complex simulation codes.

Introduction to MC Methods

The MC method is a broad class of computational algorithms that rely on repeated random sampling to obtain numerical results. MC is useful for simulating phenomena with significant uncertainty in inputs and systems with many coupled degrees of freedom (*34*, *35*), especially for high-dimensional problems because the convergence and convergence rate are not concerned with the dimensions of the problems. The error of MC is defined by $\epsilon = \frac{\lambda_\alpha \sigma}{\sqrt{n}}$, where n is the number of samples, σ is the standard deviation, λ_α is standard normal deviate relative with the confidence level α. MC involves random sampling from the distribution of inputs, and successive model runs until a statistically significant distribution of outputs is obtained. It can be used to solve problems with probability structures, or non-probabilistic problems such as finding the area under a curve. However, this kind of method requires many samples. To achieve computational efficiency, methods that sample the input distribution in an efficient manner have been introduced. One such variant of the standard MC method is the Latin Hypercube (LH) sampling method. In this method, the range of probable values for each input parameter is divided into ordered segments such that the parameter space, consisting of all uncertain parameters, is partitioned into cells with equal probabilities. Thus, parameter estimates are sampled in an efficient manner, since each parameter is sampled only once from each of its possible segments. The advantage of this approach is that it allows representation of the extremes in the probability distribution of the outputs.

Introduction to BM Method

Sevcikova et al. (*2*) presented a figure to show the basic concept of BM developed for deterministic models. There is a prior distribution of model inputs $q(X)$ from which one draws input values X_i for $i = 1, \ldots, I$. The model runs I times from the base year to the present year and for each input X_i. It produces as output the quantity of interest, Φ_i. The model can be viewed as a mapping, M, from the space of inputs to the space of outputs, which is denoted by $\Phi = M_\Phi(X)$. The 'present' time is defined as a time point with observed data available. The observed data is denoted by ψ and used to compute a weight ω_i for each input X_i: $\omega_i = L(\Phi_i)$, where $\Phi_i = M(X_i)$. Here, $L(\Phi_i)$ is the likelihood of the model outputs given the observed data ψ, $L(\Phi_i) = p(\psi|\Phi_i) \propto p(\psi|X_i) = \sum_{k=1}^{K} p(\psi_k|X_i)$. For each of the I runs, the model is run forward until a future time when making a prediction. The results, calculated by running the ith model, are denoted by Y_i. The posterior distribution of Y is approximated by a discrete distribution with values Y_i having probabilities proportional to ω_i.

The primary BM stages or steps are as follows (*9*):

1. Draw a sample $\{X_1, X_2, \ldots, X_I\}$ of values of the inputs from the prior distribution $q(X)$.
2. For each X_i, run the model to obtain Φ_i.
3. Compute weights $\omega_i = L(\Phi_i)$. Here, an approximate posterior distribution of inputs with values $\{X_1, X_2, \ldots, X_I\}$ and probabilities proportional to $\{\omega_1, \omega_2, \ldots, \omega_I\}$ are obtained.
4. The posterior distribution of Φ is no longer approximated by the set $\{\omega_i\}$ but now has a finite mixture distribution of the following form:

$$\pi(\Phi) = \prod_{i=1}^{I} w_i p(\Phi|X_i) \tag{7}$$

In Equation 7, the conditional distribution $p(\Phi|X_i)$ has an assumed parametric form that reflects the additional sources of variation. The posterior distribution of Y has a similar form, namely:

$$\pi(Y) = \prod_{i=1}^{I} w_i p(Y|X_i) \tag{8}$$

Two Test Examples

To illustrate these different methods for uncertainty propagation in transportation and LUMs, Example 1 offers LSAI and MC applications and Example 2 offers a BM application.

Example 1. Figure 1's small test network enables application of a simple travel demand model (TDM), with three nodes (with node 1 as the origin, and 2 and 3 as destinations) and four links. $(*, *)$ denotes the free-flow travel time and capacity on each link.

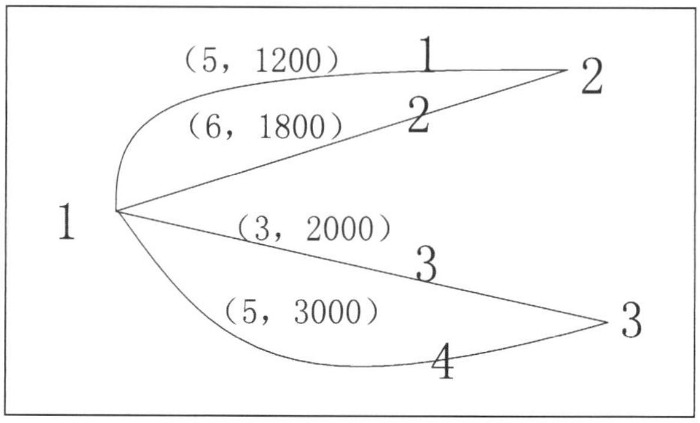

Figure 1. The test network.

This TDM's equations are as follows:

$$Z = x_1 x_2 + x_3 \tag{9}$$

$$Z_{12} = Z^* \frac{e^{x_4 * GC_{12}}}{e^{x_4 * GC_{12}} + e^{x_4 * GC_{13}}} \tag{10}$$

$$Z_{13} = Z - Z_{12} = Z^* \frac{e^{x_4 * GC_{13}}}{e^{x_4 * GC_{12}} + e^{x_4 * GC_{13}}} \tag{11}$$

$$GC_{12} = min\{LC_1, LC_2\} \tag{12}$$

$$GC_{13} = min\{LC_3, LC_4\} \tag{13}$$

$$LC_i = \gamma^* time^i + \text{Toll}^* t_f{}^i \tag{14}$$

where Z denotes number of trips generated, Z_{1j} denotes number of trips going from origin node 1 to destination node j (j = 2, 3), GC_{1j} denotes generalized cost going from node 1 to destination nodes j (j = 2,3), and LCi denotes generalized cost of using link i (i=1,2,3,4). This assumes the travelers have the same value of time (VOT), which is γ = \$6/hour (*36*). Tolls vary by link, with a toll of \$ 0.55/mile on links 1 and 3, and tolls of \$ 0.20/mile on links 2 and 4, as used in (*36*). The four model inputs x_1, x_2, x_3, x_4 follow lognormal distributions, and their summary statistics are as follows:

x_1 = population of origin zone 1, with mean = 2000 persons, standard deviation (SD) = 200 persons, and thus coefficient of variation (CoV = SD/mean) = 0.10;

x_2 = average trip-making rate per person per day, with mean = 2.303 trips/day/person, SD = 0.2303, and CoV = 0.10;

x_3 = trips generated by visitors, with mean = 1000 trips/day, SD = 100, and CoV = 0.10;

x_4 = impedance parameter for generalized cost of accessing each destination, with mean = -0.02, SD = 0.002, and CoV = 0.10 (and negative values simply obtained after drawing from the non-negative lognormal distribution).

Equation 9 describes trip generation per zone, and Equations 10 and 11 describe trip distribution. There is no mode split because only one mode is used in this example. Route choice is obtained via traffic assignment based to the network, assuming shortest-path user equilibrium (*37*) and feedback of chosen routes and TT to trip distribution.

The Bureau of Public Roads (BPR) function is used here for TT on each link i, with $time^i = t_f{}^i\left(1 + \alpha^*\left(\frac{v_i}{c_i}\right)^{\beta}\right)$, where $t_f{}^i$ is link i's free-flow travel time (i=1,2,3,4), v_i is the assigned traffic flow on link i, c_i is link i's capacity flow rate, and α and β are link-congestion parameters. The National Cooperative Highway Research Program (NCHRP) Report 365 (*38*) suggests values of α = 0.84 and β = 5.5, when using true capacity values (c_i).

LSAI for Example 1

LSAI is applied on the test example by increasing all inputs by 10%. The change of each model's outputs (traffic flow on links) from $\mathbf{x}^0$ to $\mathbf{x}^1$ can be decomposed into 15 terms that account for the individual changes in x_1, x_2, x_3, x_4, to their interactions in pairs, and in the residual term that contains their overall and residual interactions. Here, 2^4 = 16 model simulations are needed. If we compute the first-order indices, the total-order indices and the effect of interactions associated with every exogenous variable, only 2*4 = 8 simulations are needed. Figure 2 illustrates the ith-order indices as well as their sum and the total-order indices. Figure 3 shows the elasticity of the first-order indices and the total-order indices, where elasticities are computed as $\frac{\Delta v/v}{\Delta x/x}$.

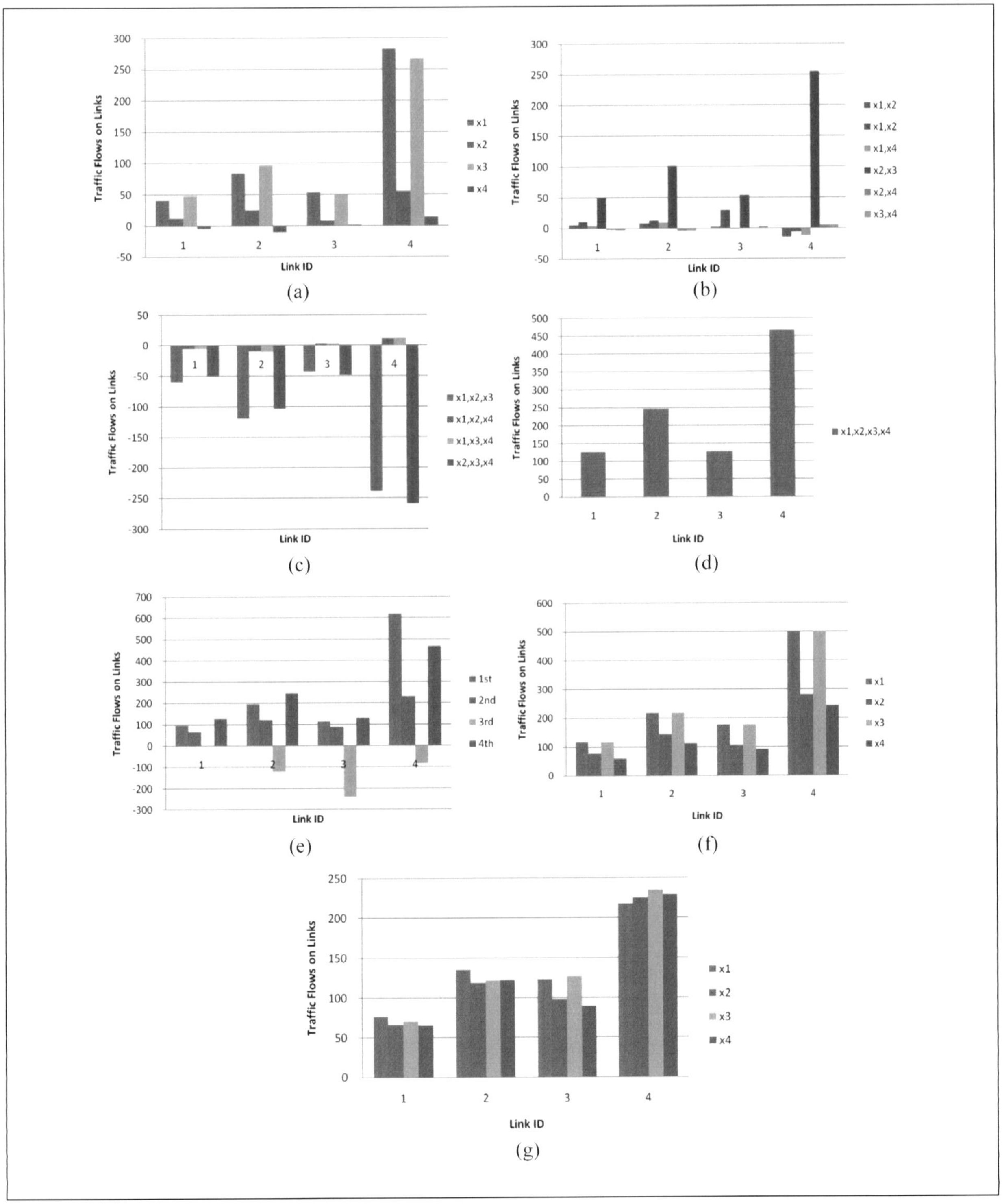

Figure 2. The *i*th-order indices as well as their sum and the total-order indices using LSAI: (*a*) first-order indices; (*b*) second-order indices; (*c*) third-order indices; (*d*) fourth-order indices; (*e*) sum of *i*th-order indices; (*f*) total-order indices; (*g*) effect of interactions associated with inputs.

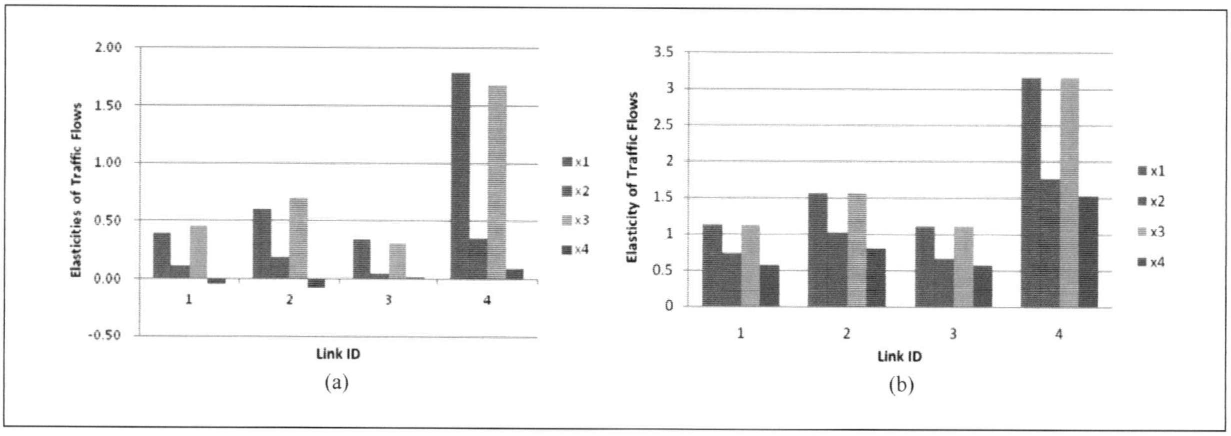

Figure 3. Elasticity of first-order and total-order indices of link flows using LSAI: (*a*): elasticity of first-order index; (*b*): the elasticity of total-order index.

Figure 2*a* shows that increasing inputs x_1, x_2 and x_3 by 10% will lead to the largest increase of traffic flows on link 4. Moreover, increasing x_4 by 10% will lead to a decrease of traffic flows on links 1 and 2, and an increase of traffic flows on links 3 and 4. The individual effect of x_1 is the most important for link 3 and 4 traffic flows. The individual effect of x_4 is the most important for link 1 and 2 traffic flows. Figure 2*b* shows how interaction effects between inputs x_2 and x_3 deliver the largest increase of traffic flows on all links among all second-order interactions. Other interactions (with the exception of the interaction between x_1 and x_3, which is 29.2) are all very small (with all absolute values under 15). Signs of some second-order indices are negative, such as the interactions between x_2 and x_4, and x_3 and x_4 on links 1 and 2, the interaction between x_1 and x_4 on link 3, the interactions between x_1 and x_2, x_1 and x_3, and x_1 and x_4 on link 4.

Figure 2*c* shows that almost all third-order indices deliver negative effects on all links' traffic flows with the exception of third-order interactions among x_1, x_2 and x_4, and among x_1, x_3 and x_4, which have positive effects on link 3 and 4 traffic flows. Interactions among x_1, x_2 and x_3, and among x_2, x_3 and x_4 have almost the same effect on all links' traffic flows. The interactions among x_1, x_2 and x_4, and among x_1, x_3 and x_4 have very small (positive or negative) effects on all links' traffic flows.

Figure 2*d* shows how the fourth-order index has positive effects on all links, with its largest effect on link 4's traffic flow and its smallest effect on the traffic flow of links 1 and 3. Figure 2*e* shows that the sum of the third-order index has non-positive effects on all links (zero for link 1, negative for other links). The sums of other order indices positively affect all links. Among all sums of the order indices with positive effects on all links, the sum of the fourth-order has the biggest effect on the traffic flow on links 1, 2, and 3, while the sum of the first-order has the biggest effect on the traffic flow on link 4. Figure 2*f* shows that the total-order indices positively affect all links. Inputs x_1 and x_3 have almost identical and the most impactful effects on all links' traffic flows, with x_2 coming next and x_4 last.

Figure 2*g* shows the effect of interactions associated with x_1, x_2, x_3, and x_4. The effect of interactions associated with every input is almost the same on the traffic flows on each link. Compared with the first-order indices in Figure 2*a*, the effect of first-order indices is less than that of interaction on the traffic flows on links 1, 2, and 3, while the effect of first-order indices is less than that of interaction on the traffic flows on link 4. The effect of interactions associated with x_2 and x_4 is almost same as that of the total-order indices on traffic flows on all links.

Finally, Figure 3 provides elasticities of first-order and total-order indices. They have the same trend with the first-order index and the total-order index in Figure 2*a* and *f*, which resulted from the same increasing of inputs of this test example.

MC for Example 1

MC was applied with the test example by randomly generating 16 different sets of model inputs and parameter values, and then solving the TDM for its equilibrium. In general, the number of simulation runs needs to be large enough to obtain robust and accurate results. Here, 16 different sets of inputs were randomly chosen, using the Example 1 distributions described earlier (for x_1 through x_4). Final link flows were obtained from the converged user equilibrium (UE) assignment results. The mean, SD and CoV for four links' traffic flows are computed as follows:

Link 1's mean (veh/day), SD and CoV values are 1006, 58.83 and 0.0584.

Table 1. OLS Parameter Estimates for Traffic Flows on Links (n = 16 Simulations)

	v_1		v_2		v_3		v_4	
Traffic flow	Coefficients	t Stat	Coefficients	t Stat	Coefficients	t Stat	Coefficients	t Stat
Intercept	285.1	12.3	−775.5	−14.6	1290	37.3	−3815	−93.9
x_1	0.1796	31.5	0.5088	38.9	0.0608	7.14	1.199	120.0
x_2	0.0916	9.76	0.2507	11.7	0.0549	3.92	0.6049	36.8
x_3	156.5	31.5	443.3	38.9	52.95	7.14	1045	120.0
x_4	−2934	−6.25	−7267	−6.76	−493.8	−0.705	−3899	−4.74
Adjusted R^2	0.9925		0.9950		0.8789		0.9995	

Link 2's mean (veh/day), SD and CoV values are 1279, 165.6 and 0.1294.

Link 3's mean (veh/day), SD and CoV values are 1569, 21.87 and 0.0139.

Link 4's mean (veh/day), SD and CoV values are 1307, 386.4 and 0.2957.

The CoVs of link 2 and 4 traffic flows are larger than the inputs' starting CoV values (of 0.10), suggesting that final flow uncertainties/variations can be compounded and end higher than input uncertainties. CoV values for link 1 and 3 flows are smaller than 0.06, which is lower than any inputs' uncertainty (as measured using CoV, or SD/mean). Moreover, the ratios of traffic flow versus capacity (v/c ratios) on links 1 through 4 are computed to be 0.84, 0.71, 0.78 and 0.44, respectively. The flow uncertainty appears not to have a strong relation with congestion, which is consistent with the result in (*24*).

For better understanding and interpretation of the results, ordinary least squares (OLS) regression was used to identify model inputs that are key contributors to uncertainty in model output (Table 1), assuming simple linear relationships between inputs (or combinations and transformations of model inputs, if one desires) and outputs. Only 16 input samples were used here, yet t-statistics and R2 values are very high in this toy network, as shown in Table 1.

BM for Example 2

Example 2. To illustrate how BM works, an integrated land-use and transportation model is presented, based on the following model equations:

$$x_1^{t_j} = (1+\delta)^{\Delta t} \quad x_1^{t_{j-1}} (j = 1, 2, \ldots), \text{where } \Delta t = t_j - t_{j-1} \tag{15}$$

$$Z^{t_j} = x_1^{t_j} x_2 + x_3 \tag{16}$$

$$Z_{12}^{t_j} = Z^{t_j} * \frac{e^{x_4 * GC_{12}}}{e^{x_4 * GC_{12}} + e^{x_4 * GC_{13}}} \tag{17}$$

$$Z_{13}^{t_j} = Z^{t_j} - Z_{12}^{t_j} = Z^{t_j} * \frac{e^{x_4 * GC_{13}}}{e^{x_4 * GC_{12}} + e^{x_4 * GC_{13}}} \tag{18}$$

$$GC_{12}^{t_j} = min\left\{LC_1^{t_j}, LC_2^{t_j}\right\} \tag{19}$$

$$GC_{13}^{t_j} = min\left\{LC_3^{t_j}, LC_4^{t_j}\right\} \tag{20}$$

$$LC_i^{t_j} = \gamma^* time^i + \text{Toll}^* t_f{}^i \tag{21}$$

where Z^{t_j} denotes number of trips generated at time t_j, $Z_{1k}^{t_j}$ denotes number of trips going from origin node 1 to the destination node k (k = 2,3) at time t_j, $GC_{1k}^{t_j}$ denotes generalized cost going from the node 1 to k at time t_j, $LC_i^{t_j}$ denotes generalized costs of each link i (i = 1,2,3,4) at time t_j, and $x_1{}^{t_j}$ denotes zone 1's population at time t_j. $x_1{}^{t_0}$ denotes the population of origin zone 1 at time t_0, with mean = 1000 persons, SD = 100 persons, and CoV = 0.10. δ denotes the annual percentage increase in population, with a mean of 2%, SD = 0.2% and CoV = 0.1. Other variables and value assumptions are the same as those provided above, in Example 1.

In this Example 2, Equation 15 is a simple LUM, while Equations 16–21 describe a simple TDM. The LUM reacts to travel network changes, and vice versa. However, the link from TDM to LUM is not as strong as the link from LUM to TDM because it is difficult to move one's home (and/or business) and costly to construct new buildings. Thus, here we investigate only the forward effect of land-use changes on the TDM outputs, rather than allow a feeding back of TDM outputs to LUM decisions. In other words, TT from the TDM's traffic assignment step feed forward into the subsequent year's LUM equation.

Here, 16 samples of random input values were chosen, and the model was run three times for each set of inputs (including all uncertain parameters). Therefore, I = 16, J = 3, and K = 4. Starting back in year t_0 = 2010, t_1 = 2015 served as the "present" year (for output validation, or comparisons to measured/regionally observed values),

Table 2. $\hat{\sigma}_i^2$ and ω_i Over All i (n = 16)

i	$\hat{\sigma}_i^2$	ω_i	i	$\hat{\sigma}_i^2$	ω_i
1	5,832	0.0017	9	3,433	0.0052
2	6,336	0.0014	10	3,748	0.0042
3	6,032	0.1630	11	1,0140	0.0014
4	5,419	0.3250	12	9,270	0.0197
5	2,182	0.0226	13	1,947	0.2650
6	2,303	0.0163	14	1,862	0.1490
7	13,883	0.0045	15	20,352	0.0014
8	12,827	0.0058	16	19,022	0.0017

Note: $\hat{\sigma}_i^2$ = the variance of model errors ε_{ik}; ω_i = the weight of conditional distribution.

and t_2 = 2020 served as the final prediction year. According to BM, ω_i can be computed using $p(\psi_k|X_i)$, where $(\psi_k|X = X_i) = \mu_{ik} + a + \varepsilon_{ik}$, and where ε_{ik} is independent and identically distributed (i.i.d.) as normal variable $N(0, \sigma_i^2)$. Here, μ_{ik} denotes the expected output, ε_{ik} denotes model errors, and a denotes the model's overall bias. μ_{ik}, $\hat{\sigma}_\delta^2$, $\hat{\sigma}_i^2$, and a can be estimated using approximate maximum likelihood methods (*39*). Thus, $\psi_k|X_i \sim N(\hat{a} + \hat{\mu}_{ik}, v_i)$ with $v_i = \hat{\sigma}_i^2 + \frac{\hat{\sigma}_\delta^2}{J}$, where $\hat{\mu}_{ik} = \frac{1}{J}\sum_j \Phi_{ijk}$, $\hat{\sigma}_\delta^2 = \frac{1}{IJK}\sum_{ijk}(\Phi_{ijk} - \hat{\mu}_{ik})^2$, $\hat{a} = \frac{1}{IK}\sum_{ik}(\psi_k - \hat{\mu}_{ik})$, and $\hat{\sigma}_i^2 = \frac{1}{K}\sum_k(\psi_k - \hat{a} - \hat{\mu}_{ik})^2$. Therefore, ω_i can be computed as follows:

$$\begin{aligned} \omega_i \propto p(\Phi|X_i) &= \prod_{k=1}^{K} p(\psi_k|X_i) \\ &= \prod_{k=1}^{K} \frac{1}{\sqrt{2\pi v_i}} exp\left[-\frac{1/2(y_k - \hat{a} - \widehat{\mu_{ik}})^2}{v_i}\right] \end{aligned} \quad (22)$$

where $\hat{a}$ and $\hat{\sigma}_\delta^2$ are computed to be 0.18 and 6.65, respectively. Table 2 provides estimates of ω_i and $\hat{\sigma}_i^2$. To compute the posterior distribution of link-level traffic flows, the propagation factors b_a and b_v were set to 5 years and 5 years, or $\frac{2015-2010}{2020-2015}$.

Posterior distributions of link-level traffic flows (Y_k values) are given by a mixture of normal distributions, as follows:

$$\begin{aligned} \pi(Y_k) &= \sum_{i=1}^{I} \omega_i N\left(\hat{a}b_a + m_{ik}, \left(\hat{\sigma}_i^2 + \frac{\hat{\sigma}_\delta^2}{J}\right)b_v\right) \\ &= N\left(\sum_{i=1}^{I} \omega_i(\hat{a}b_a + m_{ik}), \sum_{i=1}^{I} \omega_i\left(\hat{\sigma}_i^2 + \frac{\hat{\sigma}_\delta^2}{J}\right)b_v\right) \end{aligned} \quad (23)$$

where $m_{ik} = \frac{1}{J}\sum_{j=1}^{J} Y_{ijk}$.

Therefore, link k's traffic flow outputs will follow normal distribution: $N(\mu_k, \sigma^2)$, where $\mu_k = \sum_{i=1}^{I} \omega_i(\hat{a}b_a + m_{ik})$ and $\sigma^2 = \sum_{i=1}^{I} \omega_i\left(\hat{\sigma}_i^2 + \frac{\hat{\sigma}_\delta^2}{J}\right)b_v$. Using these values, one can obtain the following results: $\sigma^2 = 1110, \mu_1 = 912, \mu_2 = 862, \mu_3 = 1540$, and $\mu_4 = 646$.

In this way BM methods can provide model outputs (future-year predictions) by using present-year inputs. BM assesses output uncertainties for a land-use transport model by combining all available information on model inputs and outputs in a Bayesian way, to provide a posterior distribution of outputs.

Conclusion

In this paper, two small travel demand modeling examples (one with a land-use equation) were used to demonstrate how LSAI, MC, and BM methods of uncertainty characterization function and to illuminate their strengths and weaknesses. LSAI can provide the sign of change implied by the changes of inputs, the relative importance of change of inputs and the decomposition of change of output into the individual and interaction changes of inputs. Although LSAI needs just about 2 $\times K$ simulations to obtain first-order impact indices, a model's total-order indices and interactive effects associated with K (groups of) inputs require 2^K simulations, thus rising exponentially in respect to the number of exogenous (input) variables, to obtain the individual effects and all interaction effects.

Thus, LSAI will be effective only for problems with a small number of exogenous variables because it needs thousands of simulations for models with more than 10 inputs. Moreover, LSAI provides only point estimates, while MC and BM methods can provide distributions of needed outputs. MC involves random sampling of the distribution of inputs and successive model runs until obtaining a statistically significant distribution of outputs. MC can be used to solve problems with probability structures or non-probabilistic problems (such as finding the area under a curve), and can be used with time-series models (like land-use and travel changes, forward in time) or one-time-point models, where no validation data are needed. It is straightforward to obtain output distributions via MC random sampling of the distribution of inputs and parameters. However, such activities require hundreds of samples, even for small scale problems, to obtain accurate results. MC is especially useful for high-dimensional problems because its convergence and convergence rate do not depend on the problem's dimensionality and prediction errors depend only on the number of samples and the standard deviations of inputs.

While LSAI can only address deterministic problems, MC and BM methods are useful in solving both deterministic and stochastic problems. BM is a way of putting analysis of simulation models on a solid statistical basis. Its advantage is that it can obtain the posterior distribution of all model outputs from prior probability distributions and likelihoods of all model inputs (which include model parameters). Users are provided with probability intervals around forecasts with calibrated uncertainty statements, which add value to model validation, scenario comparison and external review and comment procedures. However, BM can be extremely computationally expensive, since it requires several hundred runs of the model. Moreover, intermediate outputs must also be known for intermediate validation. As a result, MC appears to be the best way for complex system modelers and planners to anticipate uncertainty in our urban systems' futures.

Acknowledgments

This paper was financially supported by the National Natural Science Foundation of China (71471167). The authors of this paper wish to thank Scott Schauer-West for his editing and administrative support, and several anonymous reviewers for their helpful suggestions.

Author Contributions

The authors confirm paper contributions as follows: study conception and design: KK and GW; data collection/model specifications: GW; analysis and interpretation of results: KK and GW; draft manuscript preparation: GW. All authors reviewed the results and approved the final version of the manuscript.

References

1. Rodier, C. J., and R. A. Johnston. Uncertain Socioeconomic Projections Used in Travel Demand and Emissions Models: Could Plausible Errors Result in Air Quality Nonconformity? *Transportation Research Part A: Policy and Practice*, Vol. 36, No. 7, 2002, pp. 613–631. https://doi.org/10.1016/S0965-8564(01)00026-X.
2. Sevcikova, H., M. Simonson, and M. Jensen. Assessing and Integrating Uncertainty into Land-Use Forecasting. *Journal of Transport & Land Use*, Vol. 8, No. 3, 2015, pp. 57–70. http://dx.doi.org/10.5198/jtlu.2015.614.
3. Pradhan, A., and K. Kockelman. Uncertainty Propagation in an Integrated Land Use-Transport Modeling Framework: Output Variation via UrbanSim. *Transportation Research Record: Journal of the Transportation Research Board*, 2002. 1805: 128–135.
4. Waddell, P. UrbanSim: Modeling Urban Development for Land Use, Transportation, and Environmental Planning. *Journal of the American Planning Association*, Vol. 68, No. 3, 2002, pp. 297–314. http://dx.doi.org/10.1080/01944360208976274.
5. Waddell, P., A. Borning, M. Noth, N. Freier, M. Becke, and G. Ulfarsson. Microsimulation of Urban Development and Location Choices: Design and Implementation of UrbanSim. *Networks and Spatial Economics*, Vol. 3, No. 1, 2003, pp. 43–67. https://doi.org/10.1023/A:1022049000877.
6. Miller, E. J., D. S. Kriger, and J. D. Hunt. Research and Development Program for Integrated Urban Models. *Transportation Research Record: Journal of the Transportation Research Board*, 1999. 1685: 161–170.
7. Dubus, I. G., C. D. Brown, and S. Beulke. Sources of Uncertainty in Pesticide Fate Modelling. *Science of the Total Environment*, Vol. 317, No. 1–3, 2003, p. 53. https://doi.org/10.1016/S0048-9697(03)00362-0.
8. Papoulis, A., and S. U. Pillai. *Probability, Random Variables, and Stochastic Processes*. McGraw-Hill Europe, 2002.
9. Sevcikova, H., A. E. Raftery, and P. A. Waddell. Assessing Uncertainty in Urban Simulations using Bayesian Melding. *Transportation Research Part B Methodological*, Vol. 41, No. 6, 2007, pp. 652–669. https://doi.org/10.1016/j.trb.2006.11.001.
10. Moore, R. E. *Interval Analysis*. Prentice Hall, Englewood Cliffs, NJ, 1966.
11. Klir, J. G., and B. Yuan. *Fuzzy Sets and Fuzzy Logic: Theory and Applications*. Prentice Hall, Englewood Cliffs, NJ, 1995.
12. Saltelli, A., M. Ratto, T. Andres, F. Campolongo, J. Cariboni, D. Gatelli, M. Saisana, and S. Tarantola. *Global Sensitivity Analysis: The Primer*. John Wiley & Sons Ltd, New York, NY, 2008.
13. Hamby, D. M. A Review of Techniques for Parameter Sensitivity Analysis of Environmental Models. *Environmental Monitoring & Assessment*, Vol. 32, No. 2, 1994, pp. 135–154. https://doi.org/10.1007/BF00547132.
14. Hamby, D. M. A Comparison of Sensitivity Analysis Techniques. *Health Physics*, Vol. 68, No. 2, 1995, pp. 195–204.
15. Kockelman, K. *Gravity-Based Land Use Model (G-LUM) Website*. http://www.ce.utexas.edu/prof/kockelman/G-LUM_Website/homepage.htm. Accessed July 10, 2017.
16. Borgonovo, E., and G. E. Apostolakis. A New Importance Measure for Risk-Informed Decision Making. *Reliability Engineering & System Safety*, Vol. 72, No. 2, 2001, pp. 193–212. https://doi.org/10.1016/S0951-8320(00)00108-3.
17. Borgonovo, E. Differential Importance and Comparative Statics: An Application to Inventory Management. *International Journal of Production Economics*, Vol. 111, No. 1, 2008, pp. 170–179. https://doi.org/10.1016/j.ijpe.2007.01.008.
18. Saltelli, A., and S. Tarantola. On the Relative Importance of Input Factors in Mathematical Models: Safety Assessment for Nuclear Waste Disposal. *Journal of the American Statistical Association*, Vol. 97, No. 459, 2002, pp. 702–709. http://dx.doi.org/10.1198/016214502388618447.
19. Borgonovo, E. Sensitivity Analysis with Finite Changes: An Application to Modified EOQ Models. *European*

Journal of Operational Research, Vol. 200, No. 1, 2010, pp. 127–138. https://doi.org/10.1016/j.ejor.2008.12.025.
20. Borgonovo, E., and L. Peccati. Managerial Insights from Service Industry Models: A New Scenario Decomposition Method. *Annals of Operations Research*, Vol. 185, No. 1, 2011, pp. 161–179. https://doi.org/10.1007/s10479-009-0617-1.
21. Borgonovo, E., M. Percoco, R. Polizzi, K. Kockelman, and L. Cavalli. *Sensitivity Analysis of a Gravity-Based Land Use Model: The Importance of Scenarios*. marcopercoco.files.wordpress.com/2014/11/glum-final.pdf. Accessed July 10, 2017.
22. Wang, G. M., and K. Kockelman. Local Sensitivity Analysis of Forecast Uncertainty in a Random-Utility-Based Multi-Regional Input-Output Model. *Journal of the Transportation Research Forum*, Vol. 55, No. 2, 2016, pp. 49–70.
23. Cheng, H. *Uncertainty Quantification and Uncertainty Reduction Techniques for Large-scale Simulations*. PhD dissertation. Department of Computer Science at Virginia Polytechnic Institute and State University, Blacksburg, VA, 2009.
24. Zhao, Y., and K. M. Kockelman. The Propagation of Uncertainty through Travel Demand Models: An Exploratory Analysis. *Annals of Regional Science*, Vol. 36, No. 1, 2002, pp. 145–163.
25. Krishnamurthy, S., and K. Kockelman. Propagation of Uncertainty in Transportation Land Use Models - Investigation of DRAM-EMPAL and UTPP Predictions in Austin, Texas. *Transportation Research Record: Journal of the Transportation Research Board*, 2003. 1831: 219–229.
26. Harvey, G., and E. Deakin. *Description of the STEP Analysis Package*. UC Berkeley Transportation Library, Berkeley, CA, 1996.
27. Thompson, D., M. Baker, and D. Wade. Conformity: Long-Term Prognoses for Selected Ozone Nonattainment Areas in California. *Transportation Research Record: Journal of the Transportation Research Board*, 1997. 1587: 44–51.
28. Saadi, I., A. Mustafa, J. Teller, and M. Cools. Forecasting Travel Behavior using Markov Chains-Based Approaches. *Transportation Research Part C: Emerging Technologies*, Vol. 69, 2016, pp. 402–417. https://doi.org/10.1016/j.trc.2016.06.020.
29. Clay, M. J., and R. A. Johnston. Multivariate Uncertainty Analysis of an Integrated Land Use and Transportation Model: MEPLAN. *Transportation Research Part D: Transport and Environment*, Vol. 11, No. 3, 2006, pp. 191–203. https://doi.org/10.1016/j.trd.2006.02.001.
30. Clay, M. J., A. Valdez, A. Norr, and S. M. Otterstrom. Uncertainty Analysis of the Large Zone Economic Module of the Simple, Efficient, Elegant, and Effective Model (SE3M) of Land Use and Transportation. *Transportation Planning and Technology*, Vol. 38, No. 5, 2015, pp. 855–874. http://dx.doi.org/10.1080/03081060.2015.1039231.
31. Pontius, R. G. Jr., and J. Spencer. Uncertainty in Extrapolations of Predictive Land-Change Models. *Environment and Planning B: Planning and Design*, Vol. 32, No. 2, 2005, pp. 211–230. http://journals.sagepub.com/doi/abs/10.1068/b31152.
32. Sevcikova, H., A. E. Raftery, and P. A. Waddell. Uncertain Benefits: Application of Bayesian Melding to the Alaskan Way Viaduct in Seattle. *Transportation Research Part A: Policy and Practice*, Vol. 45, No. 6, 2011, pp. 540–553. https://doi.org/10.1016/j.tra.2011.03.009.
33. Saltelli, A., S. Tarantola, F. Campolongo, and M. Ratto. *Sensitivity Analysis in Practice: A Guide to Assessing Scientific Models*. John Wiley & Sons, New York, NY, 2004.
34. Kalos, M. H., and P. A. Whitlock. *Monte Carlo Methods*. John Wiley & Sons, New York, NY, 2008.
35. Kroese, D. P., T. Brereton, T. Taimre, and Z. I. Botev. Why the Monte Carlo Method Is So Important Today. *Wiley Interdisciplinary Reviews: Computational Statistics*, Vol. 6, No. 6, 2014, pp. 386–392.
36. Chen, T. D., K. M. Kockelman, and Y. Zhao. What Matters Most in Demand Model Specifications: A Comparison of Outputs. *Journal of the Transportation Research Forum*, Vol. 54, No. 2, 2015, pp. 71–89.
37. Sheffi, Y. *Urban Transportation Networks: Equilibrium Analysis with Mathematical Programming Methods*. Prentice-Hall, Inc., Englewood Cliffs, NJ, 1985. https://doi.org/10.1016/0191-2607(87)90038-0.
38. Martin, W. *NCHRP Report 365: Travel Estimation Techniques for Urban Planning*. HRB, National Research Council, Washington, D.C., 1998.
39. Ziegler, A. *Generalized Estimating Equations*. Springer, New York, NY, 2011. https://doi.org/10.1007/978-1-4614-0499-6.

The Standing Committee on Transportation Demand Forecasting (ADB40) peer-reviewed this paper (18-00488).

Article

TRR
JOURNAL OF THE TRANSPORTATION RESEARCH BOARD

Transportation Research Record
2018, Vol. 2672(49) 23–33

DOI: 10.1177/0361198118776524
journals.sagepub.com/home/trr
SAGE

Modeling Car Drivers' On-Street Parking Decisions using the Integrated Hierarchical Information Integration Approach

Annum Khaliq[1], Peter van der Waerden[2], and Davy Janssens[1]

Abstract

Rising issues in urbanization and transportation urge municipalities to optimize the use of on-street parking spaces in order to meet local needs and complement the role of available off-street parking. In this paper car drivers' parking decisions have been investigated using a stated choice experiment, based on the method of integrated hierarchical information integration. According to this approach, a large set of parking related attributes and attribute levels are grouped into two higher order decision constructs, which are presented as hypothetical street segments. The respondents were asked if they would park their car in the street segment with the listed attributes. The collected data is used to estimate the parameters of a standard multinomial logit model. This study differs from previous studies as a large range of attributes is examined, including the parking situation and the road conditions in a street segment along with some features of off-street parking facilities present in the vicinity of the street segment. The results indicate that the contextual variables such as 'walking distance to destination' and 'parking cost' are key attributes that car drivers consider while making on-street parking decisions, while street-level attributes such as 'occupancy,' 'security,' and 'surrounding activities' seem to have only a minor impact. The study concludes with an outlook of how these insights into car drivers' parking choice process can be used by local authorities to reduce cruising in urban areas. Moreover, these findings can be integrated in multi-agent systems to investigate car drivers' movements in urban areas.

Local area parking management policies affect the amount and location of parking supply, access to parking spaces, and parking tariffs (*1*). Parking spaces are increasingly becoming a premeditated commodity and an over-consumed public good in urban areas (*2*). The challenge of managing parking spaces in the urban areas of a multi-modal transportation system is a critical issue for transport analysts and urban planners. The use of the urban road network for parking purposes is a very convenient mean for parking supply. With a variety of planning measures, local governments are trying to optimize the use of the urban road network for extending the parking supply in their cities. These measures are also used to manage on-street parking, as it affects the traffic situation of the city directly. In addition, dealing with underutilization of off-street parking facilities due to high levels of cruising for cheap on-street parking is a big challenge for local authorities concerning parking management in metropolitan cities (*3*, *4*). To achieve these goals, the factors that influence car drivers' on-street parking choice behavior need to be identified.

The relationship between parking supply and demand has been investigated in the existing literature using various approaches (*5–13*). In behavioral research significant attention has been paid to car drivers' parking choice behavior, as parking choice has a high impact on travel behavior of individuals (such as mode choice and route choice). It is considered crucial for parking management, which is an element of broader transport demand management strategy. Awareness of parking choice process enables implementation of measures to address issues created by car drivers cruising for parking (*14*). In the past, several studies of car drivers' preferences regarding parking in the urban transport system have been presented. The decision whether or not to choose on-street parking is quite complex. However, the collection of data to infer the decision making process of car drivers is a challenge. Several previous studies have

[1]UHasselt-Hasselt University, Transportation Research Institute (IMOB), Agoralaan, Diepenbeek, Belgium
[2]Faculty of the Built Environment, Eindhoven University of Technology, Eindhoven, MB, Netherlands

Corresponding Author:
Address correspondence to Annum Khaliq: annum.khaliq@uhasselt.be

used the stated choice approach for identifying parking related attributes to both on-street and off-street parking facilities (*15–24*). Similar to off-street parking, there are several factors that influence car drivers' on-street parking choice decisions. Only a few studies investigated on-street parking decisions of car drivers using stated choice approach: (*24*) and (*25*). The major focus of the above-mentioned studies was to investigate a limited set of factors related to the parking context as the traditional stated choice experiments can handle only a limited number of attributes. Therefore, the integrated hierarchical information integration (HII) approach has been employed to investigate the effect of a larger set of parking and street related attributes on parking decisions. This large set of parking related attributes enhances the possibilities for the local authorities to design parking policies efficient enough to cater to the problems related to on-street parking such as illegal parking and double parking (*26*).

This paper presents an on-street parking choice analysis designed using the integrated HII method proposed by Louviere (*27*). In this approach, it is assumed that people make complex decisions by cognitively grouping decision attributes into rational subsets. These subsets of attributes can be categorized into decision constructs. The decision constructs depict descriptive summary labels for coherent sets of attributes that define alternatives in the study. In principle, each decision construct is evaluated separately. For the current study, the evaluation of each decision construct is integrated in a single choice model. The integrated choice model estimated from concatenated sub-experiments can be used to predict the utility of alternatives. Thus, the overall preference or choice is determined by integrating the evaluations of each decision construct. In this way, the respondents use the high-order decision constructs to simplify the evaluation of multi-attribute alternatives. The main advantage of this analysis is that participants are forced to make trade-offs in their decisions; that is, respondents are forced to make inferences about omitted attribute levels to analyze whether any attribute level would lead to rejection of an alternative. Therefore, it enables the researcher to model the human decision-making process in a realistic manner. HII has also been applied to other disciplines, such as telecommunications (*27*), out-door recreation (*28*), residential choice (*29, 30*), health-care (*31*) and freight mode choice (*32*). In the field of transportation, this approach has been applied for mode choice modeling in public transport (*33*) and choice of park-and-ride facilities (*34*). To the best of our knowledge, this is the first application of HII to model car drivers' on-street parking choice behavior. The practical implication of this approach is that many attributes related to street parking can be used by the city managers to reduce parking-search traffic by identifying preferable on-street parking conditions and stimulate use of off-street parking. This increased knowledge of factors influencing car drivers' parking decisions can enhance urban parking policy applications with associated environmental and economic benefits (*14*). However, the current study aims to detail the process of car drivers' parking decisions in multi-agent systems (such as *4*) to make them more sophisticated as it is difficult to predict drivers' parking choices. To achieve the stated goal, the remainder of this paper is organized as follows. First, attention is paid to previous parking studies that include parking related attributes. Next, the adopted research approach is outlined. This section is followed by a brief description of the data collection and sample composition. The analyses are described in the subsequent section. The paper ends with the conclusions and suggestions for future research.

Related Studies

Individuals' parking choice decisions are complex and multi-faceted; therefore, the process of selecting a parking space has been studied through several decades using different choice modeling approaches. Numerous studies have been conducted to analyze off-street parking choice behavior. These studies highlight the most important factors that influence car drivers' parking decisions (parking costs, parking type, search time, parking duration, trip purpose, walking distance to destination, etc.). Stated choice experiments have been used in several parking choice studies to highlight the significant attributes/factors that influence off-street parking choice behavior of car drivers. Axhausen and Polak (*15*) used stated choice approach and identified the significance of access time, search time, egress time, parking fee, expected illegal parking fine using a multinomial logit (MNL) model. Van der Waerden and Oppewal (*16*) used stated choice technique to identify car drivers' parking preferences with respect to off-street parking and found that parking cost and walking distance are the most significant attributes in the context of shopping trips. MuConsult (*17*) investigated the relationship between walking distance and willingness to pay for parking using stated choice approach. Hensher and King (*18*) investigated factors related to off-street parking such as hours of operation, a tariff schedule, and access time to the final destination using stated choice approach. Bonsall and Palmer (*19*) investigated factors such as parking fee, walking time, number of spaces, number of car parks passed, last car park used, income, and gender using stated choice approach in the context of business and shopping trips. Hess and Polak (*20*) used stated choice approach to determine car drivers' parking choice in the context of work and shopping trips. The factors included in this study were access time, search time, egress time, parking fee, and expected fines for illegal parking. Golias et al. (*21*) identified that most time related factors such as search time for a parking space, duration of parking, and walking time from the parking space to the final destination have a significant impact on parking choice. Van der Waerden et al. (*22*) investigated attitudes and preferences of car drivers with respect to a set of possible parking measures.

A limited number of studies focused on car drivers' on-street parking choice decisions. Clinch and Kelly (*24*) used stated choice to identify car drivers' parking preferences in the context of on-street parking. The factors studied include walking distance between parking and home, visibility of the car, motorized traffic in residential street, and security. A similar approach was adopted by Chaniotakis and Pel (*25*). They designed a stated choice experiment to examine the significance of factors such as parking type, walking distance to destination, travel time, and parking fees related to on-street parking decisions of car drivers. Findings of all these studies show that major factors affecting choice of a parking space are parking cost and walking distance to destination, but all of these studies focused on investigating a limited number of attributes related to parking space rather than attributes related to the parking situation and road conditions of the street, such as speed limit, surrounding activities, available parking spaces, and so forth. This is important because a limited set of attributes provide limited possibilities for urban planners to devise effective policy measures, therefore a large set of parking related attributes are required to widen the range of parking policies. This lack of insights resulted in the present study.

Research Design

The methodology adopted for this research can be explained as follows. First, the relevant attributes and attribute levels were identified related to a regular street segment. The selection of attributes was done through rigorous effort. Initially, a list containing 25 attributes related to on-street parking was prepared based on an extensive literature review. In order to reduce the burden on respondents, the number of identified attributes was lowered to 15. The identified attributes were then grouped into two separate decision constructs (categories): parking situation and road conditions. The attribute levels were defined keeping in view the literature and common practice in Belgium. The structure of HII is presented in Figure 1.

Different experimental designs were developed for each construct such that each design included a detailed description of one decision construct while the other decision construct was added to the design as a 'global attribute.' This process makes each choice task quite detailed, consisting of 11 attributes in total: seven general parking related attributes, three attributes of detailed construct, and one global attribute (high-order description of decision construct treated as an additional attribute, replacing three detailed attributes). The first five attributes are related to on-street parking in general, and include on-street parking costs, walking distance between parking location and final destination, expected parking duration (the length of time expected by the driver to park the car), modes of payment available for paying on-street parking costs, and the number of streets visited by the driver before entering the street segment. The other two attributes are related to the closest off-street parking facility: off-street parking tariffs and walking distance between off-street parking facility and final destination. The decision construct 'road conditions' includes attributes that explain in more detail the condition of the street segment in terms of average number of parking spaces available in the street segment, surrounding activities indicating the type of land-use present in the street segment, and the allowed speed limit in the street segment. The decision construct 'parking situation' includes the attributes occupancy rate, presence of security features (i.e., CCTV cameras and guards), and the maximum parking duration allowed in the street segment. The global attribute was listed as 'level of convenience' represented by three levels (high, medium, and low). These levels depict the convenience (ease) of parking in the street segment. Level of convenience explains the contribution of decision construct in overall judgement of a given situation (i.e., the respondent has to consider if level of convenience provided by favorable road conditions affect parking decisions significantly or not). For example, in the construct 'parking situation' a 'high convenience' means that the parking situation is very favorable for parking, meaning that the street segment has a low occupancy rate with multiple payment options. A level of 'low convenience' indicates that the street segment is not favorable for parking referring to limited parking duration allowance or high occupancy rate. A similar reasoning is valid for the other decision construct. The basic logic to list the global attributes as decision variables is to assess if 'road conditions' or 'parking situation' as a whole affect parking decisions. Some studies in the literature include a similar approach (*31*, *33*). The combinations of attribute levels and decision construct levels are presented as choice profiles (tasks) to the respondents. The responses obtained from the experiment are analyzed using MNL regression analysis. Furthermore, the parameters of all attributes are estimated simultaneously.

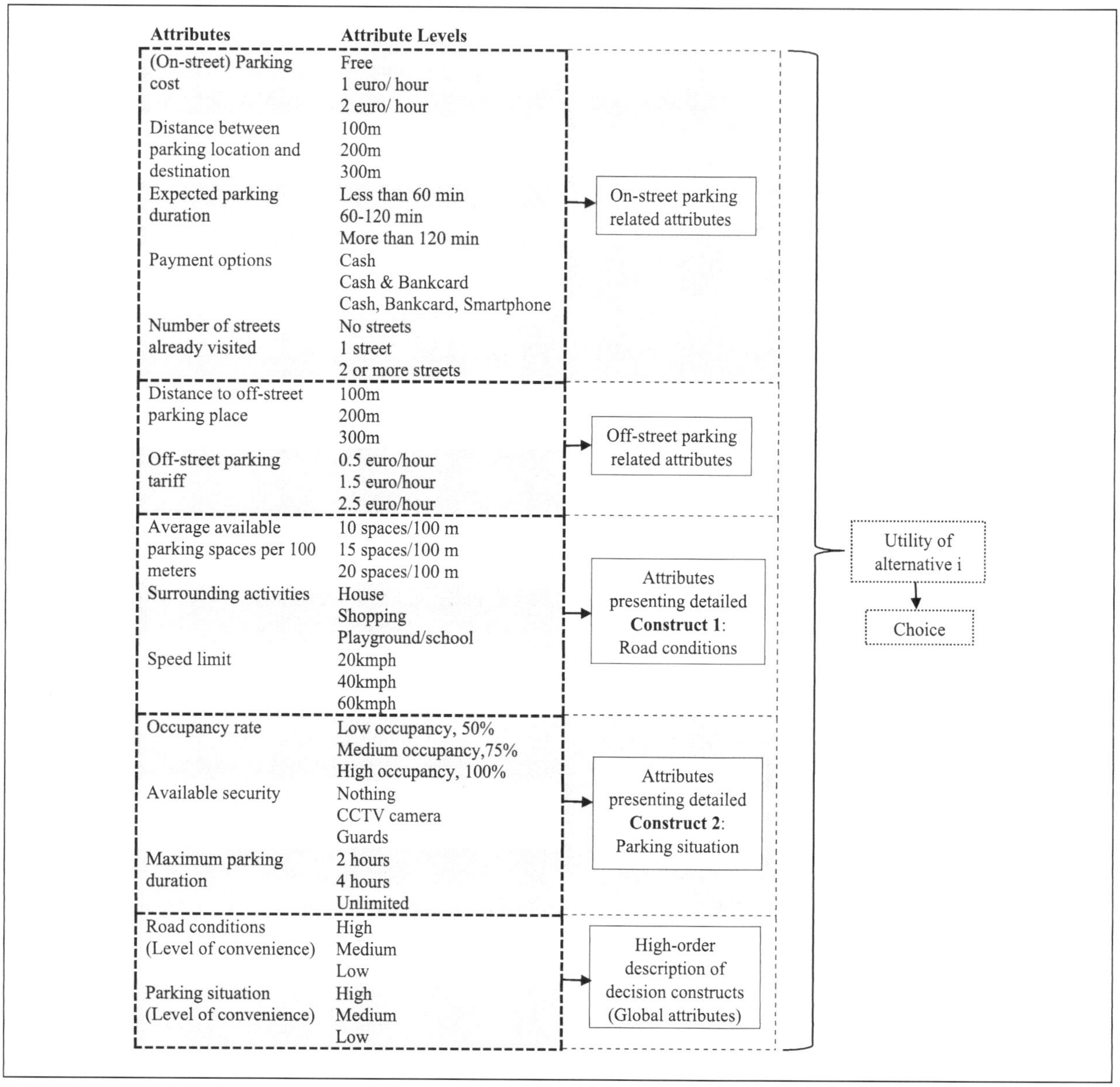

Figure 1. Overview of attributes and constructs.

Questionnaire Design

An online questionnaire was designed for collecting the stated choice data. The questionnaire consisted of three parts and was constructed using an online system developed by the Eindhoven University of Technology (see Figure 2). Two screening questions were asked at the beginning of the questionnaire, to select respondents who owned one or more cars and had some experience with on-street parking in urban areas. The first section of the questionnaire was related to car drivers' experiences with on-street parking. This section included questions about on-street parking issues faced by the respondents, search time, parking purpose, parking frequency, walking time between parking place and final destination, parking information sources, preference for on-street and off-street parking, and which street they find most difficult to park at. The second section consists of the hypothetical choice situations. At the beginning of each section a detailed explanation was provided to prepare the respondents for the forthcoming questions. The context of parking decision was mentioned in the explanation of section 2. The respondents were asked to assume that he/she is driving from his/her home to a destination (generic)

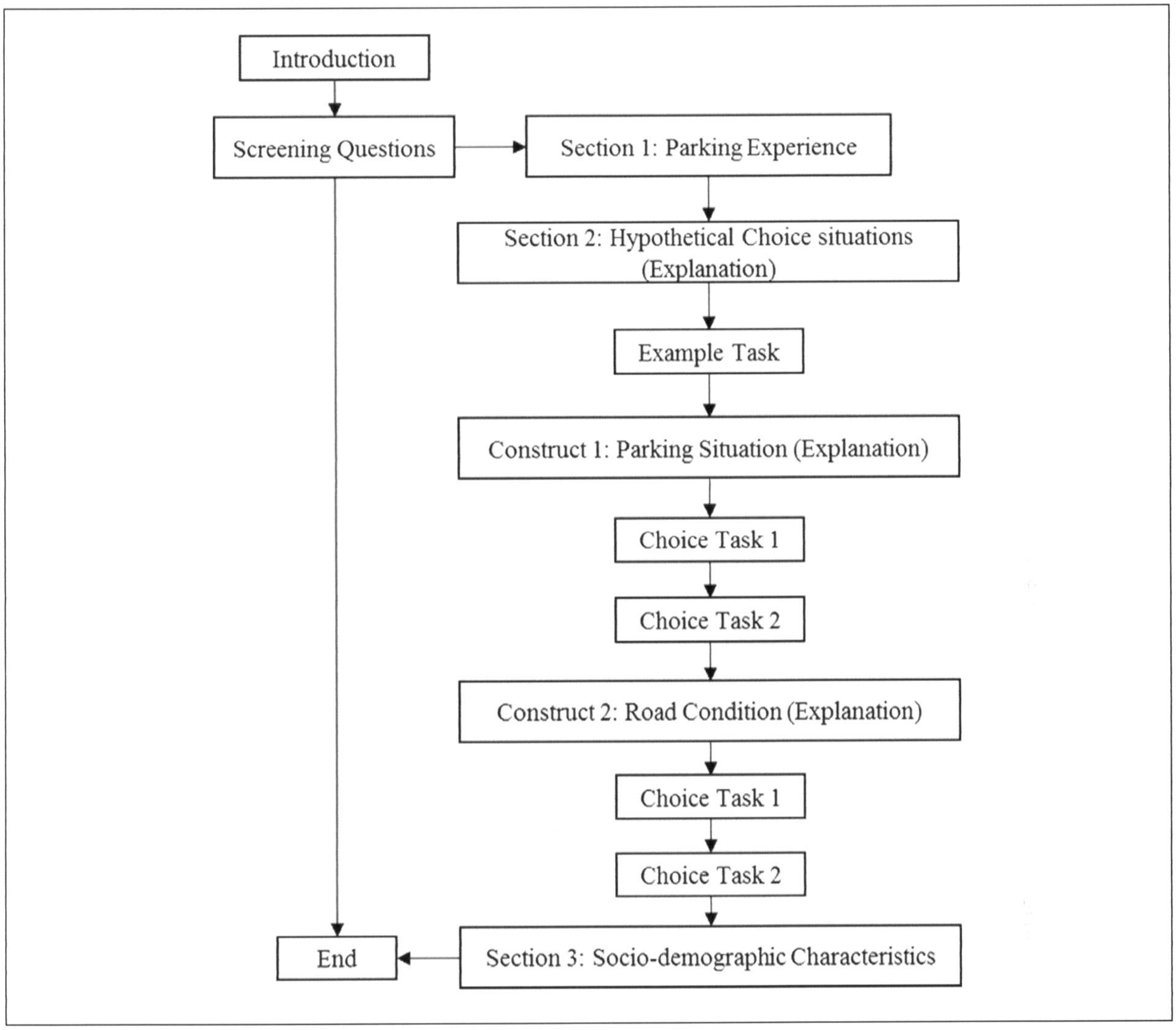

Figure 2. Questionnaire structure and routing.

located in the inner city, while driving he/she enters a street segment to search for a suitable parking space (see Figure 3). The choice tasks were designed using a 3^{11} fractional factorial design. Only main effects were considered in this experiment to limit the number of generated hypothetical profiles. In total, 27 efficient profiles were prepared using SAS %Mktex macros (%Mktex macros are used to create efficient factorial designs in SAS) (*35*). These profiles show a well-balanced subtraction from the full factorial design. The choice task consisted of the presented on-street parking context attributes, a detailed description of one decision construct, and a global attribute representing other decision construct. Per construct, two choice tasks were presented to each respondent; this makes a total of four choice tasks per respondent. The details of the hypothetical choice situation are presented in the form of choice tasks (profile consisting attributes and corresponding attribute levels). For making the choice task more understandable for the respondents (*36*), it was decided to include some pictures and colored symbols in the task. The basic aim of adding pictures/symbols was to reduce the complexity of the task for the respondents. Before presenting the choice tasks to the respondents, a detailed explanation of attributes, attribute levels, and constructs along with the pictures used in the choice tasks (see Figure 3) was included in the questionnaire. A hypothetical scenario is described at the beginning of the experiment. A small description of the context in which the parking decision has to be made is provided in the explanation section of the questionnaire. In the explanation section, it is mentioned that the focus of this research is on short-term trips to an inner city (city

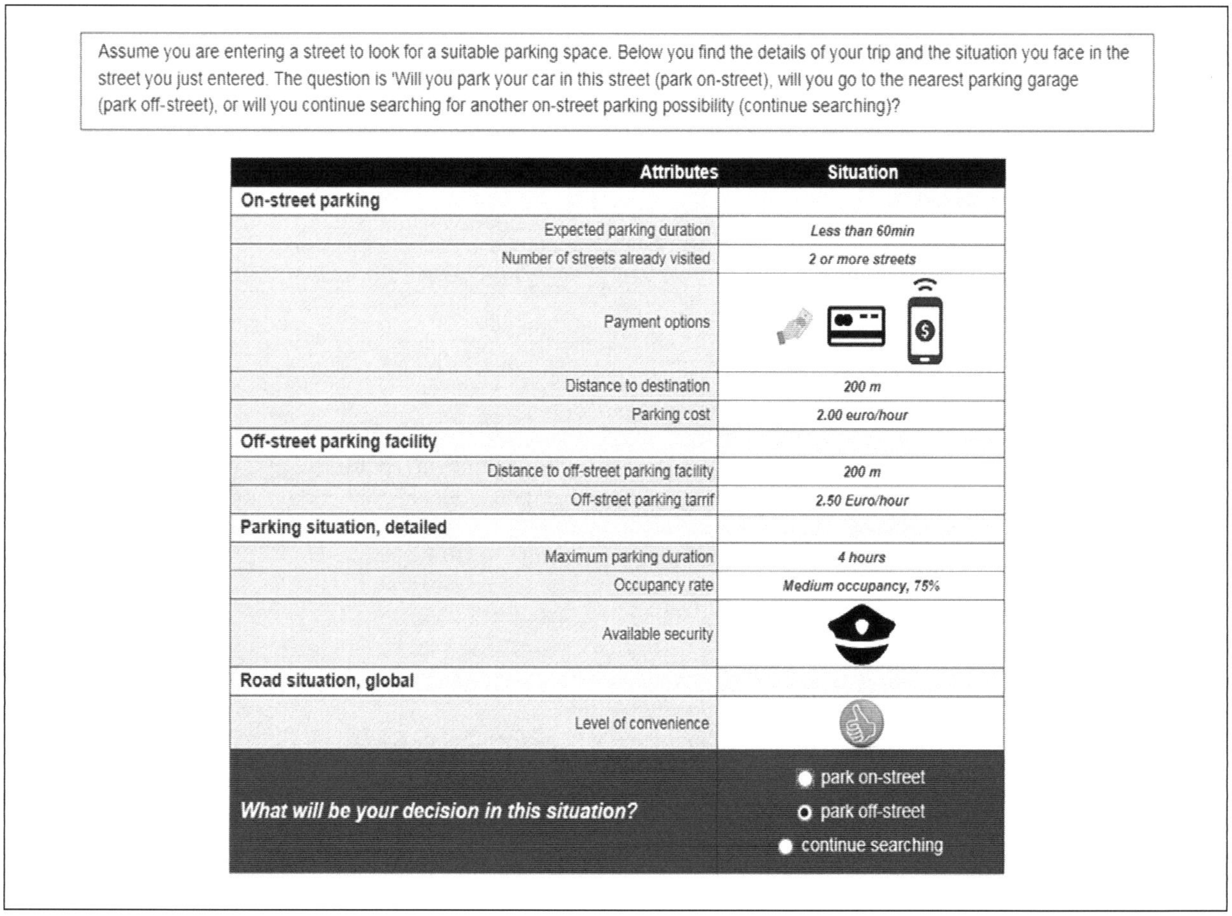

Assume you are entering a street to look for a suitable parking space. Below you find the details of your trip and the situation you face in the street you just entered. The question is 'Will you park your car in this street (park on-street), will you go to the nearest parking garage (park off-street), or will you continue searching for another on-street parking possibility (continue searching)?

Attributes	Situation
On-street parking	
Expected parking duration	*Less than 60min*
Number of streets already visited	*2 or more streets*
Payment options	
Distance to destination	*200 m*
Parking cost	*2.00 euro/hour*
Off-street parking facility	
Distance to off-street parking facility	*200 m*
Off-street parking tarrif	*2.50 Euro/hour*
Parking situation, detailed	
Maximum parking duration	*4 hours*
Occupancy rate	*Medium occupancy, 75%*
Available security	
Road situation, global	
Level of convenience	
What will be your decision in this situation?	park on-street park off-street continue searching

Figure 3. Example of presented choice task.

center) without considering any time pressure. At the end of the task the respondent has to indicate her/his decision: park on-street in the presented street segment (parking on-street), go to the nearest off-street parking facility (park off-street), or continue searching for another on-street parking possibility (continue searching). The last section of the questionnaire was related to personal characteristics (age, income, gender, education) of the respondents.

Model Specification

The car drivers' choices are analyzed using the standard MNL model. Traditionally, the MNL model has been used to analyze discrete choices. The MNL model assumes that the random components are independently and identically double exponential (or Gumbel) distributed. The double exponential distribution is convenient because, in contrast to the normal distribution, it leads to a closed, therefore tractable, model form as presented below, for example. (*37*).

$$P_{qi} = \frac{e^{\mu V_{qi}}}{\sum_{j}^{J} e^{\mu V_{qj}}} \tag{1}$$

where:

P_{qi} is the probability that alternative i is chosen by individual q from a set of J alternatives;

V_{qi} is the systematic (or representative) utility of alternative i (see Equation 2);

μ is a scale parameter, usually assumed to be equal to 1.0.

$$V_{qi} = \sum_{k} \beta_k . X_{qik} \tag{2}$$

where:

x_{qik} represents the value of attribute k of alternative i for individual q,

β_k is a parameter indicating the contribution of attribute k to the utility of each alternative.

The model parameters are estimated using NLOGIT version 5. The choices of car drivers (park on-street, park

Table 1. Personal Characteristics of Research Sample (N = 548)

Characteristics	Level	Frequency	Percentage
Gender	Male	332	61
	Female	216	39
Age	Less than 45 years	261	48
	More than 45 years	287	52
Education	Medium	289	53
	High	259	47
Income	Less than 5,000 euros	285	52
	More than 5,000 euros	263	48
Total		548	100

off-street, and continue searching) are used as a dependent attribute. The alternative 'continue searching' was used as a base alternative. In this study, dummy coding is used to represent all effects of the attribute levels. Therefore, part-worth utilities of all base attribute levels is zero.

Data Collection

The data used to estimate the model were collected using members of the marketing panel of PanelClix (www.panelclix.be). The aim of the panel is to create a platform for members willing to participate in different marketing related studies. In April 2017, Belgian members of the panel were invited to fill out the online questionnaire. In total, 548 respondents completed the questionnaire. The characteristics of the respondents are presented in Table 1. Most of the respondents in the sample have a medium education level and an income lower than 5,000 euros. The sample characteristics show that the group of respondents is representative of the reference population.

Analysis and Results

Table 2 presents the results of the model estimation. The log-likelihood ratio statistic (LRS) was used to test the estimated model against the null model (a model with all coefficients equal to zero resulting in equal probabilities for all included alternatives). The test of the LRS-value indicates that the estimated model performs significantly better than the null model (*38*). The LRS-value is equal to 999.70 while the critical chi-square value for 62 degrees of freedom is equal to 499.852 at the confidence level of 95 percent. The model has a McFadden's Rho-square value of 0.275, which indicates that the estimated model is well able to predict the observed choices. Only a few attributes are significant at the conventional level (95 percent), such as 'on-street parking cost,' 'off-street parking tariff,' 'expected parking duration,' 'maximum parking duration,' 'surrounding activities,' and 'speed limit.' The significant parameters are in anticipated direction. This result might indicate that car drivers do not consider 'available security' or 'payment options' while looking for a parking space on-street. The negative signs of the estimated parameters (e.g., in the case of parking cost and occupancy rate), suggest that low cost and lower occupancy rate stimulate car drivers to park on-street. Similarly, 'average available parking spaces per 100 meters,' 'number of streets already visited,' and 'off-street parking tariff' have negative signs. Therefore, the probability of considering a street segment for parking on-street increases with more spaces available per 100 meters, fewer number of streets visited, and at a higher off-street parking tariff. The model also predicts that the decision to park off-street depends on the 'parking cost' of the off-street parking facility, 'maximum parking duration,' and the 'surrounding activities' in the street segment. According to the parameter estimates, if shops are located in the street segment, the maximum parking duration is 2 hours, and the off-street parking tariff is low, then the probability that a car driver parks off-street increases. Moreover, the global attribute representing decision constructs do not show any influence on car drivers' parking choice. All the significant attributes are presented in bold. In general, it can be concluded that all part-worth utilities are in the anticipated direction, giving face validity to the estimated model.

Conclusion

In this paper, the car drivers' preferences for on-street parking are outlined by applying an integrated HII approach. A stated choice experiment was formulated to investigate the influence of various attributes related to both parking situation and road conditions of street segments on car drivers' on-street parking choice decisions. An integrated choice model (both constructs estimated jointly) was estimated to identify the part-worth utilities and significance levels of attributes using a standard MNL model. A large set of insignificant attributes show that respondents are focusing on a limited set of attributes (the significant ones) while making parking

Table 2. Estimated Part-Worth Utilities

Choice	Attributes	Level values	Part-worth utilities	Significance
Park on-street		On-street parking facility		
	Intercept		1.2577	0.000
	On-street parking cost	Free	0.4047	0.039
		1.00 euro/hour*	0**	
		2.00 euro/hour	−0.6400	0.001
	Distance between parking location and destination	100 m	0.1283	0.488
		200 m	0.2594	0.186
		300 m*	0	
	Expected parking duration	Less than 60 min	−0.4156	0.033
		60–120 min*	0	
		More than 120 min	−0.1976	0.3275
	Payment options	Cash	0.0237	0.900
		Cash & bankcard	0.1454	0.447
		Cash, bankcard, & smartphone *	0	
	Number of streets already visited	No streets	−0.0686	0.536
		1 street	0.1197	0.719
		2 or more streets*	0	
		Off-street parking facility		
	Distance to off-street parking place	100 m	0.0536.	0.777
		200 m	0.1224	0.522
		300 m*	0	
	Off-street parking tariff	0.50 euro/hour	0.0834	0.668
		1.50 euro/hour	−0.0510	0.770
		2.50 euro/hour*	0	
		Road conditions		
	Average available parking spaces per 100 meters	10 spaces/100 m*	0	
		15 spaces/100 m	−0.459	0.857
		20 spaces/100 m	0.0054	0.983
	Surrounding activities	House	−0.0426	0.865
		Shopping	0.3628	0.163
		Playground/school*	0	
	Speed limit	20 km/h	0.5208	0.042
		40 km/h	0.3825	0.123
		60 km/h*	0	
		Parking situation		
	Occupancy rate	Low occupancy, 50%	0.4080	0.102
		Medium occupancy, 75%*	0	
		High occupancy, 100%	−0.1182	0.625
	Available security	Nothing	0.3690	0.144
		CCTV camera*	0	
		Guards	0.0019	0.994
	Maximum parking duration	2 hours	−0.0937	0.175
		4 hours	0.3496	0.698
		Unlimited*	0	
		Decision construct levels		
	Level of convenience for parking situation	Low	−0.3366	0.158
		Medium*	0	
		High	0.2948	0.255
	Level of convenience for road condition	Low	−0.2709	0.265
		Medium*	0	
		High	0.1427	0.603
		Off-street parking facility		
Park off-street	Intercept		0.4177	0.267
	On-street parking cost	Free	−0.3718	0.071
		1 euro/hour*	0	
		2 euro/hour	−0.0174	0.926
	Distance between parking location and destination	100 m	−0.0916	0.627
		200 m	0.0948	0.633
		300 m*	0	

(continued)

Table 2. Continued

Choice	Attributes	Level values	Part-worth utilities	Significance
	Expected parking duration	Less than 60 min	−0.5102	0.010
		60–120 min*	0	
		More than 120 min	−0.2991	0.145
	Payment options	Cash	0.0891	0.644
		Cash & bankcard	0.0475	0.809
		Cash, bankcard, & smartphone*	0	
	Number of streets already visited	No streets	−0.1215	0.532
		1 street	0.0471	0.812
		2 or more streets*	0	
		Off-street parking facility		
	Distance to off-street parking place	100 m	0.2407	0.212
		200 m	0.1809	0.357
		300 m*	0	
	Off-street parking tariff	0.5 euro/hour	1.3776	0.000
		1.5 euro/hour	0.7505	0.000
		2.5 euro/hour*	0	
		Road conditions		
	Average available parking spaces per 100 meters	10 spaces/100 m*	0	
		15 spaces/100 m	0.0109	0.967
		20 spaces/100 m	−0.0475	0.857
	Surrounding activities	House	0.4438	0.085
		Shopping	0.5200	0.054
		Playground/school*	0	
	Speed limit	20 km/h	0.1672	0.524
		40 km/h	0.0029	0.991
		60 km/h*	0	
		Parking situation		
	Occupancy rate	Low occupancy, 50%	0.2207	0.393
		Medium occupancy, 75%*	0	
		High occupancy, 100%	0.1234	0.615
	Available security	Nothing	0.1622	0.532
		CCTV camera*	0	
		Guards	0.0642	0.793
	Maximum parking duration	2 hours	0.0945	0.700
		4 hours	0.4627	0.079
		Unlimited*	0	
		Decision construct levels		
	Level of convenience for parking situation	Low	−0.1014	0.679
		Medium*	0	
		High	0.2604	0.330
	Level of convenience for road condition	Low	0.0938	0.707
		Medium*	0	
		High	0.1892	0.507
Goodness-of-fit				
Log-likelihood of the null model, LL(0)		–2,416.947		
Log-likelihood of the optimal model, LL(B)		–1,917.095		
LRS=-2[LL(0)-LL(B)]		999.704 (df = 62)		
McFadden's Rho-Square MNL		0.275		
McFadden's adjusted Rho-Square MNL		0.264		

*Base attribute level. **Part-worth utilities of dummy base levels.

decisions. This means that the insignificant attributes do not actually affect car drivers' parking decisions. It is important to mention the insignificant attributes because the part-worth utilities are necessary to describe the effect of each attribute level on choice. The results of applying the HII method show that if speed limit is decreased (e.g., 20 km/h) car drivers will tend to park on-street rather than cruising for parking. Other attributes (e.g., security and payment options) seem to be less important in respondents' evaluation of road conditions.

The results also indicate that besides parking cost, expected parking duration, maximum parking duration, speed limit, and surrounding activities also play a considerable part when determining car drivers' parking preferences. The part-worth utilities of the attributes in the choice model show that car drivers are more willing to park on-street if the expected parking duration is low (e.g., less than 60 minutes). Car drivers will prefer to park off-street if surrounding activities in the street include shops or residences.

Moreover, this approach highlights preferable on-street parking conditions that can help planners to reduce cruising for parking. A large set of parking related attributes broaden the range of possible parking policies, as parking choice has wide implications for parking and traffic policy. Nowadays, municipalities are employing parking policies that restrict streets for on-street parking and direct car drivers to other streets or off-street parking facilities. Such policies are inducing search traffic as more time is required by car drivers to find an optimal parking place and also more time is required to walk from selected parking space to final destination. Therefore, this approach can identify which other measures cities can take to reduce such issues. Moreover, this approach reduces respondents' bias as a large set of attributes divides respondents' attention to several parking related attributes. It is recommended to set up more studies where this approach can be employed to produce better results. The results of the study can also be used to develop more sophisticated agent-based parking models.

Author Contributions

The authors confirm contributing to the paper as follows: study conception and design: Annum Khaliq, Peter van der Waerden, Davy Janssens; data collection: Annum Khaliq, Peter van der Waerden; analysis and interpretation of results: Annum Khaliq, Peter van der Waerden, Davy Janssens; draft manuscript preparation: Annum Khaliq, Peter van Waerden. All authors reviewed the results and approved the final version of the manuscript.

References

1. McShane, M., and M. D. Meyer. Parking Policy and Urban Goals: Linking Strategy to Needs. *Transportation*, Vol. 11, No. 2, 1982, pp. 131–152.
2. Broaddus, A. No Such Thing as a Free Spot: Pricing a Demand Management Strategy for Parking. Presented at the Association of European Schools of Planning (AESOP), Naples, Italy, 2007.
3. Shoup, D.C. Cruising for Parking. *Transport Policy*, Vol. 13, No. 6, 2006, pp. 479–486.
4. Martens, K., I. Benenson, and N. Levy. The Dilemma of On-Street Parking Policy: Exploring Cruising for Parking Using an Agent-Based Model. In *Geospatial Analysis and Modelling of Urban Structure and Dynamics* (B. Jiang, and X. Yao, eds.), Springer, New York, 2010, pp. 121–138.
5. Arnott, R., and J. Rowse. Modeling Parking. *Journal of Urban Economics*, Vol. 45, No. 1, 1999, pp. 97–124.
6. Eldin, N., and H. K. Ismail. A Combined Parking and Traffic Assignment Model. *Traffic Engineering and Control*, Vol. 22, No. 10, 1981, pp. 524–530.
7. Thompson, R. G., and A. J. Richardson. A Parking Search Model. *Transportation Research Part A: Policy and Practice*, Vol. 32, No. 3, 1998, pp. 159–170.
8. Leephakpreeda, T. Car-Parking Guidance with Fuzzy Knowledge-Based Decision Making. *Build Environment*, Vol. 42, No. 2, 2007, pp. 803–809.
9. Benenson, I., K. Martens, and S. Birfir. PARKAGENT: An Agent-Based Model of Parking in the City. *Computers, Environment and Urban Systems*, Vol. 32, No. 6, 2008, pp. 431–439.
10. Steenberghen, T., K. Dieussaert, S. Maerivoet, & K. Spitaels. SUSTAPARK: An Agent-based Model for Simulating Parking Search. *Journal of the Urban & Regional Information Systems Association*, Vol.24, No.1, 2012, pp. 63–76.
11. Waraich, R., & Axhausen, K. Agent-Based Parking Choice Model. *Transportation Research Record: Journal of the Transportation Research Board*, 2012. 2319: 39–46.
12. Gallo, M., L. D'Acierno, and B. Montella. A Multilayer Model to Simulate Cruising for Parking in Urban Areas. *Transport Policy*, Vol. 18, No. 5, 2011, pp. 735–744.
13. Guo, L., S. Huang, J. Zhuang, and A. W. Sadek. Modeling Parking Behavior Under Uncertainty: A Static Game Theoretic versus a Sequential Neo-additive Capacity Modeling Approach. *Networks and Spatial Economics*, Vol. 13, No. 3, 2013, pp. 327–350.
14. Brooke, S., S. Ison, and M. Quddus. 2014. "Parking Choice." In *Transport and Sustainability, Volume 5: Parking Issues and Policies*, edited by S. Ison and C. Mulley, 115–35. Emerald Group Publishing Limited, Bingley, UK.
15. Axhausen, K. W., and J. W. Polak. Choice of Parking: Stated Preference Approach. *Transportation*, Vol. 18, No. 1, 1991, pp. 59–81.
16. van der Waerden, P., and H. Oppewal. Modelling the Combined Choice of Parking Lot and Shopping Destination. Presented at 7th World Conference on Transport Research, Sydney, Australia, 1995, pp. 16–21.
17. MuConsult. *Parkeeronderzoek Zuid-Limburg*. Synthese, Amersfoort, The Netherlands, 1997. (in Dutch)
18. Hensher, D. A., and J. King. Parking Demand and Responsiveness to Supply, Pricing and Location in the Sydney Central Business District. *Transportation Research Part A: Policy and Practice*, Vol. 35, No. 3, 2001, pp. 177–196.
19. Bonsall, P., and I. Palmer. Modelling Drivers' Car Parking Behaviour Using Data from a Travel Choice Simulator. *Transportation Research Part C: Emerging Technologies*, Vol. 12, No. 5, 2004, pp. 321–347.
20. Hess. S., & J. W. Polak. 2009. "Mixed Logit Modelling of Parking Type Choice Behaviour." In *Transportation Statistics*, edited by B. W. Sloboda, 77–102. J. Ross Publishing, Fort Lauderdale, FL.

21. Golias, J., G. Yannis, and M. Harvatis. Off-Street Parking Choice Sensitivity. *Transportation Planning and Technology*, Vol. 25, No. 4, 2002, pp. 333–348.
22. van der Waerden, P., A. Borgers, and H. Timmermans. Attitudes and Behavioral Responses to Parking Measures. *European Journal of Transport and Infrastructure Research*, Vol. 6, No. 4, 2006, pp. 301–312.
23. van der Waerden, P., H. Oppewal, and H. Timmermans. Adaptive Choice Behaviour of Motorists in Congested Shopping Centre Parking Lots. *Transportation*, Vol. 20, No. 4, 1993, pp. 395–408.
24. Clinch, J. P., and J. A. Kelly. Influence of Varied Parking Tariffs on Parking Occupancy Levels by Trip Purpose. *Transport Policy*, Vol. 13, No. 6, 2004, pp. 487–495.
25. Chaniotakis, E., and A. J. Pel. Drivers' Parking Location Choice Under Uncertain Parking Availability and Search Times: A Stated Preference Experiment. *Transportation Research Part A: Policy and Practice*, Vol. 82, 2015, pp. 228–239.
26. Arnott, R. Spatial Competition Between Parking Garages and Downtown Parking Policy. *Transport Policy*, Vol. 13, No. 6, 2006, pp. 458–469.
27. Louviere, J. J. Hierarchical Information Integration: A New Method for the Design and Analysis of Complex Multi-Attribute Judgement Problems. *Advances in Consumer Research*, Vol. 11, 1984, pp. 148–155.
28. Louviere, J. J., and H. J. P. Timmermans. Using Hierarchical Information Integration to Model Consumer Responses to Possible Planning Actions: Recreation Destination Choice Illustration. *Environment and Planning A*, Vol. 22, No. 3, 1990, pp. 291–308.
29. Molin, E. J. E. *Conjoint Modelling Approaches for Residential Group Preferences*. Dissertation, Faculty of Architecture, Building and Planning, Eindhoven University of Technology, 1999.
30. Molin, E. J. E., H. Oppewal, and H. J .P. Timmermans. Conjoint Modelling of Residential Group Preferences: A Comparison of Hierarchical Information Integration Approaches. *Journal of Geographical Systems*, Vol. 4, No. 4, 2003, pp. 343–358.
31. van Helvoort-Postulart, D., B. G. C. Dellaert, T. van der Weijden, M. F. von Meyenfeldt, and C. D. Dirksen. Discrete Choice Experiments for Complex Health-Care Decisions: Does Hierarchical Information Integration Offer a Solution? *Health Economics*, Vol. 18, No. 8, 2009, pp. 903–920.
32. Norojono, O., and W. Young. A Stated Preference Freight Mode Choice Model. *Transportation Planning and Technology*, Vol. 26, No. 2, 2003, pp. 195–212.
33. Richter, C., and S. Keuchel. Deriving Constructs for Hierarchical Information Integration Experiments in Regional Public Transport. *Procedia - Social and Behavioral Sciences*, Vol. 54, 2012, pp. 666–674.
34. Bos, D. M., E. J. E. Molin, H. J. P. Timmermans, and R. E. C. M. van der Heijden. Cognitions and Relative Importance Underlying Consumer Valuation of Park and Ride Facilities. *Transportation Research Record: Journal of the Transportation Research*, 2003. 1835: 121–127.
35. Kuhfeld, W. F. *The Macros*. MR-2010I-Experimental Design and Choice Modelling Macros, 2010, pp. 1017–1089.
36. He, C., and Z. Gao. Do Picture Labels Give Better Idea to Customers? A Comparison of Picture Labels to Traditional Text Describe Labels in Choice Experiments. Presented at Agricultural and Applied Economics Association & Western Agricultural Economics Association. Joint Annual Meeting, San Francisco, Calif., 2015.
37. Hensher, D. A., and L. W. Johnson. *Applied Discrete Choice Modelling*. Croom Helm, New York, 1981.
38. Hensher, D. A., J. M. Rose, and W. H. Greene. *Applied Choice Analysis: A Primer*. Cambridge University Press, 2005.

The Standing Committee on Transportation Demand Forecasting (ADB40) peer-reviewed this paper (18-01623).

Article

TRR
JOURNAL OF THE TRANSPORTATION RESEARCH BOARD

Transportation Research Record
2018, Vol. 2672(49) 34–45

DOI: 10.1177/0361198118791911
journals.sagepub.com/home/trr
SAGE

Comparing Survival Analysis and Discrete Choice Specifications Simulating Dynamics of Vehicle Ownership

Milad Ghasri[1], Taha Hossein Rashidi[1], and Meead Saberi[1]

Abstract

Modeling travel-related decisions of transport system users is the core of many behavioral travel demand models. It is of great significance to use these models for planning purposes in which decisions of individuals are simulated for specific time intervals or on a continuous dimension until the target year. Discrete choice and survival analysis methods are two popular econometric structures to model and forecast time-dependent outcomes. This paper elaborates the conceptual and practical differences between these two methods in the context of vehicle ownership modeling. There are meaningful differences between these two methods including data preparation approaches, interpretations of the variable of interest in the model, and the simulation procedures. Further, this paper shows how negligent application of discrete choice methods for modeling time-to-event variables results in specification bias. This discussion paves the path for using hazard-based models in travel demand modeling, as the application of these models have been quite limited compared with their capacity.

Vehicle ownership has been a topic of interest in a variety of fields including energy, environment, and transportation, with the intention of forecasting demand for oil, vehicle demand, energy consumption, emission levels, and transportation demand (*1*). In the field of travel behavior, vehicle ownership substantially affects households' long-term and short-term decisions (*2*). Vehicle ownership influences travel decisions such as mode choice and destination choice in day-to-day decisions (*3*, *4*), whereas in long-term decisions, availability of a private vehicle provides the opportunity to access amenities and activities, and allows individuals to select home and job locations that otherwise would not be included in their choice sets (*2*, *5*, *6*).

Discrete choice modeling specification based on the utility maximization theory is one of the most commonly used methods for vehicle ownership modeling. Discrete choice models, as the name implies, were initially developed to model the behavior of decision makers when facing a finite set of alternatives. In the vehicle ownership context, this method is used to model fleet size, fleet type, or the decision of vehicle transaction (*7–9*). Vehicle transaction refers to the decision to purchase a new vehicle, to dispose of a currently owned vehicle, or to dispose of a current vehicle and purchase a new one in a predefined time interval.

Modeling vehicle transaction is different from modeling fleet size or fleet type because it requires an extra assumption about the time interval. When modeling fleet size or fleet type, defining an exhaustive and finite choice set does not require any additional presumption. However, in modeling vehicle transaction, the modeler needs to decide how to address time in the model. The most common approach for considering time is discretizing time into intervals and modeling whether the decision of making a transaction takes place in each time interval. The methodology section of this paper shows how assuming a predetermined time interval often results in losing modeling accuracy, and sometimes it causes specification bias.

This paper discusses potential risks of using discrete choice methods for time-dependent outcomes such as vehicle transaction. Then it introduces survival analysis (aka hazard-based modeling) and elaborates on its capacities in modeling time-dependent outcomes, as the

[1]Research Centre for Integrated Transport Innovation (rCITI), School of Civil and Environmental Engineering, University of New South Wales, Kensington, Australia

Corresponding Author:
Address Correspondence to Milad Ghasri: m.ghasri@unsw.edu.au

literature of transport modeling is quite slim when it comes to practical survival analyses. Further, survival analysis and discrete choice methods are compared and the interpretation of time, as an essential element in dynamic modeling, is illustrated. This paper also discusses the special condition where a discrete choice formulation collapses into a hazard-based formulation.

The rest of the paper is organized as follows. First, the literature of vehicle ownership studies is reviewed. Then in the methodology section, the shortcomings of discrete choice method in modeling time-dependent outcomes are explained. The mathematical formulation of survival analysis method is also introduced in this section. After that, the data section discusses the process of preparing a real case dataset to validate the models. Then, the outcome of modeling practice is discussed in the results section. Finally, the highlights of the study are summarized in the conclusions section.

Literature Review

Vehicle ownership is studied from different angles in the literature. In a broad categorization, previous studies can be labeled as dynamic or static, depending on their approach toward the concept of time. In a static modeling approach, the variable of interest is either fleet size or fleet type. Early studies addressing fleet size utilize an aggregate model to estimate the number of vehicles within a zone (*10*), or owned by a household (*11*, *12*). Along with rapidly growing computational improvements and easier access to individual-level datasets, studies on fleet size also evolved to adapt disaggregate and behavioral models (*8*, *9*). Studies on fleet type investigate the impact of socio-demographic attributes on the type of vehicle that people drive (*7*, *13*). The most recent approach in this category models fleet type and their usage in a single formulation, using multiple discrete–continuous extreme value (MDCEV) specification (*14*).

In the dynamic category, the time of vehicle transaction, or the duration of owning a vehicle is targeted (*15*, *16*). A dynamic vehicle ownership model can estimate the timing of acquiring, disposing of, or trading a vehicle (*2*, *17*); thereby, it can show the progression of households' fleet type and size over time, which eventually can improve policy analysis accuracy (*18*).

As noted in the introduction, the discrete choice and survival analysis methods are the most commonly used techniques in dynamic vehicle ownership modeling. Survival analysis deals with the occurrence timing of an event. This method considers time as a stochastic term and the model's output is the probability distribution of the vehicle transaction timing (*13*, *18–21*). Traditionally used in medical studies (*22*), survival analysis has other applications in the transportation engineering and planning fields (*23*), such as modeling housing relocation (*24*, *25*), job relocation (*26*), activity duration modeling (*27–29*), incident duration (*30*), and pavement analyses (*31*). Furthermore, technical details of hazard-based models such as capturing unobserved heterogeneity (*32*) are also discussed in the literature, which is beyond the scope of this study.

Applying the discrete choice method in dynamic vehicle ownership context is possible through devising a pseudo dynamic framework (*33*). In this approach, time is discretized into arbitrary intervals and the probability of a vehicle transaction occurring at each interval is modeled using discrete choice models. Critically speaking, the discrete choice method answers the question of whether a transaction takes place in a predefined time interval, whereas survival analysis answers the question of when the transaction takes place.

Although both approaches have been discussed in the literature with heavier weight on the discrete choice modeling side, comparison between these approaches is rare in the transportation literature. Some studies in this area investigated the relationship between survival analysis and discrete choice models (*21*, *34*). These studies show that under some specific assumption, a discrete hazard-based model (HBM) is equivalent to a discrete choice model. Specifically, Cox shows that a proportional discrete HBM is equivalent to a logistic regression specification (*34*). Similarly, Bhat shows a proportional discrete competing HBM can be converted into an ordered logit (*28*).

Methodology

This section starts with hypothetical data on vehicle trade. Then in the next two subsections, the discrete choice method and survival analysis are introduced to model the vehicle trade decision.

Illustrative Example

Consider Table 1, showing the result of a hypothetical vehicle trade data, where four individuals are observed over 2 years. In this table, "duration" shows the time since purchasing the vehicle, and the "trade" indicates whether a trade has taken place or not. According to this table, person 1 and person 3 have traded their vehicles after 1 and 2 years, respectively, but person 2 and 4 did not trade their vehicle during the survey. Note that, person 2 has exited the study after 1 year.

Binary Logit Model (BLM)

Basic Introduction to Logit Model. Logit is the most widely used discrete choice modeling format (*35*). Previous

Table 1. Hypothetical Survey for Vehicle Trade

Person	Income	Duration (years)	Trade
1	Low	1	1
2	High	1	0
3	High	2	1
4	High	2	0

studies have widely used this approach to model households' vehicle transactions (*36, 37*). In its original definition, logit is used to model the scenario of decision makers facing two (or multiple) alternatives, in which they receive some utility by selecting an alternative. Assuming a rational decision maker, the alternative with the highest utility should be selected. Equation 1 shows the utility of selecting alternative j by decision maker n. This utility consists of an observed part of V_{nj} and a random variable measuring unobservable attributes of alternative and decision maker (ϵ_{nj}). The logit formulation (Equation 2) is obtained based on the assumption of ϵ_{nj} being independent and identically distributed extreme values:

$$U_{nj} = V_{nj} + \epsilon_{nj} \tag{1}$$

$$P_{ni} = \frac{e^{V_{ni}}}{\Sigma_j e^{V_{nj}}} \tag{2}$$

Regarding the illustrative example, the choice is whether to trade the vehicle or not. Assume the utility function for the trading decision as a linear function of income, as shown in Equation 3, where U_n denotes the utility of choosing to trade a vehicle, x_n is a binary variable which equals to 1 for high-income people and θ is the parameter of the model:

$$U_n = \theta x_n + \epsilon_n \tag{3}$$

Negligent Discrete Choice Specification. Using the maximum likelihood method, a likelihood function is needed for a point estimate of θ. Imagine a modeler defines the likelihood function based on the last decision of each subject in the survey (Equation 4). Considering a binary logit specification, the likelihood function converts to Equation 5, and MLE results in $\theta = -0.69$:

$$L = P(y_1 = 1) \times P(y_2 = 0) \times P(y_3 = 1) \times P(y_4 = 0) \tag{4}$$

$$L = \frac{e^{0\theta}}{e^{0\theta}+1} \times \frac{1}{e^{\theta}+1} \times \frac{e^{\theta}}{e^{\theta}+1} \times \frac{1}{e^{\theta}+1} \tag{5}$$

A noteworthy fact in this likelihood function is that person 3 and person 4 have not traded their vehicle in the first year of the survey, but there is no trace of this decision in the likelihood function. This is why this model is called a negligent specification, because it does not reflect the entire available information in the survey. As Shumway states, when the history of decision is ignored, the likelihood function provides biased estimates (*38*).

Appropriate Discrete Choice Specification. To overcome the shortcoming of the negligent specification, one needs to add the time dimension to the model and use conditional probabilities. In this case, the binary variable of y_i changes to y_{it} which indicates whether a trade has taken place by individual i in time interval of t. The likelihood function in this case is revised to Equation 6:

$$\begin{aligned} L = {} & P(y_{11} = 1) \times P(y_{21} = 0) \times [P(y_{31} = 0) \\ & \times P(y_{32} = 1 | y_{31} = 0)] \\ & \times [P(y_{41} = 0) P(y_{42} = 0 | y_{41} = 0)] \end{aligned} \tag{6}$$

Assuming unchanged income (no time-varying coefficients) and using a logit specification, the likelihood function of Equation 6 can be simplified into Equation 7. The appropriate likelihood function has two extra terms compared with the negligent likelihood function, which reflects the first year decisions of person 3 and 4. The estimated value for θ in this case is -1.39 which is considerably different from the negligent estimate:

$$\begin{aligned} L = {} & \frac{e^{0\theta}}{e^{0\theta}+1} \times \frac{1}{e^{\theta}+1} \times \left(\frac{1}{e^{\theta}+1} \times \frac{e^{\theta}}{e^{\theta}+1}\right) \\ & \times \left(\frac{1}{e^{\theta}+1} \times \frac{1}{e^{\theta}+1}\right) \end{aligned} \tag{7}$$

In some studies, it is recommended to convert the data into "person-period" format when a discrete choice specification is applied to a time-dependent outcome (*21, 38*). In the person-period format, multiple records are generated corresponding to discrete time intervals over which the subject is surveyed. Although converting data to person-period format solves the bias in the parameter estimation, the arbitrary definition of the interval can cause practical issues. In the illustrative example, it was arbitrarily decided to discretize time into yearly intervals. This decision was intuitive and did not cause any loss of data in this example. However, if the duration in the survey was collected in finer resolution, discretizing time could be more challenging. For instance, if the duration vector is reported as (370, 355, 740, 72) days then selecting yearly intervals is not definite. The next section shows how survival analysis can be helpful in these situations.

HBM

Basic Introduction to HBM. HBM is usually used to model time-to-event or failure time (*21, 39*). The event of failure in this paper is vehicle transaction. In HBM, four primary functions are defined: (i) *Failure function* is defined as the probability of failure before a specific time, (ii) *Survival function* is defined as the probability of not failing before a specific time, (iii) *Failure probability density function (PDF)* is defined as the probability of failure in a minuscule time interval, which is the slope of failure function as expressed in Equation 8, and (iv) *Hazard rate*, (aka baseline hazard), which is defined as the conditional probability of failure at time t, given the fact that the failure has not occurred before this time. By definition, hazard rate is formulated as in Equation 9 where F_0 denotes the failure function, S_0 shows the survival function, f_0 represents the failure PDF and h_0 denotes the hazard rate:

$$f_0(t) = \lim_{\Delta t \to 0} \frac{\Pr(t \leqslant T \leqslant t + \Delta t)}{\Delta t} = \frac{\partial F_0(t)}{\partial t} = -\frac{\partial S_0(t)}{\partial t} \tag{8}$$

$$h_0(t) = \lim_{\Delta t \to 0} \frac{\Pr(\mathrm{T} + \Delta t > \mathrm{t} \geqslant \mathrm{T} | \mathrm{T} \leqslant \mathrm{t})}{\Delta t} = \frac{f_0(t)}{S_0(t)} = \frac{f_0(t)}{1 - F_0(t)} \tag{9}$$

A non-parametric function of survival and hazard rate can be estimated for any dataset using Kaplan–Meier estimators (*21*). Non-parametric cumulative hazard rate is commonly used to measure the goodness-of-fit when estimating a parametric distribution in HBM.

To allow the hazard rate to vary between individuals, another dimension should be added to the specification. There are two defined methods to incorporate the impact of exogenous variables, called proportional hazard, and accelerated failure time (AFT) (*21*). It was found that Gamma provides the best parametric distribution fit. This distribution belongs to AFT category. In AFT, covariates are incorporated in the definition of failure PDF. Equation 10 shows the definition of $f(t,x)$ where x_{ij} denotes the explanatory variable j for instance i. Using Equation 10, the hazard function in AFT can be expressed as in Equation 11:

$$f(t, x_i) = f_0(t) \exp\left(-\sum_{j=1}^{J} \beta_j x_{ij}\right) \tag{10}$$

$$h(t, x_i) = h_0\left(t \exp\left(-\sum_{j=1}^{J} \beta_j x_{ij}\right)\right) \exp\left(-\sum_{j=1}^{J} \beta_j x_{ij}\right) \tag{11}$$

The Relationship between Survival Analysis and Discrete Choice Methods. Equations 8 and 9 show the definition of continuous failure and hazard functions. Survival analysis is applicable to discrete outcomes as well. By definition, discrete failure function is the probability of failure in a time interval, discrete survival function is the probability of not observing failure until a time interval, and discrete hazard function is the probability of failure in a time interval conditional on surviving until the previous time interval.

When modeling a time-dependent outcome, discrete HBM and discrete choice method are similar. The relationship between survival analysis and discrete choice methods is established using the definition of hazard function. As shown in Equation 12, the likelihood function in survival analysis consists of failure function for the failure records (f_{it}), and survival function for survived records (S_{it}). It can be shown that $f_{it} = h_{it} S_{it-1}$ and $S_{it} = \prod_{k=1}^{t} (1 - h_{ik})$ (*21*). Substituting f_{it} and S_{it} in the likelihood function results in Equation 13 (for more details refer to Jenkins [*21*]). Therefore, a discrete HBM collapses into a discrete choice model. Note than, the arbitrary selection of intervals can be problematic for discrete HBM as well. Course intervals result in reduction of accuracy and fine intervals result in computational difficulties:

$$L = \prod_i f_{iT}^{y_i} S_{iT}^{(1-y_i)} \tag{12}$$

$$L = \prod_i \prod_{k=1}^{T} (h_{ik})^{y_{ik}} (1 - h_{ik})^{y_{ik}} \tag{13}$$

Benefits of HBM. The main strength of the hazard-based method over the discrete choice method originates from the fact that the variable of interest in HBM is time. This obviates the need for arbitrary discretization of time. Note that, when independent variables are time varying, then time should be discretized according to the variations in the independent variables (*21*, *28*). The second benefit of HBM over discrete choice method is that for a wide range of parametric stochastic term of the model, HBM has a closed form formulation; whereas, in the discrete choice method, only if the stochastic term follows the extreme value distribution is a close form formulation available.

Exchanging Outputs between HBM and BLM

Considering time as the variable of interest, a connection can be made between BLM and HBM outputs. The models' output can be compared in terms of "mean failure time" and "probability of failure in time intervals."

Regarding the mean failure time, in HBM, the integral of the failure function provides the mean failure time, (Equation 14). However, logit does not directly address time of failure, but the probability of its occurrence in intervals. Instead of addressing the question of when the event would occur, logit addresses the question of whether the event occurs in a specific time interval. Therefore, the mean value of failure time can be estimated from the weighted average of time intervals' median, where weights are the probability of failure in time intervals. Equation 15 shows the mean transaction time in BLM, where $\bar{t}_i$ represent the median of time interval t_i and $\Pr(t_i)$ is the probability of failure in interval t_i:

$$\mu_{\text{HBM}} = \int f(t,x)t\,dt \tag{14}$$

$$\mu_{\text{BLM}} = \sum \bar{t}_i \left\{ \Pr(t_i = 1) \prod_{j=1}^{i-1} (1 - \Pr(t_j = 1)) \right\} \tag{15}$$

From a different point of view, one can convert the output of HBM in a comparable format with the BLM output and compare "probability of failure in time intervals". When using BLM as a pseudo dynamic approach, time is discretized into intervals denoted by $t_i = [t_i^1, t_i^2]$, and the BLM output is the probability of transaction in each interval. To link HBM and BLM, the probability of transaction in each time interval resulting from HBM should be calculated. Equation 16 shows that the probability of transaction in time interval t_i is the integral of developed hazard function in HBM over the boundaries of the interval:

$$\Pr(t_i = 1) = \int_{t_i^1}^{t_i^2} f(t,x)dt = F(t_i^2) - F(t_i^1) \tag{16}$$

Data

Anonymized data from 50,000 individuals during a 10-year period from 2005 to 2015 were obtained from an insurance company. The data included comprehensive information on members' demographics, vehicle(s) owned by each member, insurance type I contracts, breakdown history, insurance type II contracts, and vehicle insurance claims.

This study aims to model the time to vehicle acquisition and trade. As the information about actual vehicle acquisitions and disposals is not directly observed, a surrogate measure is used to infer such information from existing data. To do so, the insurance type I and type II contracts are combined and categorized by vehicle IDs. The earliest time of policy commencement for a vehicle is assumed to be its purchasing time and the latest policy cancelation is assumed to represent its disposal. Although transferring between insurance companies is overlooked, this approach seems to be a suitable to create vehicle transaction dataset. The pseudo code of data preparation algorithm is shown below. This pseudo code extracts the latest transaction for each individual and determines the type of transaction.

```
A ← Insurance type I contracts data
B ← Insurance type II contracts data
UNION A and B
SORT BY vehicle registration ID and policy start time
For each vehicle registration ID and member ID
compute the followings
   LC ← The most recent policy cancelation time
   (if no value, set to M infinity)
   LS ← The most recent policy start time
   FS ← The oldest policy start date
IF LC>LS THEN
   Acquisition date (AD) ← FS
   Disposal date (DD) ← LC
ELSE
   Acquisition date (AD) ← FS
   Disposal date (DD) ← M (infinity)
```

After extracting vehicle transaction records for every individual, the duration of a transaction (d) is defined as in Equation 17, where s_n is the time of the last vehicle purchase/change and s_{n-1} is the time of the $n - 1$ vehicle purchase/change:

$$d = s_n - s_{n-1} \tag{17}$$

In cases where only one transaction is recorded for an individual, transaction duration is calculated as the duration between the last vehicle purchase/change time s_n and the end of survey time T_e as expressed in Equation 18:

$$d = T_e - s_n \tag{18}$$

If a disposal and an acquisition occur within a month, the vehicle transaction is assumed as a trade or change of vehicle. Otherwise, the new policy start time is assumed as an acquisition or new vehicle purchase.

Censoring is a form of the missing data problem which can be handled in the survival analysis. Right censoring occurs when a subject leaves the study before an event occurs, or the study ends before the event has occurred (*21*). In this study censorship occurs when a member purchases a vehicle without trading or purchasing a new car, or when the study ends after the member has conducted a vehicle transaction (acquisition or trade).

Figure 1 shows vehicle transaction records for an arbitrary example. In this case, three vehicle purchases are recorded (green arrows) in which two of them have been disposed of (V1 and V2). At time T_2 a transfer occurred, as a vehicle disposal is coupled with a vehicle purchase. Further, the datum of V3 is censored because neither a trade nor a new acquisition is observed by the end of the survey.

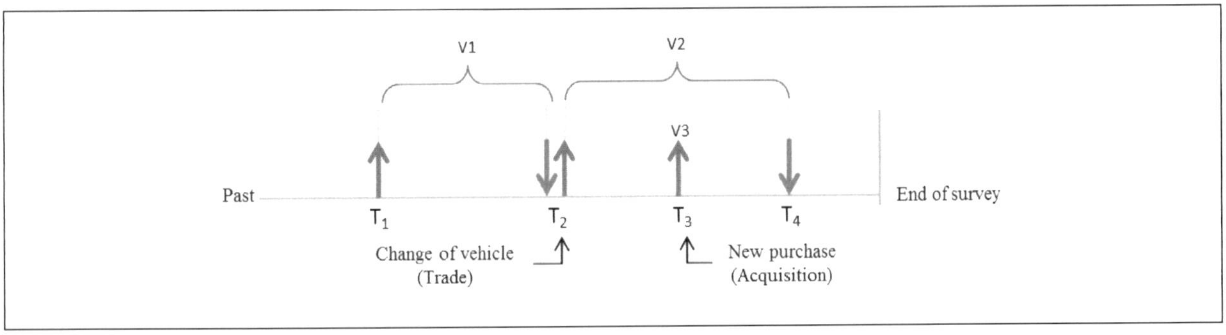

Figure 1. Schematic illustration of vehicle transaction.

Table 2. Descriptive Statistics of the Explanatory Variables

Variable	Description	Mean	St. dev.
Open contract	Number of currently open contracts	0.679	0.608
Completed contract	Number of terminated contracts	0.406	0.828
1-year breakdown	Number of breakdown claims during past 1 year	0.390	0.970
5-year breakdown	Number of breakdown claims during past 5 years	1.232	2.526
1-year insurance	Number of insurance claimed during past 1 year	0.050	0.243
5-year insurance	Number of insurance claimed during past 5 years	0.131	0.470
Inflation	Inflation rate when the transaction occurred	1.732	0.599
Price index	Customer price index when transaction occurred	105.733	4.054
Female	Binary variable indicating if the member is female	0.484	0.500
Brand 1	Binary variable indicating vehicle's brand	0.037	0.188
Brand 2	Binary variable indicating vehicle's brand	0.035	0.184
Duration (year)	Observed duration from a vehicle acquisition to the next acquisition/trade	1.998	2.468

After collation, the dataset of this this study includes nearly 18,000 records of vehicle transactions. Table 2 presents the descriptive statistics of the examined explanatory variables in the modeling practice. Some 80% of the records are randomly selected for developing the models and the rest are used in testing the models. In this table, two binary variables are defined to represent three different brands of vehicle (*13*).

Results

BLM and HBM models are developed using the Stata software. This section presents and discusses the estimated coefficients and validation of models.

Baseline Hazard

The first step in developing a parametric HBM is deciding about the parametric distribution of hazard rate. The distribution of baseline hazard (h_0) determines how the probability of a transaction changes as time passes. Six well-known parametric distributions of Weibull, exponential, Lognormal, Gompertz, log-logistic, and generalized gamma are tested, and the results are presented in Figure 2. The horizontal axis of this figure represents time. The exponential distribution has a constant hazard rate, and Weibull and Gompertz have monotonic hazard functions (*21*). The rest of the distributions can produce non-monotonic hazard functions (*21*). In Figure 2, the values of log likelihood (LL) and Akaike information criterion (AIC) indices are provided to help with selecting the most suitable distribution for the dataset of this study. The two distributions of lognormal and generalized gamma are similar and better than the rest of the distributions according to the AIC. This study uses the generalized gamma distribution because of its lower LL value and the more flexible structure of this distribution.

As shown in Figure 2, non-monotonic distributions provide better fits to the data. Based on the shape of the hazard function, the probability of a vehicle transaction increases by time in the first 2 years, and then starts a decreasing pattern.

Model Estimation

Estimated coefficients for both models are presented in Table 3. The first part of Table 3 shows the coefficients of BLM utility function. The estimated parameters are

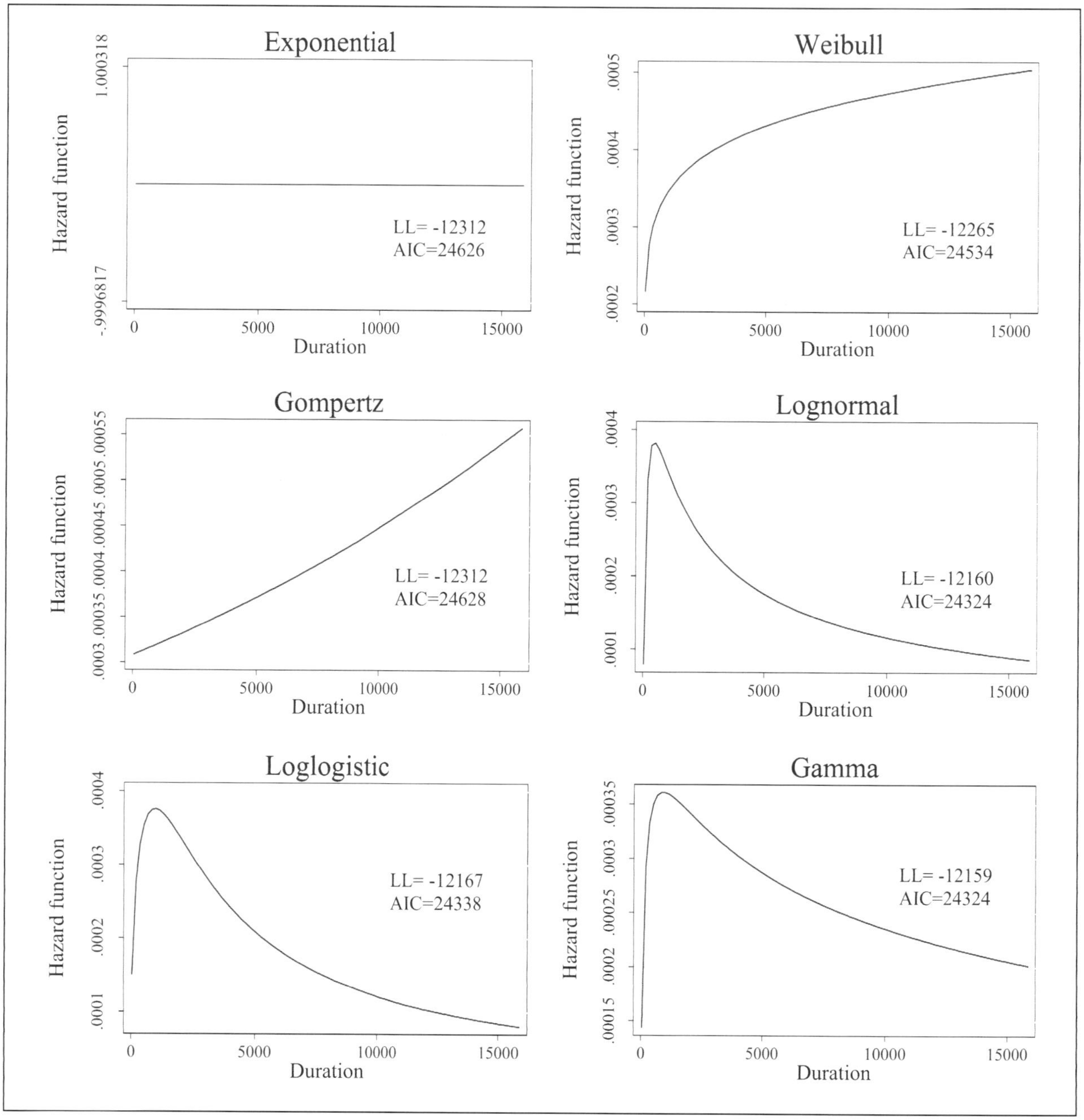

Figure 2. Parametric baseline-hazard function.

for the utility function of making a transaction, where the utility function of not making a transaction is set to zero to cope with identifiability issues (*35*). To convert the data into the person-period format, time is discretized into yearly intervals. As shown, all the included coefficients are statistically significant at the 95% confidence level.

The values in the second part of Table 3 show the coefficients of included variables in Equation 10. Note that, by definition, a negative sign is multiplied to all coefficients in this equation (*21*), therefore, the interpretation of estimated coefficients in Table 3 should be carefully discussed. For instance, the coefficient for "completed contracts" has a negative sign, which means higher values of this variable *increase* the hazard rate and consequently the probability of making a transaction.

The captured impacts from explanatory variables in HBM agree with their corresponding estimated impacts in BLM. Completed contracts both have positive impact on the probability of transaction. Higher number of open

Table 3. Estimated Coefficients

	Coefficient	Std. err.	z	Pr() > |z|
Logit model				
Variable				
Constant	−2.07	0.03	−67.89	0.000
Duration	0.128	0.01	12.14	0.000
Completed contracts	0.27	0.01	25.50	0.000
1-year insurance	0.72	0.08	8.95	0.000
5-year insurance	−0.31	0.05	−6.52	0.000
1-year breakdown	0.39	0.03	14.67	0.000
5-year breakdown	−0.20	0.01	−15.04	0.000
Female	−0.09	0.03	−2.63	0.008
Brand 1	0.00	0.05	−0.01	0.993
Brand 2	0.12	0.05	2.59	0.010
AIC	23404			
Hazard-based model				
Variable				
Gamma distribution parameters				
Sigma	1.15	0.03	4.86	0.000
Kappa	0.43	0.05	8.57	0.000
Constant	8.38	0.04	233.75	0.000
Completed contracts	−0.25	0.01	−23.22	0.000
1-year insurance	−0.59	0.07	−8.51	0.000
5-year insurance	0.14	0.04	3.52	0.000
1-year breakdown	−0.34	0.02	−15.46	0.000
5-year breakdown	0.13	0.01	12.36	0.000
Female	0.11	0.03	3.35	0.001
Brand 1	0.03	0.05	0.52	0.604
Brand 2	−0.11	0.05	−2.42	0.015
AIC	16422			

contracts results in higher vehicle ownership rate. Other variables with a positive impact on the probability of transaction are number of insurance claims during last year, and number of breakdown claims during last year. This observation implies customers' satisfaction with the provided service in the past.

Although age is found to have a significant impact on vehicle transaction (*2, 18–20*), this study did not include age of the customer in the model. The reason behind it is that age is a time-varying variable, and including it in a HBM without discretizing time is associated with the assumption that age does not change with time. Therefore, this variable was excluded.

Duration is one of the explanatory variables in BLM and based on its coefficient, 1% increase in the duration increases the probability of transaction by 0.13%. This finding confirms the findings of previous studies. Previous studies on vehicle ownership also show significant impact of the duration of owning a vehicle on the probability of vehicle transaction (*2, 18–20*). The specifications of these studies only allow for a constant impact from the duration and the impact is often found to be positive, which means the longer the duration of owning a vehicle, the higher the probability of trading it.

One of the strengths of survival analysis is its flexibility for investigating the impact of time on the outcome. The structure of the baseline hazard reveals the variations in the duration impact on the probability of making a transaction. The selected Gamma distribution in this study suggests an initially increasing impact from duration of owning a vehicle on the probability of trading it. However, after a few years the relationship takes a reverse direction, and longer duration of owning a vehicle results in lower probability of trading it. The cutoff point is not constant and is determined based on other influencing factors.

Validation

Validity of the estimated models is explored using the test dataset including 20% of the records. Validity of the modes is examined based on two measurements: the time of transactions and probability of transaction in intervals.

Time of Transaction. The mean value of time of transaction is one of the standard outputs of HBM. Equation 14 shows how mean time of transaction is calculated in

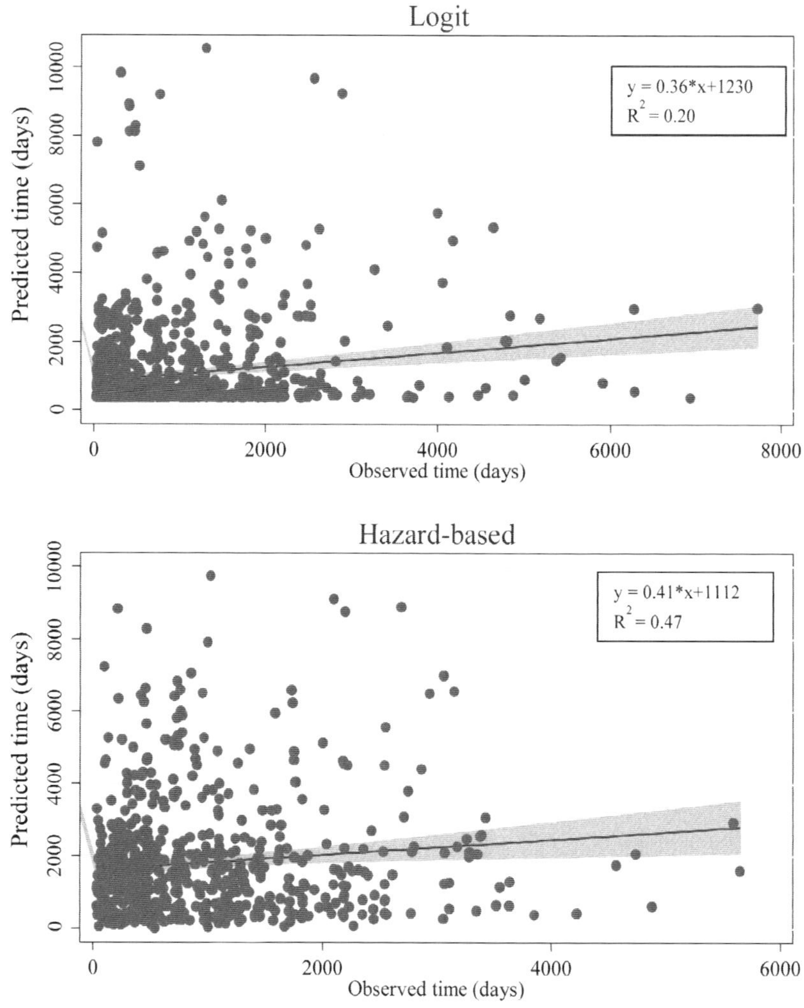

Figure 3. Observed–predicted time of transaction.

HBM. Using this equation, mean transaction time is calculated for each record in the test dataset for which a transaction is observed. Equation 15 provides the value of the mean transaction time in BLM. Mathematically the chain of intervals in Equation 15 has infinite terms. However, $\prod_{j=1}^{i-1}(1-\Pr(t_j))$ asymptotically converges to zero. This study considers $t^* = 15$ years to be the longest considered period between two consecutive transactions. Note that, the longest observed duration in the dataset was 12 years.

The estimated transaction times from each model are compared against their corresponding observed ones. Figure 3 illustrates the differences between the two models in estimating the mean transaction time. In this figure, the slope of the best regression in HBM is closer to 1, and the goodness-of-fit measurement of R^2 is higher for HBM. Therefore, HBM outperforms BLM in predicting time of transaction.

Probability of Transaction. In contrast to the first validation measurement, the probability of transaction in time intervals is the standard output of BLM. In BLM, Equation 2 calculates this value and in HBM Equation 16 provides the probability of transaction in a time interval. This validation measurement includes both censored and uncensored instances. For instances with an observed transaction, the probability of transaction is estimated for the time interval in which the transaction has occurred. For censored instances the probability of

Table 4. Cross-Table of Probability of Transaction

Model	Predicted outcome	Observed outcome: Transaction	Observed outcome: No transaction	Marginal probabilities
BLM	Transaction	17.0%	9.4%	26.4%
	No transaction	8.2%	65.4%	73.6%
HBM	Transaction	14.9%	12.8%	27.7%
	No transaction	10.3%	62.0%	72.3%
Marginal probabilities		25.2%	74.8%	100.0%

survival (no transaction) is calculated for the last observed time interval. Table 4 shows the cross-table of transaction/no-transaction probabilities.

Based on this metric, BLM outperforms HBM. The expected value of correctly predicting transaction and no-transaction in BLM are 17.0% and 65.4%, respectively (the diagonal in the table). That is while their corresponding generated values by HBM are 14.9% and 62.0%, which are slightly smaller than BLM results. Besides, the marginal estimated probabilities in BLM (26.4% and 73.6% for transaction and no-transaction) are closer to the observed marginal probabilities (25.2% and 74.8%), compared with what HBM produces (27.7% and 72.3%).

Conclusions

This paper investigated the merits and demerits of two of the most commonly used models in dynamic vehicle ownership modeling. Utilizing a rich dataset of 50,000 individuals from an insurance companay, a HBM is developed that directly models time of transition. On the other hand, a pseudo dynamic model is developed using a binary logit formulation (BLM).

This paper discusses the required condition for BLM to collapse into HBM, then explains the limitations of BLM in modeling time-dependent outcomes. The appropriate specification of BLM requires converting the available dataset into a person-period format, in which time should be discretized. In the person-period dataset, each record in the original dataset is repeated as many times as the number of intervals. If the intervals are too coarse, then this process reduces the accuracy of analysis. This study assumed yearly intervals to be a practical discretization for time in the context of vehicle trade modeling. Therefore, durations are aggregated into the annum format. In this fashion, the variations in the intervals will be overlooked. The results show that the increasing or decreasing impacts of explanatory variables on the transaction time are similar in both models; however, the magnitude of the impact is different.

This study proposed a converting formulation to estimate the expected time of transaction based on the BLM outputs. Similarly, the probability of transaction in specified time intervals can be estimated using the HBM output. These formulations are utilized to connect the output of HBM and BLM and to compare their performance. Moreover, this discussion enhances our knowledge about modeling dynamic decisions.

Finally, two measurements are defined in order to validate the developed models. In the first measurement, the expected transaction time from the models were compared again the observed transaction time. In the second measurement, the expected probability of transaction in the interval that a transaction has been observed is examined. As expected, HBM performs better in the first measurement and BLM in the second one. The main reason is that expected transaction time is the standard output of HBM whereas probability of transaction is the standard output of BLM.

The most fundamental advantage of using the HBM specification is its capacity to capture the varying impact of time on the probability of making a transaction. The best parametric distribution for the hazard function in HBM was Gamma with a non-monotonic pattern. This implies an initially increasing impact from time on the probability of transaction, which then follows a decreasing pattern.

Author Contributions

The authors confirm contribution to the paper as follows: study conception and design: MG, THR, MS; data collection: MG, THR, MS; analysis and interpretation of results: MG, THR, MS; draft manuscript preparation: MG, THR, MS. All authors reviewed the results and approved the final version of the manuscript.

References

1. Jong, G. D., J. Fox, A. Daly, M. Pieters, and R. Smit. Comparison of Car Ownership Models. *Transport Reviews*, Vol. 24, No. 4, 2004, pp. 379–408.
2. Rashidi, T., and A. K. Mohammadian. A Dynamic Hazard-Based System of Equations of Vehicle Ownership with Endogenous Long-Term Decision Factors Incorporating Group Decision Making. *Journal of Transport Geography*, Vol. 19, No. 6, 2011, pp. 1072–1080.

3. Bhat, C. R., and F. S. Koppelman. An Endogenous Switching Simultaneous Equation System of Employment, Income, and Car Ownership. *Transportation Research Part A: Policy and Practice*, Vol. 27, No. 6, 1993, pp. 447–459.
4. Dissanayake, D., and T. Morikawa. Household Travel Behavior in Developing Countries: Nested Logit Model of Vehicle Ownership, Mode Choice, and Trip Chaining. *Transportation Research Record: Journal of the Transportation Research Board*, 2002. 1805: 45–52.
5. Giuliano, G., and J. Dargay. Car Ownership, Travel and Land Use: A Comparison of the US and Great Britain. *Transportation Research Part A: Policy and Practice*, Vol. 40, No. 2, 2006, pp. 106–124.
6. Zhang, J., B. Yu, and M. Chikaraishi. Interdependences Between Household Residential and Car Ownership Behavior: A Life History Analysis. *Journal of Transport Geography*, Vol. 34, 2014, pp. 165–174.
7. Choo, S., and P. L. Mokhtarian. What Type of Vehicle Do People Drive?: The Role of Attitude and Lifestyle in Influencing Vehicle Type Choice. *Transportation Research Part A: Policy and Practice*, Vol. 38, No. 3, 2004, pp. 201–222.
8. Bhat, C. R., S. Sen, and N. Eluru. The Impact of Demographics, Built Environment Attributes, Vehicle Characteristics, and Gasoline Prices on Household Vehicle Holdings and Use. *Transportation Research Part B: Methodological*, Vol. 43, No. 1, 2009, pp. 1–18.
9. Bhat, C. R., and V. Pulugurta. A Comparison of Two Alternative Behavioral Choice Mechanisms for Household Auto Ownership Decisions. *Transportation Research Part B: Methodological*, Vol. 32, No. 1, 1998, pp. 61–75.
10. Button, K. J., A. Pearman, and A. Fowkes. *Car Ownership Modelling and Forecasting*, 1982.
11. Bates, J., M. Roberts, S. Lowe, and P. Richards. *The Factors Affecting Household Car Ownership*. Gower Press, London, UK, 1981.
12. Sanko, N., D. Dissanayake, S. Kurauchi, H. Maesoba, T. Yamamoto, and T. Morikawa. Household Car and Motorcycle Ownership in Bangkok and Kuala Lumpur in Comparison with Nagoya. *Transportmetrica A: Transport Science*, Vol. 10, No. 3, 2014, pp. 187–213.
13. Mannering, F., and C. Winston. Brand Loyalty and the Decline of American Automobile Firms. Brookings Papers on Economic Activity. *Microeconomics*, Vol. 1991, 1991, pp. 67–114.
14. Bhat, C. R., and S. Sen. Household Vehicle Type Holdings and Usage: An Application of the Multiple Discrete-Continuous Extreme Value (MDCEV) Model. *Transportation Research Part B: Methodological*, Vol. 40, No. 1, 2006, pp. 35–53.
15. Rashidi, T., and A. Mohammadian. A Competing Hazard Model of Household Vehicle Transaction Behavior with Discrete Time Intervals and Unobserved Heterogeneity. *Transportation Letters*, Vol. 3, No. 3, 2013, pp. 219–229.
16. Hossein Rashidi, T., and K. Mohammadian. Application of a Nested Trivariate Copula Structure in a Competing Duration Hazard-Based Vehicle Transaction Decision Model. *Transportmetrica A: Transport Science*, Vol. 12, No. 6, 2016, pp. 1–18.
17. Bunch, D. S., D. Brownstone, and T. F. Golob. *A Dynamic Forecasting System for Vehicle Markets with Clean-Fuel Vehicles*. University of California Transportation Center, Berkeley, CA, 1995.
18. Mohammadian, A., and T. Rashidi. Modeling Household Vehicle Transaction Behavior, Competing Risk Duration Approach. *Transportation Research Record: Journal of the Transportation Research Board*, Vol. 2014, No. 1, 2007, pp. 9–16.
19. Yamamoto, T., R. Kitamura, and S. Kimura. Competing-Risks-Duration Model of Household Vehicle Transactions with Indicators of Changes in Explanatory Variables. *Transportation Research Record: Journal of the Transportation Research Board*, 1999. 1676: 116–123.
20. Yamamoto, T., J.-L. Madre, and R. Kitamura. An Analysis of the Effects of French Vehicle Inspection Program and Grant for Scrappage on Household Vehicle Transaction. *Transportation Research Part B: Methodological*, Vol. 38, No. 10, 2004, pp. 905–926.
21. Jenkins, S. P. *Survival Analysis*. Unpublished Manuscript. Institute for Social and Economic Research, University of Essex, Colchester, UK, 2005.
22. Cox, D. R., and D. Oakes. *Analysis of Survival Data*. CRC Press, Boca Raton, FL, 1984.
23. Hensher, D. A., and F. L. Mannering. Hazard-Based Duration Models and Their Application to Transport Analysis. *Transport Reviews*, Vol. 14, No. 1, 1994, pp. 63–82.
24. Deurloo, R. C., F. M. Dieleman, and W. A. V. Clark. Tenure Choice in the German Housing Market: A Competing Risks Model. *Tijdschrift voor economische en sociale geografie*, Vol. 88, No. 4, 1997, pp. 321–331.
25. Ghasri, M., and T. H. Rashidi. A Residential Relocation Timing and Tenure Choice Modeling Structure: Competing Hazard Based or Cause-Specific Formulation? Presented at the 94th Annual Meeting Transportation Research Board, Washington, D.C., 2015.
26. Ghasri, M., and T. H. Rashidi. Modeling Reason and Time of Residential and Job Relocations: Application of a Simultaneous Nested Hazard Based Modeling Structure. Presented at 95th Annual Meeting of the Transportation Research Board, Washington, D.C., 2016.
27. Hamed, M. M., and F. L. Mannering. Modeling Travelers' Postwork Activity Involvement: Toward a New Methodology. *Transportation Science*, Vol. 27, No. 4, 1993, pp. 381–394.
28. Bhat, C. R. A Hazard-Based Duration Model of Shopping Activity with Nonparametric Baseline Specification and Nonparametric Control for Unobserved Heterogeneity. *Transportation Research Part B: Methodological*, Vol. 30, No. 3, 1996, pp. 189–207.
29. Mannering, F., E. Murakami, and S.-G. Kim. Temporal Stability of Travelers' Activity Choice and Home-Stay Duration: Some Empirical Evidence. *Transportation*, Vol. 21, No. 4, 1994, pp. 371–392.
30. Nam, D., and F. Mannering. An Exploratory Hazard-Based Analysis of Highway Incident Duration. *Transportation Research Part A: Policy and Practice*, Vol. 34, No. 2, 2000, pp. 85–102.

31. Anastasopoulos, P. C., and F. L. Mannering. Analysis of Pavement Overlay and Replacement Performance Using Random Parameters Hazard-Based Duration Models. *Journal of Infrastructure Systems*, Vol. 21, No. 1, 2014, p. 04014024.
32. Anastasopoulos, P., J. Haddock, M. Karlaftis, and F. Mannering. Analysis of Urban Travel Times: Hazard-Based Approach to Random Parameters. *Transportation Research Record: Journal of the Transportation Research Board*, 2012. 2302: 121–129.
33. Hensher, D. A. *Dimensions of Automobile Demand: A Longitudinal Study of Household Automobile Ownership and Use*. Elsevier, Duivendrecht, the Netherlands, 2013.
34. Cox, D. R. Regression Models and Life-Tables. In *Breakthroughs in Statistics*, Springer, New York, NY, 1992, pp. 527–541.
35. Train, K. E. *Discrete Choice Methods with Simulation*. Cambridge University Press, Cambridge, UK, 2009.
36. Brownstone, D., D. S. Bunch, T. F. Golob, and W. Ren. A Transactions Choice Model for Forecasting Demand for Alternative-Fuel Vehicles. *Research in Transportation Economics*, Vol. 4, 1996, pp. 87–129.
37. Hensher, D. A., and V. Le Plastrier. Towards a Dynamic Discrete-Choice Model of Household Automobile Fleet Size and Composition. *Transportation Research Part B: Methodological*, Vol. 19, No. 6, 1985, pp. 481–495.
38. Shumway, T. Forecasting Bankruptcy More Accurately: A Simple Hazard Model. *The Journal of Business*, Vol. 74, No. 1, 2001, pp. 101–124.
39. Rashidi, T., and A. Mohammadian. Parametric Hazard Functions: Overview. *Transportation Research Record: Journal of the Transportation Research Board*, 2011. 2230: 48–57.

The Standing Committee on Transportation Demand Forecasting (ADB40) peer-reviewed this paper (18-02489).

Article

TRR
JOURNAL OF THE TRANSPORTATION RESEARCH BOARD

Transportation Research Record
2018, Vol. 2672(49) 46–57

DOI: 10.1177/0361198118787090
journals.sagepub.com/home/trr
SAGE

Incorporation of Travel Time Reliability in Regional Travel Model

Surabhi Gupta[1], Peter Vovsha[2], Arup Dutta[3], Vladimir Livshits[3], Wang Zhang[3], and Haidong Zhu[3]

Abstract

The paper presents a practical method for incorporation of travel time reliability in a regional travel model. The discussion includes five consecutive steps. First it describes how the vehicle speed dataset for the metropolitan area of Phoenix, AZ, termed HERE, was processed and link-level volume–delay–reliability functions were estimated, and then how link-level reliability measures can be applied for network path building. The third step describes how trip origin–destination (OD) reliability measures can be constructed out of the link-level reliability measures. The fourth step involves implementation of the link-level and OD-level reliability measures in highway assignment, mode choice, and other travel models. The fifth step includes model validation and sensitivity tests. The paper addresses several long-standing issues associated with incorporation of travel time reliability in operational travel models in practice. These issues include construction of OD reliability measures with the recognition that the core link-level reliability measures such as standard deviation or variance are not additive in a general case, accounting for a partial correlation between travel time distributions for different links, and incorporation of travel time reliability in a standard static assignment.

Travel time reliability represents an important aspect of transportation level of service that is highly valued by travelers. According to previous research, the attractiveness to travelers of toll roads and managed lanes relates to their more reliable and predictable travel times rather than to savings in average travel times (*1–5*). Unreliability of highway travel times is also one of the primary factors that explain the advantages of rail transit modes. However, the majority of travel models in practice still rely solely on average travel times (*4, 6*).

The purpose of this research is to incorporate travel time reliability measures such as standard deviation (SD) of travel times into the travel model in addition to average travel time and cost. In this project, the extensive body of research on travel time reliability, including SHRP 2 Projects such as C04 "Improving Our Understanding of How Highway Congestion and Pricing Affect Travel Demand" (*6, 7*), L03 "Analytical Procedures for Determining the Impacts of Reliability Mitigation Strategies" (*4*), and L04 "Incorporating Reliability Performance Measures in Operations and Planning Modeling Tools" (*4*) were effectively used. This research provides prototype functions for SD at the highway link level and estimates for value-of-reliability for different travel segments. The focus of this project is on the less explored areas of correlation of travel times between different links, construction of measures for trip origin–destination (OD) reliability such as SD out of the link-level reliability measures, and operational implementation of highway assignment and skimming procedures with reliability measures. The development utilizes a region-wide commercial speed dataset (HERE).

Incorporation of travel time reliability in a regional model includes the following steps which are described in the subsequent sections:

- Estimation of link-level volume–delay–reliability functions (VDRFs)
- Incorporation of the link-level reliability measures in highway assignment
- Construction of trip OD reliability measures out of the link-level reliability measures
- Model implementation schema
- Model validation and sensitivity tests

[1]WSP, San Francisco, CA
[2]WSP, New York, NY
[3]Maricopa Association of Governments (MAG), Phoenix, AZ

Corresponding Author:
Address correspondence to Surabhi Gupta: Surabhi.Gupta@wsp.com

Link-Level Analysis of Reliability

Previous Research on Link-Level Reliability

Several studies have established a relationship between mean and SD of time per unit distance. Scaling both mean and SD per unit distance allows for bringing links of different length to a common denominator. Pioneering research undertaken by Mahmassani at the University of Maryland in the 1990s, described in (*4*), established a linear relationship between the mean travel time and SD per unit distance. However, this research was very limited in terms of the available data and it required special speed measurement surveys. Subsequently, the linear model was re-estimated by Mahmassani based on GPS traces from the Seattle Traffic Choices Study, 2007 (*4*).

Another body of empirical evidence that travel time mean is a good predictor of variance came from the SHRP 2 Project L03 led by Cambridge Systematics (*4*). This study found that concave logarithmic functions or convex exponential functions may slightly out-perform the linear function. Most of the functions used congestion index (CI), that is, a ratio of mean travel time to free-flow time as the explanatory variable, which is very close to mean time per unit distance. The research also showed that different road types (freeways, arterials, etc.) are characterized by different functions.

The research led by Resource Systems Group as part of the application of the SHRP 2 C04 Project (*7*) extended the link-level analysis of SD in two important directions. First, this research was based on much more extensive INRIX data for one month aggregated to 15-minute bins for more than 2,000 links. Second, regression functions for SD were not only segmented by four road types (freeways, arterials, collectors, and ramps) but also included a rich set of explanatory variables in addition to CI, such as number of lanes, lane drop/increase indicator, and control type, as well as traffic dynamics.

Analysis of Link-Level Reliability with HERE Data

The HERE data included travel times and speeds for more than 18,000 links in the metropolitan area of Phoenix, AZ, collected during the entire year of 2015. Speed outliers were identified and filtered out from raw HERE speed data. This step eliminated the potential chance of enlarging reliability measures caused by outliers in the analysis. Travel times and speeds were averaged within 15-minute intervals (epochs) during the day. To eliminate the travel time variability that can be perceived by drivers as systematic (such as seasonal or between weekdays and weekends), only data from regular weekdays for six representative months in fall and spring were used (more than 100 days).

In the link-level database, mean travel times and SDs were calculated for each link and epoch across multiple days. No averaging or SD calculation was done across epochs since it is assumed that the drivers are familiar with travel time variation from epoch to epoch and consider it systematic. Each link–epoch combination represented an independent observation for the estimation of VDRFs (more than 1.2 million observations). This database was joined with the regional planning network used by Maricopa Association of Governments (MAG) for modeling purposes which provided multiple additional variables.

VDRF functions were estimated separately by four road types (freeways, arterials, collectors, and ramps). The dependent variable was SD of travel time per mile. The main independent variable was CI which is always greater than or equal to 1.0. Additional explanatory variables included number of lanes, lane change (difference between the maximum number of lanes across all downstream links and the number of lanes on the current link), and control types (none, signal, stop, roundabout). The best VDRF regression specifications in terms of statistical fit and logical properties are summarized in Table 1. All regressions for different facility types were estimated with a reasonable level of fit. Most of the variables obtained meaningful signs indication of logical impacts on travel time reliability although some variables proved insignificant for certain road types.

In order to account for a non-linear impact of congestion, the regressions were estimated with an intercept and several terms including linear CI, log of CI, and squared CI. Combination of these terms always yields a monotonically increasing function but some of the terms can have a negative sign. A negative sign on the CI squared for freeways indicates concavity of function while a positive sign on CI squared for arterials and ramps indicates convexity of the function.

Additional variables were explored in interaction with CI to account for the fact that in free-flow conditions (CI close to 1.0) travel time variation should be minimal and only when the congestion grows (CI becomes substantially greater than 1.0) do the other factors come into play. Number of lanes and lane increase from the current link to the downstream link logically reduce SD for freeways and arterials. Presence of signs or signals of three different types (traffic light, stop sign, or roundabout intersection) was analyzed separately for the downstream node and upstream node of the link. For arterials, both downstream and upstream signals and signs proved to be beneficial for reliability (reduced SD). More complex reliability effects with a mix of positive and negative impacts of traffic signals and signs were found for ramps.

Table 1. Regression Models for Link SD per Mile

Variable	Freeways		Arterials		Collectors		Ramps	
	Coeff.	*T*-stat	Coeff.	*T*-stat	Coeff.	*T*-stat	Coeff.	*T*-stat
Observations (link × epoch)	152,528		909,093		49,787		124,921	
R^2	0.663		0.606		0.463		0.523	
Intercept	0.026	0.8	−0.016	−0.4	−3.750	−64.8	−0.1000	−0.7
CI	0.092	2.6	0.054	0.8	4.267	71.6	0.241	1.5
LN(CI)	1.201	36.9	0.677	16.0	−3.760	−46.2	0.322	2.4
CI^2	−0.017	−3.9	0.243	29.8			0.202	8.1
CI × Lanes	−0.030	−98.6	−0.015	−46.5				
CI × LaneChange	−0.003	−7.3						
CI × SignalDown			−0.004	−8.3			−0.022	−14.3
CI × StopDown			−0.008	−2.0			0.034	7.6
CI × RoundAbtDown			−0.020	−3.6			−0.067	−4.2
CI × SignalUp			−0.018	−39.9			−0.046	−32.3
CI × StopUp			−0.013	−3.4			0.088	23.2
CI × RoundAbtUp			−0.035	−6.4			0.031	1.9

Incorporation of Reliability in Path Building in Static Assignment

Previous Research on Incorporation of Reliability in Network Model

Multiple approaches have been suggested for incorporation of travel time reliability in network models (*4*, *7*, *9–12*) and specifically for route choice. Either a link SD or link variance was suggested as the path building attribute representing reliability that corresponds to two extreme cases—either a perfect correlation between link travel times (which would make SD additive by link [*4*, *9*, *10*]) or complete independence of travel times between different links (which would make variance additive by link [*7*, *12*]). Unfortunately, neither of the extreme cases provides a good approximation for a trip OD SD. In this regard, the research work of Horowitz and Granato (*11*) is conceptually the closest to the approach suggested in the current research since it recognizes a partial correlation of travel times between different links on the same route. However, it is based on a simplified assumption where correlation was considered only between adjacent links on a route.

Huang et al. (*13*) explored a correlation pattern between travel times on five consecutive freeway links for spatial and temporal (hour-by-hour) proximity. That study showed that, for the same hour, levels of correlation between pairs of links vary between 0.2 and 0.8; higher correlation was logically observed for the adjacent links while lower correlation was observed between the most distant links. Another important observation was that the correlation increased with growing congestion levels.

Suggested Approach for Path Building and Analysis

A new method for path building and subsequent construction of OD reliability measures is proposed that is based on the following principles:

- Travel time reliability is first captured at the link level by VDRFs that predict link SD. It is assumed that SD is the reliability measure perceived by travelers as part of the generalized cost of travel along with average travel time and cost. Consequently, SD is used in both route choice (path building) and travel models.
- Path building and corresponding construction of OD SD out of link SDs recognizes the correlation of travel times between different links on the same path.
- The path is built by using a marginal SD measure that is the closest approximation implemented with a standard static assignment. Marginal SD is defined for a given link as an increment of the route SD after traversing this link (see below).

Construction of Origin–Destination (OD) Reliability Measures

Previous Research on OD Trip Reliability Measures

Direct Methods for Calculation of OD Trip Reliability. Several attempts have been made to build a direct model for OD reliability that operates with the entire-route characteristics. This approach does not provide a linkage between link-level reliability measures and route-level reliability measures explicitly; thus, it is not useful for network path building with link-based assignment algorithms, although it can be used in a framework where paths can be enumerated for each OD pair. However, models of this type provide important independent statistical evidence and insights into the construction principles for OD reliability. One notable example developed by Arup et al. (*9*) is a model for OD reliability similar to the link-level functions discussed earlier in the way it relates SD to the mean travel time in a monotonically increasing way. The suggested formula is convex with respect to CI, that is, it assumes that the time variability grows faster than the average travel time when congestion grows. The formula included a unique additional term that expresses a diminishing OD SD with growing route length. In general, the longer the route (and consequently the lower the level of correlation between the links on the route) the lower will be the route-level SD.

Construction of OD Reliability Out of Link Reliabilities (Reliability "Skims"). There are just a few methods of this type suggested in the literature which either require a complex path building procedure (*11*) or assume an extreme correlation pattern (either perfect correlation or complete independence) between the links (*7*). However, path building still leaves open the question of how the OD reliability measure (reliability "skim") can be calculated for use in mode choice and other travel models. In this regard, the method that is most consistent but difficult to implement, with partial correlations between adjacent links (*11*), assumes skimming the same route measure of SD. However, imperfection is introduced because it accounts only for correlation between the adjacent links and ignores route characteristics and length, which would affect both route choice and other travel choices. The simplified approach adopted by the Resource Systems Group (*7*) uses link SD as an addition to the link VDRF for path building but the link variances are skimmed. A square root of the skim of link variances is fed to the travel model. This creates a principal discrepancy between the assignment and travel model. With a path-based method for calculation of reliability (*11*) it is possible to achieve a complete consistency between route choice and other travel choices. However, a complete path enumeration is largely impractical for large-scale networks. Additionally, a reliability measure built at the entire-route level is bound to be simplistic.

Suggested Approach to OD Trip Reliability and Analysis

Theoretical Considerations and Statistical Evidence on OD Trip Reliability. The cornerstone of the proposed approach is that OD SD should be between the two extreme cases—perfect correlation (modeled by an OD skim of link SD) and complete independence (modeled as square root of the OD skim of link variance). Figure 1 shows the three corresponding measures, skim of link SDs, square root of the skim of variances, and constructed actual OD SD, mapped against each other pair-wise to support this central hypothesis. For this statistical analysis, an OD reliability database was created by skimming contemporaneous HERE link data on travel times. One hundred representative traffic analysis zones uniformly distributed across the region were selected which produced almost 10,000 OD pairs. For each OD pair, contemporaneous link travel times (for the same day and epoch) were skimmed and subsequently OD mean and SD were calculated across multiple days.

Figure 1 shows that the defined lower and upper bounds for OD SD perform properly for most OD pairs and departure time epochs. One interesting exception relates to a relatively small number of cases of short trips and trips in the off-peak period (0.7% of all cases) where the observed OD SD proved to be higher than the OD skim of link SDs (theoretical upper bound)—see Figure 1*b*. This can be explained by an inherent imperfection of the data itself when it comes to a small level of variation. HERE queries devices in the vehicle at fixed time intervals for their position and speed, and spot observations are averaged to obtain mean speed. Hence, slower vehicles on facilities such as ramps will generate more HERE observations than higher speed vehicles. Another interesting exception relates to the cases (3.5%) where the observed OD SD proved to be lower than the square root of the OD skim of link variances (theoretical lower boundary)—see Figure 1*c*. This can be a manifestation of possible negative correlations between travel times on different links due to certain links, like freeway ramps, acting as a buffer for the downstream main links. This is an interesting topic for future research.

Demand Commonality Measure. From the previous research, it can be concluded that trip length and overall congestion level (route CI), as well as the route composition in terms of proportion of different road types, can be important measures to define the level of correlation. Another measure that can potentially affect the correlation of travel times along the route is demand commonality, that is, a proportion of mutual flow carried from link to link in the total vehicle volume. The measure is calculated as a ratio of the sum of turn volumes between two consecutive links ($v_{a,a+1}$) to the sum of total link volumes (v_a) skimmed along the links (a) of the route. This measure is bound to be within the unit interval. A value close to 1 corresponds to the flow of the same vehicles between the links causing a higher correlation of travel times. Conversely, if the measure is close to 0, this means that the flow of vehicles is different for different links and travel times on the links would be independent.

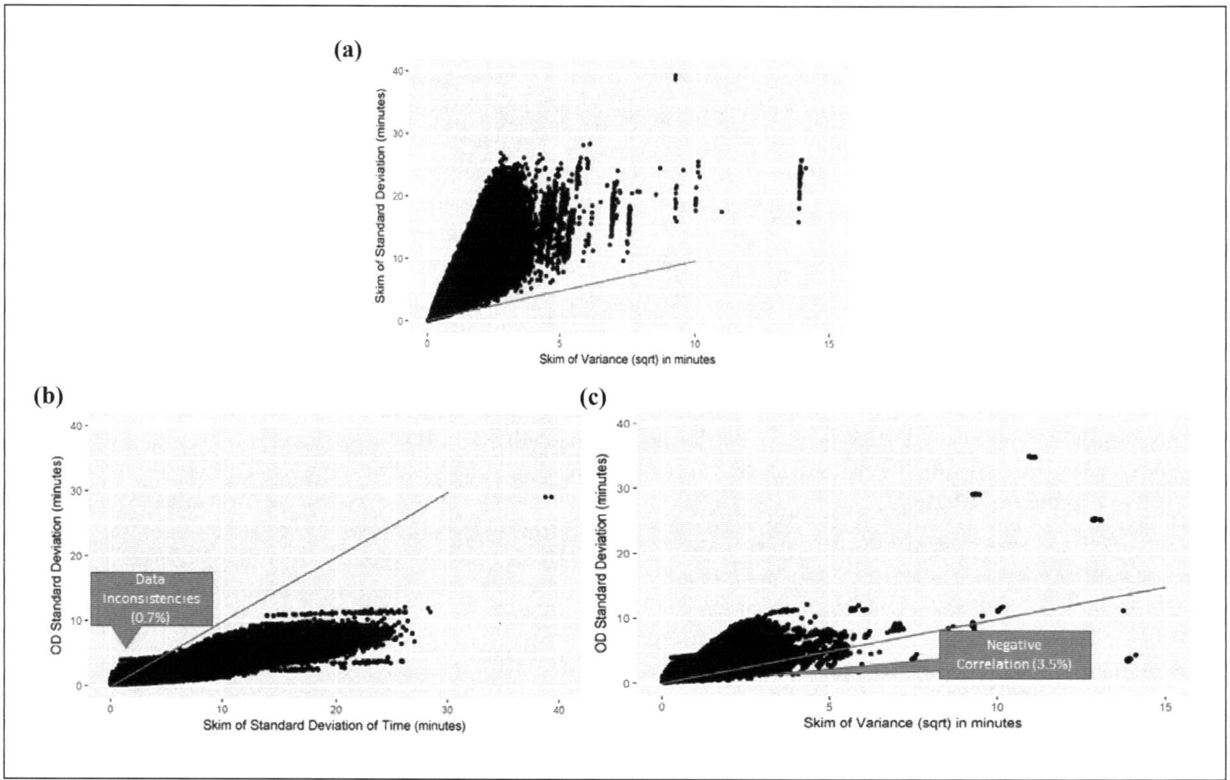

Figure 1. Statistical evidence on OD measures of SD: (*a*) OD skim of link SD (upper bound) should be greater than square root of skim of link variance (lower bound); (*b*) is OD skim of link SD indeed the upper bound for OD SD? (*c*) is square root of the skim of link variance indeed the lower bound for OD SD?

Suggested Approach to Construction of OD Reliability. The developed method includes the following key components:

- Calculation of SD of travel times for each link using VDRFs
- Skimming of link SD to generate the maximum possible SD for the OD trip
- Skimming of link variance of travel times and taking a square root of it to generate the minimum possible SD for the OD trip
- Skimming of OD trip length (L) along the path
- Calculation of the proportion of the freeway distance in the total OD trip length (F)
- Calculation of the demand (traffic flow) commonality factor (D)
- Approximation of OD SD by a blend of the skim of SDs and square route of the skim of link variances where the proportion of each component was estimated statistically as a function of trip length (L), proportion of freeway distance (F), and demand commonality factor (D)

The model for OD SD can be written as follows:

$$\sigma_{od} = \alpha_{od} \times \bar{\sigma}_{od} + \beta_{od} \times \underline{\sigma}_{od}, \quad (1)$$

where

σ_{od} = "Blended" OD SD

$\bar{\sigma}_{od}$ = OD skim of link time SDs as the upper bound for OD SD

$\underline{\sigma}_{od}$ = Square root of OD skim of link time variances as the lower bound for OD SD

α_{od}, β_{od} = weights for skim of SDs and square root of skim of variances in the blend

Two versions of the model (Equation 1) were explored: unconstrained and constrained. In the unconstrained version, weights α_{od} and β_{od} are estimated independently and the logical condition $\alpha_{od} + \beta_{od} = 1$ does not necessarily hold. In the constrained version, only one weight α_{od} is estimated while β_{od} is constrained to be $1 - \alpha_{od}$. The unconstrained version yields a better statistical fit but does not guarantee that the blend will always be between the upper and lower bound:

$$\underline{\sigma}_{od} \leq \sigma_{od} \leq \bar{\sigma}_{od}. \quad (2)$$

Weights α_{od} and β_{od} are calculated as functions of trip length (L_{od}), proportion of freeways in the trip length ($0 \leq F_{od} \leq 1$), and demand commonality factor ($0 \leq D_{od} \leq 1$). The following logical properties of these functions can be formulated:

- Impact of trip length: if L_{od} grows, a lower correlation between travel times on different links is expected that results in smaller α_{od} and/or bigger β_{od}.
- Impact of freeway proportion: if F_{od} grows, a higher correlation between travel times on different links is expected that results in bigger α_{od} and/or smaller β_{od}.
- Impact of demand commonality if D_{od} grows, a higher correlation between travel times on different links is expected that results in bigger α_{od}, smaller β_{od}.

The most theoretically appealing specifications for weight functions are:

$$\alpha_{od} = D_{od} \times (b - c \times L_{od} + d \times F_{od}), \quad (3)$$

$$\beta_{od} = 1 - \alpha_{od}, \quad (4)$$

where:

$b>0, c>0, d>0$ are global parameters estimated across all ODs.

The specifications based on the constrained version of model (Equation 1) ensure that if the demand commonality factor tends to zero (no correlation), the resultant OD SD will be equal to the square root of skim of variances as the lower bound $\sigma_{od} = \underline{\sigma}_{od}$. Conversely, if the demand commonality factor tends to 1, trip length is relatively short, and freeway proportion is high, the resultant OD SD will approach the skim of link SDs as the upper bound $\sigma_{od} = \bar{\sigma}_{od}$. This specification is also easy to estimate as a regression linear in parameters (b, c, d).

Alternative specifications for the unconstrained version of model (Equation 1) would be:

$$\alpha_{od} = b - c \times L_{od} + d \times F_{od} + e \times D_{od}, \quad (5)$$

$$\beta_{od} = f + h \times L_{od} - k \times F_{od} - m \times D_{od}, \quad (6)$$

where:

$b>0, c>0, d>0, e>0, f>0, h>0, k>0, m>0$ are global parameters estimated across all ODs.

In both specifications (Equations 3 to 6) parameters are assumed to be positive and corresponding positive or negative impacts are reflected by the signs explicitly. In the model estimation, parameters are not constrained and the sign is defined as an outcome of estimation. In order to make the model linear in parameters, the expressions for weights (Equations 3 to 6) are substituted into the main model (Equation 1). For the constrained model, the following is obtained after the substitution:

$$\begin{aligned} \sigma_{od} = {} & D_{od} \times (b - c \times L_{od} + d \times F_{od}) \times \bar{\sigma}_{od} \\ & + [1 - D_{od} \times (b - c \times L_{od} + d \times F_{od})] \times \underline{\sigma}_{od} \end{aligned} \quad (7)$$

After an equivalent transformation an expression is obtained for the constrained model that is estimated over parameters (b, c, d):

$$\begin{aligned} \sigma_{od} = {} & 1 \times \underline{\sigma}_{od} + b \times (\bar{\sigma}_{od} - \underline{\sigma}_{od}) \times D_{od} - c \\ & \times L_{od} \times (\bar{\sigma}_{od} - \underline{\sigma}_{od}) \times D_{od} + d \times F_{od} \\ & \times (\bar{\sigma}_{od} - \underline{\sigma}_{od}) \times D_{od}. \end{aligned} \quad (8)$$

For the unconstrained model, the following is obtained after the substitution:

$$\begin{aligned} \sigma_{od} = {} & (b - c \times L_{od} + d \times F_{od} + e \times D_{od}) \times \bar{\sigma}_{od} \\ & + (f + h \times L_{od} - k \times F_{od} - m \times D_{od}) \times \underline{\sigma}_{od} \end{aligned} \quad (9)$$

After an equivalent transformation an expression is obtained for the unconstrained model that is estimated over parameters (b, c, d, e, f, h, k, m):

$$\begin{aligned} \sigma_{od} = {} & b \times \bar{\sigma}_{od} - c \times L_{od} \times \bar{\sigma}_{od} + d \times F_{od} \times \bar{\sigma}_{od} \\ & + e \times D_{od} \times \bar{\sigma}_{od} + f \times \underline{\sigma}_{od} \\ & + h \times L_{od} \times \underline{\sigma}_{od} - k \times F_{od} \times \underline{\sigma}_{od} - m \times D_{od} \times \underline{\sigma}_{od} \end{aligned} \quad (10)$$

Statistical Analysis and Model Estimation for OD Reliability with MAG Data. The OD reliability database was used to estimate linear regression models specified in Equations 8 and 10. Coefficients for the final model specifications are summarized in Table 2. Both final specifications—constrained and unconstrained—are characterized by a very good statistical fit.

The following observations can be made from the estimated constrained model. Two factors—demand commonality and demand commonality interacted with freeway proportion—work in the direction of larger OD SD. If both demand commonality and freeway proportion tend to 1 (which is the worst case for unreliability in relative terms) the OD SD will be shifted toward the skim of link SD (upper bound for OD SD) by about 0.4 (sum of 0.2438 and 0.1792) multiplied by the difference between the upper and lower boundaries $(\bar{\sigma}_{od} - \underline{\sigma}_{od})$. This means that in these worst conditions the blend would include about 0.6 of the square root of skim of variance and about 0.4 of the skim of SD. However, there is one more complex factor that affects the blend. It is represented by the product of trip length, demand commonality, and difference between the upper and lower boundaries for OD SD. This factor, with growing

Table 2. Estimated Constrained and Unconstrained Model for OD SD

Variable	Notation for variable	Notation for coefficient	Coefficient value	T-stat for coefficients and R^2 for model
Constrained model				$R^2 = 0.938$
Square root (SQRT) of OD skim of Variance (VAR)	$\underline{\sigma}_{od}$		1.0 by specification	N/A
Difference between OD skim of SD and SQRT of OD skim of VAR Interactions				
Demand commonality	$(\bar{\sigma}_{od} - \underline{\sigma}_{od}) \times D_{od}$	b	0.2438	363.4
Trip length and Demand commonality	$L_{od} \times (\bar{\sigma}_{od} - \underline{\sigma}_{od}) \times D_{od}$	$-c$	–0.0008	–127.7
Freeway proportion and Demand commonality	$F_{od} \times (\bar{\sigma}_{od} - \underline{\sigma}_{od}) \times D_{od}$	d	0.1792	208.8
Unconstrained model				$R^2 = 0.953$
OD skim of link SD and Interactions	$\bar{\sigma}_{od}$	b	Illogical sign, dropped	N/A
Trip length	$L_{od} \times \bar{\sigma}_{od}$	$-c$	–0.000225	–50.6
Freeway proportion	$F_{od} \times \bar{\sigma}_{od}$	d	0.3255	195.3
Demand commonality	$D_{od} \times \bar{\sigma}_{od}$	e	0.1903	127.8
SQRT of OD skim of link VAR and Interactions	$\underline{\sigma}_{od}$	f	2.1868	501.2
Trip length	$L_{od} \times \underline{\sigma}_{od}$	h	Illogical sign, dropped	N/A
Freeway proportion	$F_{od} \times \underline{\sigma}_{od}$	$-k$	–1.1400	–140.6
Demand commonality	$D_{od} \times \underline{\sigma}_{od}$	$-m$	–1.6138	–172.0

trip length, would shift the blend somewhat back to the lower bound. With the same example of demand commonality and freeway proportion approaching 1, and additionally, assuming a trip length of 10 miles, the share of lower bound in the blend will increase from 0.60 to 0.61 while the share of upper bound in the blend will decrease from 0.40 to 0.39. For a trip that is 100 miles long the corresponding shifts will be 10 times stronger, resulting in 0.7/0.3 shares of lower and upper bounds.

The following observations can be made from the estimated unconstrained model. The model is heavily centered on the square root of the skim of variance as the most significant explanatory variable. However, the corresponding coefficient value of 2.19 witnesses that the actual OD SD is not very close to the lower bound but rather is twice greater than it on the average. Freeway proportion and demand commonality bring the upper bound to the mix. Following the same extreme example of both these factors tending to 1, one can see that it would logically bring more than 0.5 (0.3255 + 0.1903) of the upper bound to the blend. Moreover, under these conditions, the share of lower bound would be reduced by approximately 2.75 (–1.1400–1.6138), that is, a greater value than the original coefficient of the lower bound. This means that in the unconstrained estimation for the worst unreliability conditions, the blend will be more dominated by the upper bound and the resulting OD SD can be close to half of the upper bound. Additional logical effects are associated with trip length. In a similar way to the constrained regression, the blend is shifted toward the lower bound with growing trip length. For example, for a 10-mile trip, the share of the upper bound will be reduced by 0.00225. For a 100-mile trip, the share of the upper bound will be reduced by 0.0225.

Important conclusions can be made based on comparison of the estimated models with the previously reported models for OD SD. On the one hand, certain effects were confirmed by this study, for example, overall diminishing impact of trip length on unreliability (*9*). On the other hand, the current research showed that the simplified approach adopted for OD SD by the Resource Systems Group (*7*), where it was approximated by the lower bound, is too crude and systematically biased downward. The constrained model indicated that the lower bound should be blended with the upper bound in approximately 60/40 proportion.

Implementation Schema for Reliability in MAG Travel Model

Approximation of Marginal SD

Approximation of marginal SD (π_a) for each link (a) that is used for path building is based on the following formula:

$$\pi_a = \left(\sqrt{\sum_{a=1}^{A} \sigma_a^2}\right) - \left(\sqrt{\sum_{a=1}^{A-1} \sigma_a^2}\right) = \frac{\sigma_A^2}{\left(\sqrt{\sum_{a=1}^{A} \sigma_a^2}\right) + \left(\sqrt{\sum_{a=1}^{A-1} \sigma_a^2}\right)} \geq \frac{\sigma_A^2}{2\sqrt{\sum_{a=1}^{A} \sigma_a^2}} \tag{11}$$

For simplicity of illustration, it is shown for the case of independent travel times, where link variances can be summed along the path to obtain the path variance. The first expression for π_a from the left calculates the true marginal increment of the path SD when the link is added. It was implemented this way by Horowitz and Granato (*11*) with the addition of the assumed partial correlation between the adjacent links. The second expression after the inequality sign is an approximation suggested in the current study. Per this approximation, marginal link SD is proportional to the link variance and inversely proportional to the SD of the entire route. This formula provides a conservative (lower) value of marginal SD versus the true calculation, as was shown by Horowitz and Granato (*11*). However, the advantage of this formula is that it can be implemented efficiently within the standard static user equilibrium assignment and skimming capabilities of the available transportation planning packages.

Equilibration Algorithm

The complete developed method for calculation of link-level and OD-level reliability measures was incorporated in the traffic assignment and skimming procedures of the MAG travel model. The OD SD skim was incorporated in the mode choice model and accessibility measures that are used for destination choice. The added procedures were integrated with the existing demand-network equilibration strategy in the MAG travel model. The step-by-step algorithm is outlined below.

Step 0: Set initial generalized cost attributes (g_a) for each link for path building without reliability as function of link time (t_a)and link cost (c_a):

$$g_a = t_a(v_a) + c_a \tag{12}$$

Step 1: Run assignment to get link volumes $\{v_a\}$ with demand (T_{od}).
Step 2: Calculate link SD using VDRF:

$$\sigma_a = f_a(t_a, v_a) \tag{13}$$

Step 3: Skim SD and variance to get OD measures where $\delta_a^{od} = 0, 1$ is the link-path incidence:

$$\bar{\sigma}_{od} = \sum_a \delta_a^{od} \sigma_a \text{ (maximum OD SD for perfect correlation)} \tag{14}$$

$$\underline{\sigma}_{od}^2 = \sum_a \delta_a^{od} \sigma_a^2$$

(minimum OD variance for independent travel times) (15)

Step 4: Calculation of marginal link SD for path building:

$$\left(\sigma^2\right)_a^{OD} \times T_{od}$$

(assign additional demand as product of $\underline{\sigma}_{od}^2 \times T_{od}$) (16)

$$\left(\sigma^2\right)_a^{OD} = \frac{\left(\sigma^2\right)_a^{OD} \times T_{od}}{v_a} \tag{17}$$

(average OD variance for the link)

$$\pi_a = \frac{\sigma_a^2}{2\sqrt{\left(\sigma^2\right)_a^{OD}}}$$

(additive path - building attribute as marginal SD) (18)

Step 5: Calculation of OD correlation for travel times:

$$V_{od}^1 = \sum_a \delta_{a,a+1}^{od} v_{a,a+1} \tag{19}$$

(skim direct turns for continuing demand)

$$V_{od}^2 = \sum_a \delta_a^{od} v_a \text{ (skim link volumes for total demand)} \tag{20}$$

$$D_{od} = V_{od}^1 / V_{od}^2 \text{ (demand commonality factor)} \tag{21}$$

$$\alpha_{od} = D_{od} \times (b - c \times L_{od} + d \times F_{od}) \tag{22}$$

(OD correlation measure)

Step 6: Calculation of average link correlation for OD travel times:

$$\gamma_a^{OD} \times T_{od}$$

(assign additional demand as product of $\alpha_{od} \times T_{od}$) (23)

$$\gamma_a = \frac{\gamma_a^{OD} \times T_{od}}{v_a} \text{ (average OD correlation for the link)} \tag{24}$$

Step 7: Update link generalized cost function with reliability by blending independent and correlated travel time distributions where ρ is the reliability ratio estimated as 0.8 (*3, 5, 14, 15*):

Table 3. Summary Results for Base Scenario for Traffic Assignment with Reliability Measures

Outer loop	Reliability iteration	Facility type	VMT	VHT	VRHT	VHT/VMT (min/mile)	VRHT/VMT (min/mile)	Reliability overhead ([F]/[E])	Average speed (mph)
1	1	Highways	9,874,157	247,149	247,149	1.50	1.50	1.00	39.95
1	2	Highways	9,963,405	248,795	270,188	1.50	1.63	1.09	40.05
1	3	Highways	9,959,838	248,514	270,264	1.50	1.63	1.09	40.08
1	4	Highways	9,959,426	248,428	270,321	1.50	1.63	1.09	40.09
1	5	Highways	9,959,173	248,391	270,350	1.50	1.63	1.09	40.09
1	1	Arterials	10,712,912	466,118	466,118	2.61	2.61	1.00	22.98
1	2	Arterials	10,714,075	459,234	515,678	2.57	2.89	1.12	23.33
1	3	Arterials	10,709,532	459,417	515,629	2.57	2.89	1.12	23.31
1	4	Arterials	10,708,895	459,487	515,750	2.57	2.89	1.12	23.31
1	5	Arterials	10,708,770	459,516	515,816	2.57	2.89	1.12	23.30
1	1	Collectors	740,148	32,951	32,951	2.67	2.67	1.00	22.46
1	2	Collectors	725,774	31,457	35,947	2.60	2.97	1.14	23.07
1	3	Collectors	727,561	31,662	36,041	2.61	2.97	1.14	22.98
1	4	Collectors	728,122	31,696	36,066	2.61	2.97	1.14	22.97
1	5	Collectors	728,460	31,718	36,080	2.61	2.97	1.14	22.97
1	1	Ramps	662,338	28,851	28,851	2.61	2.61	1.00	22.96
1	2	Ramps	659,786	28,292	31,672	2.57	2.88	1.12	23.32
1	3	Ramps	660,187	28,332	31,672	2.57	2.88	1.12	23.30
1	4	Ramps	660,187	28,342	31,673	2.58	2.88	1.12	23.29
1	5	Ramps	660,176	28,344	31,670	2.58	2.88	1.12	23.29
1	1	All	21,989,555	775,069	775,069	2.11	2.11	1.00	28.37
1	2	All	22,063,041	767,777	853,484	2.09	2.32	1.11	28.74
1	3	All	22,057,119	767,924	853,605	2.09	2.32	1.11	28.72
1	4	All	22,056,630	767,953	853,810	2.09	2.32	1.11	28.72
1	5	All	22,056,578	767,969	853,916	2.09	2.32	1.11	28.72

Note: VMT = vehicle miles traveled; VHT = vehicle hours traveled; VRHT = vehicle hours traveled with reliability; mph = miles per hour.

$$g_a = t_a(v_a) + c_a + \rho \times [\gamma_a \sigma_a + (1 - \gamma_a)\pi_a] \tag{25}$$

Go to Step 1 and equilibrate until link generalized cost becomes stable.

Step 8: Calculate OD SD as blended reliability measure for mode choice (see Equation 1):

$$\sigma_{od} = \alpha_{od}\bar{\sigma}_{od} + (1 - \alpha_{od})\underline{\sigma}_{od}. \tag{26}$$

For computation of reliability measures, the highway assignment must be run at least two to three times, first with zero reliability measures and then with updated (equilibrated) reliability measures. The increase in run times is directly proportional to the highway assignment run times for all time periods.

Travel Model Validation with Reliability

Base Year Validation of Model Results with and without Reliability

The first set of tests related to convergence of the traffic assignment with fixed demand and reliability measures where both average travel times and reliability terms (marginal SDs) are equilibrated with path choice and resulting link volumes. A test run with five external iterations for update of the link reliability terms (with moving successive average method applied) and 500 internal assignment iterations of the path equilibration on each external iteration is summarized in Table 3. The first external iteration corresponds to the traffic assignment without reliability measures while starting from the second iteration reliability is considered. Overall convergence and stability of the results between iterations 2 to 5 is observed. In terms of simulated Vehicle Hours Traveled (VHT) and Vehicle Miles Traveled (VMT), the addition of reliability terms did not cause a substantial shift between different road types.

The second set of tests included comparisons of the outcomes of different outer iterations with respect to local traffic effects. The map in Figure 2 portrays the differences in traffic volumes between the first external iteration (standard traffic assignment without reliability measures) and fifth iteration (with equilibrated reliability measures). Inclusion of reliability affected several key roads and overall made the model more sensitive to congestion levels. Several freeway facilities around the city core gained more traffic (green) at the expense of many arterials and local collectors (red).

It should be noted that the MAG travel model had very high validation statistics against the traffic counts

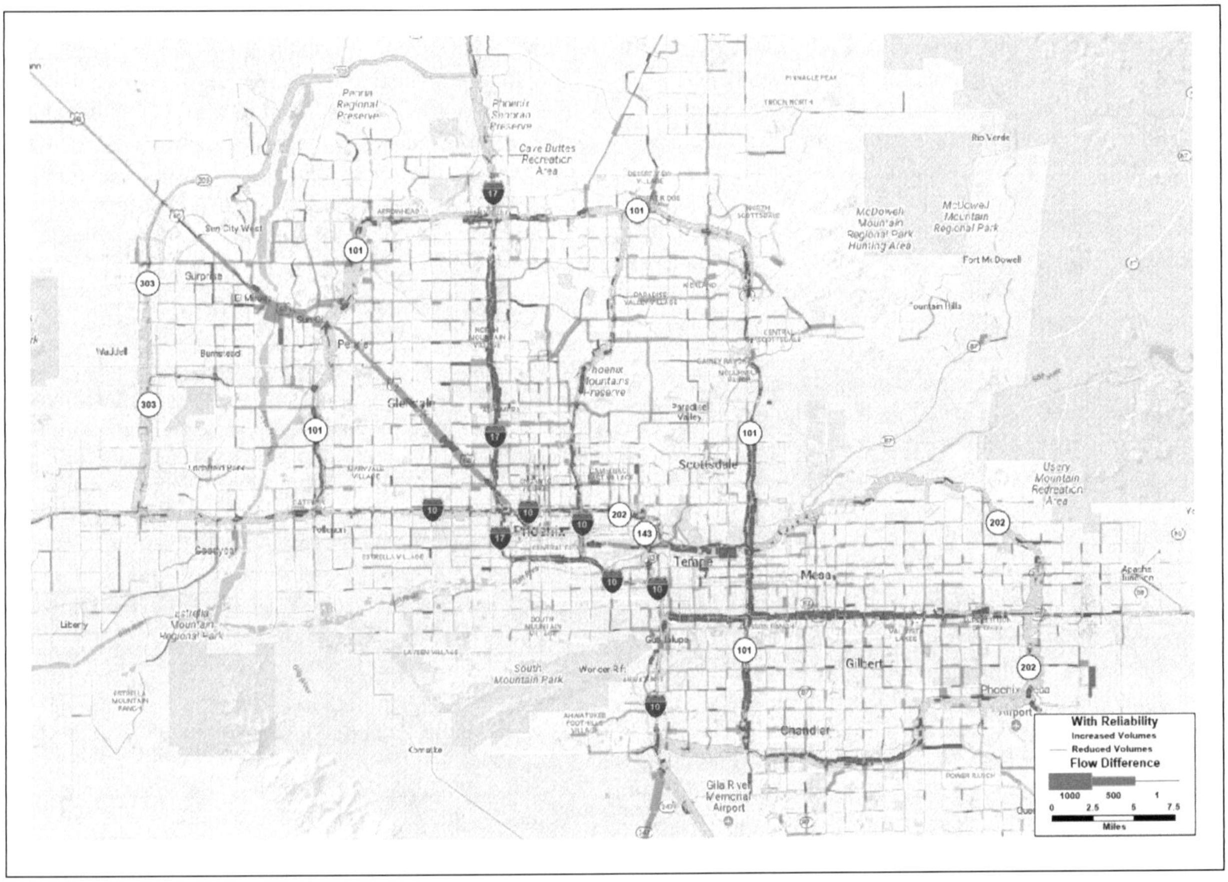

Figure 2. Map of Pheonix, AZ, showing differences in traffic assignment volumes with and without reliability.

such as R^2 = 97.20% and RMSE = 27.98% even without the reliability component added in. With reliability, we observe only a slight change: R^2 = 97.31%; RMSE = 28.41%. Thus the inclusion of reliability neither destroys nor improves the validation against traffic counts. However, it should be noted that the primary purpose of including reliability is not to match traffic counts for the base. It is to make the model more sensitive to projects and policies which benefit from reliability improvement in addition to improved average travel times.

Model Sensitivity Tests with and without Reliability Measures

The third set of tests included comparisons of the model sensitivity with and without reliability when applied to a highway pricing project. The project includes converting all high occupancy vehicle (HOV) lanes to high occupancy vehicle and toll (HOT) lanes using a distance-based toll (20 cents per mile). This analysis included a set of comparisons between four simulated scenarios: 1. Base network without reliability; 2. Base network with reliability; 3. "Build" network with pricing without reliability; 4. "Build" network with pricing and reliability. These four scenarios were simulated in two model settings: highway assignments only with fixed demand and complete travel model run with three feedback loops where travel demand itself was affected by reliability.

The first step of the analysis included comparison of scenarios 1. Base network without reliability and 3. "Build" network with pricing and reliability. Conversion of HOV lanes into HOT lanes without accounting for reliability predictably showed an additional traffic volume on all HOT lanes and parallel reduction of the traffic volumes on all general-purpose lanes (GPLs). Also, other roads were affected by allowing single-occupancy vehicles (SOVs) to use managed lanes for the toll. On the one hand, the roads (specifically arterials) that are competing alternatives to the managed lanes lost a certain amount of traffic. On the other hand, the roads that provide access or egress for managed lanes gained some traffic.

The second step of analysis included comparison of scenarios 2. Base network with reliability and 4. "Build" network with pricing for reliability. This analysis showed a similar pattern with additional traffic volume on all HOT lanes and parallel reduction of the traffic volumes on all GPLs. Other roads were also affected by allowing SOVs to use managed lanes for the toll in a similar way. This shows that reliability in general does not change the traffic impact drastically at the regional level. However, analysis with reliability revealed many local effects, such as differential shifts toward the managed lane in different situations. When the HOV lane was somewhat underutilized but GPLs were congested, incorporation of reliability produced a relatively greater shift toward the HOT lane. However, if the HOT lane was congested in the first place, incorporation of reliability produced a smaller additional volume (primarily SOVs willing to pay toll) compared with the model results without reliability.

Conclusions

The research addresses several long-standing issues associated with incorporation of travel time reliability in operational travel models in practice. These issues include a construction of OD reliability measures with the recognition that the core link-level reliability measures such as SD or variance are not additive in a general case, accounting for a partial correlation between travel time distributions for different links and incorporation of travel time reliability in a standard static assignment without a complication of its basic algorithm. New methods were suggested for incorporation of travel time reliability in traffic assignment path building through link marginal SDs and in travel models through "blended" OD reliability measures that combine a skim of link SDs and a skim of link variances.

The overall conclusion is that the methods developed are reasonable and may affect project evaluation in a logical way. The model is characterized by a very good convergence pattern for the Base year and Build scenario where HOV lanes were converted to HOT lanes although it is slightly more fluctuating compared with the Base scenario (Scenario 2) due to a greater variety of path choices for SOVs that can choose between free and tolled path in the Build scenario (Scenario 4). However, even in this case three iterations of reliability terms calculations can be recommended for practical applications.

The results for the Build scenario showed a strong pattern of additional traffic volume on all HOT lanes and parallel reduction of the traffic volumes on all GPLs. Overall, reliability does not change the traffic impact drastically at the regional level. However, there are many local effects revealed with reliability, such as differential shifts of traffic toward the managed lane in different situations. The main purpose of quantifying reliability in a travel model is to introduce sensitivity to projects and policies such as managed lanes or toll roads which benefit from improved reliability in addition to improved average travel times. It is expected that the traffic forecast for toll roads and managed lanes might be systematically higher for a model with reliability versus a standard model without reliability. This of course is yet to be explored and proved in practice. The authors plan to implement more sensitivity tests in the near future to analyze the impact of including reliability in the analysis.

Author Contributions

The authors confirm contribution to the paper as follows: study conception and design: Peter Vovsha, Surabhi Gupta; data collection: Vladimir Livshits, Arup Dutta, Haidong Zhu, Wang Zhang; implementation: Surabhi Gupta, Arup Dutta; analysis and interpretation of results: Surabhi Gupta, Arup Dutta, Peter Vovsha, Vladimir Livshits; draft manuscript preparation: Peter Vovsha, Surabhi Gupta, Arup Dutta. All authors reviewed the results and approved the final version of the manuscript.

References

1. Noland, R. B., and K. A. Small. *Travel Time Uncertainty, Departure Time Choice and the Cost of the Morning Commute*. Working Paper UCI-ITS-WP-95-1. Institute of Transportation Studies, University of California, Irvine, CA, 1995.
2. Noland, R. B., K. A. Small, P. M. Koskenoja, and X. Chu. Simulating Travel Reliability. *Regional Science & Urban Economics*, Vol. 28, 1998, pp. 535–564.
3. Small, K. A., R. Noland, X. Chu, and D. Lewis. *Valuation of Travel-Time Savings and Predictability in Congested Conditions for Highway User-Cost Estimation*. National Cooperative Highway Research Program Report 431, National Academy Press, 1999.
4. *Incorporating Reliability Performance Measures in Operations and Planning Modeling Tools*. Final Report. Prepared by Delcan Corporation, Parsons Brinckerhoff, and Northwestern University, 2013.
5. Small, K. A., C. Winston, and J. Yan. Uncovering the Distribution of Motorists' Preferences for Travel Time and Reliability. *Econometrica*, Vol. 73, No. 4, 2005, pp. 1367–1382.
6. *Improving Our Understanding of How Highway Congestion and Pricing Affect Travel Demand*. Final Report. Prepared by Parsons Brinckerhoff, Northwestern University, and Mark Bradley, Research & Consulting, 2010.
7. *Pricing and Reliability Project. Report Prepared by Resource Systems Group*. FHWA Webinar, 2017.
8. *Analytical Procedures for Determining the Impacts of Reliability Mitigation Strategies*. Final Report. Prepared by Cambridge Systematics, Texas A&M University, and University of Washington, 2012.
9. Arup. *Frameworks for Modeling the Variability of Journey Times on the Highway Network*. Arup, London, UK, 2003.
10. Bates, J., I. Black, J. Fearon, C. Gilliam, and S. Porter. Supply Models for Use in Modeling the Variability of Journey Times on the Highway Network. *AET – Proceedings of*

the European Transport Conference. Association for European Transport (AET), Cambridge, UK, 2002.
11. Horowitz, A. J., and S. Granato. Practical Considerations for Travel Time Reliability in Regional Travel Forecasting. *Transportation Research Record. Journal of the Transportation Reseach Board*, 2012. 2302: 184–191.
12. Multimodal Travel Time Variability. Final Report. Institute for Transportation Studies, University of Leeds, Imperial College London, John Bates Services, 2008.
13. Huang, H., P. Rachtan, and S. Gao. Link Correlations in Evaluating Trip Travel Time Reliability. Presented at 90th Annual Meeting of Transportation Research Board, Washington, D.C., 2011.
14. Brownstone, D., and K. A. Small. Valuing Time and Reliability: Assessing the Evidence from Road Pricing Demonstrations. *Transportation Research A*, 39A, 2005, pp. 279–293.
15. Concas, S., and A. Kolpakov. *Synthesis of Research on Value of Time and Value of Reliability*. Center for Urban Transportation Research, University of South Florida, Tampa, FL, 2009.

The Standing Committee on Transportation Demand Forecasting (ADB40) peer-reviewed this paper (18-03028).

Article

TRR
JOURNAL OF THE TRANSPORTATION RESEARCH BOARD

Transportation Research Record
2018, Vol. 2672(49) 58–68
© National Academy of Sciences:
Transportation Research Board 2018
Article reuse guidelines:
sagepub.com/journals-permissions
DOI: 10.1177/0361198118796357
journals.sagepub.com/home/trr
SAGE

A Heteroscedastic Polarized Logit Model to Investigate the Competition of Bicycle for the Bikeable Trips with the Other Modes

Khandker Nurul Habib[1]

Abstract

The paper proposes a new discrete choice model, named the Heteroscedastic Polarized Logit (HPL) to investigate choice contexts with one or more alternatives with remarkably low market shares. The proposed model is used to investigate the factors influencing the choice of a bicycle as a travel mode in the National Capital Region (NCR) of Canada. Data from the latest household travel survey of the NCR are used to investigate the mode choices of bikeable trips. Bikeable trips are defined as trips with lengths shorter than 16 km as this is the observed maximum limit of a bicycle trip in the dataset. A large dataset with over 40,000 trip records is used for empirical investigation where the bicycle has the lowest mode share of 3%. The HPL model clearly shows its appropriateness and superiority over comparable models in such a context. The choice to walk is found to be more sensitive to trip length than the choice to cycle, yet walking is found to have three times larger market share than that of cycling. Similarly, motorized modes are found to have low sensitivity to travel time and other impedances and have larger market shares. Women and students are found not to prefer the bicycle as a travel mode. Cycling infrastructure is seen to be effective in increasing the choice of the bicycle as a travel mode, but it also becomes clear that additional soft policy initiatives would be necessary to increase the popularity of cycling among young people, students, and women.

Discrete choice models are widely used in mode choice investigations as they allow capturing competitions and substitutions of alternative modes; they also provide an efficient means of economic evaluation (e.g., value of travel time savings, the value of reliability, etc.). The most commonly used assumptions of developing a standard discrete mode-choice model (e.g., Multinomial Logit [MNL] or even Nested Logit [NL]) is the 'Independence of Irrelevant Alternatives/Alternative-nests' (IIA) and Random Utility Maximizing (RUM). The IIA assumption leads to a binding restriction that the alternative modes (or nests/groups of modes) are perfect substitutes for each other, which may not always be practically correct (*1*). In reality, mode alternatives are rarely perfect substitutes for one another. Moreover, in many urban contexts often one or more alternatives dominate by a large margin, leaving one or more with very low market share (often called minor modes).

A standard discrete mode-choice model with a full set of alternative-specific-constants (ASC) and under the IIA assumption guarantees that the model predictions match the aggregate shares (of alternatives in case of an MNL and of alternative nests in the case of an NL) of the sample dataset used for the model estimation. However, this is not guaranteed for future predictions and can even be problematic as the base case ASCs will have biased influences. The presence of any inter-alternative correlations or IIA violation can lead to wrong future predictions of choice probabilities. In most of the cases, the modes with the lowest market shares in the base dataset (minor modes) are the ones that suffer from a large percentage of prediction errors (*2*).

The theories of RUM discrete choice models have been well advanced in last two decades to capture imperfect substitutions of choice alternatives to accommodate IIA violation. However, most such advances are based

[1]Department of Civil Engineering, University of Toronto, Toronto, ON, Canada

Corresponding Author:
Address correspondence to Khandker Nurul Habib: khandker.nurulhabib@utoronto.ca

on the bottom-up approach of model formulation (starting from the individual choice to match aggregate market shares). These advanced models focus mostly on choice set definition, precise specification of the systematic utility functions and the random utility error specifications. Such advanced models include Generalized Extreme Value (GEV) models, Generalized Nested Logit (GNL) models, Mixed Logit models, Logit models with choice set formation, Latent segmentation Logit and Hybrid choice models with Latent Variables and so forth. However, none of these advanced models focus on the issue of imbalance in market shares of the alternatives and resulting prediction errors of minor (or inferior) alternatives (*3*). This is a critical issue for mode choice modeling when minor modes (e.g., bicycle, walking) are the most sustainable modes and need to be promoted.

In a somewhat separate trend of research (from the bottom-up approach of developing discrete choice models), there have been some developments in discrete choice models by using a top-down approach: from market shares to disaggregate choice probability functions. Such an approach uses the concept of Entropy Maximization of market shares to derive the disaggregate choice model formulation (*4*). Anas (1983) (*5*), in fact, derived the MNL model by using such an approach. Donso and Grange (*6*) further elaborated this formulation by providing a clear interpretation of the entropy theory used to derive the MNL formulation. They also explained the equivalence of this top-down approach with the RUM theory.

However, in both of these works, the derived discrete choice models were of MNL formulation that carried the same limitations innate in the IIA assumption at the disaggregate choice levels. In a recent work, Grange et al. (*7*) extend the top-down formulation of discrete choice models by inducing an additional constraint that generates a new model called the Polarize Logit (PL) model. This is a very promising model as it forces the choice model predictions to be more realistic than those of a standard discrete choice model by accommodating choice endogeneity. Choice endogeneity pushes the probability of best alternative toward 1 and the worst alternative toward 0. This paper uses this PL model formulation and extends it further to investigate mode choices in a context of the bicycle being the minor mode for the bikeable trips, that is, trips for which the bicycle is a feasible alternative mode.

The paper proposes an extension to the PL model to accommodate different substitution patters of the bicycle with other modes and randomness in mode choice behaviour. The proposed model is a Heteroscedastic Polarized Logit (HPL) model. For the empirical investigation, the choice of using the bike for the bikeable trips in the National Capital Region (NCR) of Canada is investigated. A subset of NCR's latest household travel survey data is used. The dataset reveals that the bicycle is a minor mode sharing only around 3% of market share of all bikeable trips. The proposed HPL model is used to further our understanding of the influential factors of the choice of using a bicycle for travel in the NCR.

The paper is organized as follows: the next section presents a brief literature review on mode choice modeling in the context of the current investigation. This section is followed by a section explaining the proposed HPL model; a description of data used for empirical investigation and a discussion on an empirical model with the policy relevance of the findings. The paper concludes with a summary of key findings and recommendations for further studies.

Context and Literature Review

The purpose of this section is not to present a comprehensive literature review on discrete mode choice modeling as the area is too vast to cover in one paper and the information is available elsewhere (e.g., [*8*]). The discrete choice model has been the widely used modeling technique for its representation of the behavioral theory. Application of a discrete choice model for mode choice analysis has been an essential tool in transportation planning exercises, yet there are many issues related to the applicability of discrete choice models in a wide variety of practical situations (*9*). This section presents a review of studies that focused on the issue of over- or under-prediction of choice probabilities when the alternative modes have unequal market shares.

In the typical trip-based investigation, minor modes tend to have disproportionate market shares. In car dominated cities, private car (car driving, car passenger, etc.) related modes usually have major shares compared with those of transit and non-motorized modes. Often, standard discrete choice models do not do as expected in accurately predicting future shares of modes that have lower market shares in the base case. This, in fact, hinders, policy initiatives that tend to promote minor modes. For example, Idris et al. (*10*) used various advanced discrete choice models and found that typical discrete choice models often over-predict the shares of transit. Their investigations reveal augmentation of observed mode choice data by stated preference data can help improve the model predictions. While the recommendation to use stated preference data is intuitive, may not always be feasible for many reasons (e.g., time and cost requirements in collecting additional stated preference data).

Mahmoud et al. (*11*) investigated the competitions of alternative travel modes in downtown Toronto considering the subsets of trips (from Toronto's latest household travel survey data) for which private car, local transit, bicycle, and walking are all feasible modes. They compared an MNL model against a parameterized logit captivity (PLC) model and found that PLC outperforms the multinomial logit model for competitive modal arrangements, for example, in downtown Toronto. The PLC model is an advanced formulation of a Dogit model that parameterizes the captivity parameters of the Dogit model. The Dogit model is developed by Guadry and Degenais (*12*) to overcome the IIA limitations of the MNL model. The Dogit model adds choice set consideration constraints in the choice probability function of an MNL model that can overcome the limitation generated by its IIA assumption. It assumes that the choice of an alternative is composed of two parts: choice of considering the alternative alone (myopic choice) and choice of the same alternative considering an IIA MNL model. Mahmoud et al. (*11*) used a parametrized version of the myopic choice component and found that it improves the model's prediction accuracy of low-market-share alternatives.

Habib et al. (*13*) used a similar version of the PLC model but considered a consideration constraint only for a minor mode (carpooling). They found that a model so constrained outperforms the standard MNL model in terms of better understanding of the potential of a minor mode. In another study, Habib and Weiss (*14*) also compared the prediction capacity of the PLC models against those of the MNL models in mode choice modelling with a variety of choice alternatives and a wide range of market shares. They used repeated correctional datasets collected over two decades in the Greater Toronto Area to investigate the evolution of modal preferences over time. They also proved that by adding consideration constraints in the choice set formation, the Dogit type PLC model can outperform the MNL model.

Dogit or PLC type models explicitly recognize the fact that some of the alternatives may have underlying reasons to have low or high probabilities (vis-à-vis market shares). So, this issue is tackled by introducing a binary choice of captivity to the concerned alternatives within the discrete mode choice model. Still, this does not guarantee that endogeneity between the choice of a particular alternative and its market share is satisfied. It is understandable that, in many cases, a lower share of a choice alternative induces a lower probability of choices, which is the choice endogeneity. For example, the lower the number of people using a bicycle for trips, the lower the number of people who will be encouraged to use a bicycle as a travel mode. Similar cases can happen for walking and transit mode choices as well.

Grange et al. (*7*) are the first to recognize this is as a choice model endogeneity issue. They proposed a new approach to discrete choice modelling that forces the model on predicting realistic outcomes and it can thus further outperform even the Dogit or PLC type model. Their proposed model is named a PL model that polarizes the probability of alternative choices toward 0 or 1. The alternatives that are most likely to be chosen are pushed to have probabilities closer to 1 and the alternatives that are least likely to be chosen are pushed to have probabilities closer to 0. Such a mechanism is induced by formulating the model considering entropy maximization of the aggregate market shares under defined constraints. Although the model is derived by using the aggregate entropy theory, the formulation of choice probabilities become exactly similar to those derived from the RUM approach.

Grange et al. (*7*) compared their proposed PL model with MNL, NL, Mixed Logit, and Dogit models in their empirical mode choice investigation of a Chilean city. They found that the PL model outperforms the MNL, NL and the Dogit model in predicting choice probabilities. Their empirical investigation also proves that the PL approach is suitable for the context of mode choices when one or more alternatives have very low market shares (minor modes).

However, like the RUM based discrete choice model, the PL model proposed by Grange et al. (*7*) also considered the IIA assumption in some way. It does consider choice endogeneity in the model, but the basic tenet of IIA holds within the choice model formulation. Also, it implicitly considers that the sources of choice randomness across the alternatives are the same (homoscedasticity). It is understandable that these may limit the prediction accuracy of a PL model. So, to contribute to this area of advancement, this paper extends the formulation of the PL model into a HPL model that can accommodate choice heteroscedasticity as well as imperfect substitution between the choice alternatives. The HPL used error mixing approach to develop an error-component version of the PL model. The model is then used to investigate mode choices in the context of bikeable trips with the bicycle as the minor mode in the NCR of Canada.

The Heteroscedastic Polarized Logit Model

Consider a random sample of individuals (I), where any individual i has a total of M feasible modes to choose from. Assuming that the choice probability of any mode m of the individual i is Pr^i_m, the entropy maximization optimization problem to derive the choice probability function is:

$$\underset{(Pr^i_m)}{Minimize} \sum_{i=1}^{I} \sum_{m=1}^{M} Pr^i_m \left(\ln\left(Pr^i_m\right) - 1\right) \tag{1}$$

Subject to:

$$\sum_{m=1}^{M} Pr_m^i = 1 \tag{2}$$

$$\sum_{i=1}^{I} Pr_m^i x_m^{ik} = D_m^k \tag{3}$$

The constraint specified in Equation 2 is the first law of probability that the summation of all probabilities of discrete alternatives is unity. The constraint specified in Equation 3 defines that exogenous covariate x of category k, (x_m^{ik}) as perceived in alternative m by the individual i and $D_m^k = \sum_{i=1}^{I} \vartheta_m^i x_m^{ik}$, where ϑ_m^i takes the value 1 if the alternative m is chosen by individual i and 0 otherwise.

This is a standard optimization problem that leads to the formulation of MNL that has the underlying implicit assumption of IIA choice alternatives. As discussed previously, this assumption tends to give higher probability values of less likely alternatives. So, Grange et al. (*7*) proposed new constraints that will minimize the variance $(Pr_m^i(1 - Pr_m^i))$ of any individual choice probability, considering the choice probability as a Bernoulli variable. Since, in a random choice model, it is impossible to reduce the variance to zero, it is assumed that the variance should be lower than a very small value (υ). This results in the 3rd constraint as follows:

$$Pr_m^i(1 - Pr_m^i) \leq \upsilon \tag{4}$$

The Lagrangian function (LG) of the optimization problem in Equation 1 subjected to constraints (2), (3), and (4) is:

$$\begin{aligned} LG = & \sum_{i=1}^{I}\sum_{m=1}^{M} Pr_m^i(\ln(Pr_m^i) - 1) \\ & + \sum_{km} \beta_m^k \left(D_m^k - \sum_{i=1}^{I} Pr_m^i x_m^{ik} \right) \\ & + \sum_{i} \gamma_i \left(1 - \sum_{m=1}^{M} Pr_m^i \right) \\ & + \alpha \left(\upsilon - \sum_{im} (Pr_m^i(1 - Pr_m^i)) \right) \end{aligned} \tag{5}$$

Applying first order conditions to the Lagrangian function with respect to Pr_m^i and some manipulation, the resulting PL probability function becomes:

$$Pr_m^i = \frac{\exp(\sum_k \beta_m^k x_m^{ik} + \alpha(1 - 2Pr_m^i))}{\sum_{m'=1}^{M} \exp(\sum_k \beta_{m'}^k x_{m'}^{ik} + \alpha(1 - 2Pr_{m'}^i))} \tag{6}$$

Here, α is the polarization parameter.

This is equivalent to an MNL with an endogenous variable, $(1-2\ Pr_m^i)$. As per the Karush–Kuhn–Tucker optimality condition, this values should be negative. For the case where its value is close to zero, the PL probability function converges to that of an MNL model. The same probability function can be derived by considering $(1-2\ Pr_m^i)$ as a variable (which is endogenous) in the systematic utility function as follows:

$$U_m^i = \sum_k \beta_m^k x_m^{ik} + \alpha(1 - 2Pr_m^i) + \varepsilon_m^i \tag{7}$$

Here, U is the total utility and ε is the random error component. Sub and superscripts m of individual i are defined previously.

Deriving Equation 6 as an MNL model requires that assumption that ε_m^i has IID Type I extreme value distribution. The endogenous variable, $(1-2\ Pr_m^i)$ does not necessarily accommodate any direct correlations among the random errors of alternative utility functions to overcome IID assumptions. However, any patterns of substitution can be induced by mixing error distribution as mixed logit with error component modelling (Train (*8*)) as follows:

$$U_m^i = \sum_k \beta_m^k x_m^{ik} + \alpha(1 - 2Pr_m^i) + \varepsilon_m^i + \zeta_m^i \tag{8}$$

Here, ζ_m^i is the mixing random error that has multivariate normal distribution.

This results in the new formulation of the model as:

$$Pr_m^i = \int \frac{\exp(\sum_k \beta_m^k x_m^{ik} + \alpha(1 - 2Pr_m^i) + \zeta_m^i)}{\sum_{m'=1}^{M} \exp(\sum_k \beta_{m'}^k x_{m'}^{ik} + \alpha(1 - 2Pr_{m'}^i) + \zeta_{m'}^i)} f(\zeta_m^i) d(\zeta_m^i) \tag{9}$$

This leaves the model in non-closed form as the integration is necessary over the distribution of ζ_m^i, $f(\zeta_m^i)$.

This is a mixed version of the PL model and as we can accommodate difference in sources of randomness in the choice model along with correlations between the alternatives through a multivariate normal specification of ζ_m^i, the model is named as the HPL model. For further simplification, the probability function specified in Equation 9 can be simplified by using a simulation technique as follows:

$$Pr_m^i = \frac{1}{R}\sum_{r=1}^{R} \left(\frac{\exp(\sum_k \beta_m^k x_m^{ik} + \alpha(1 - 2Pr_m^i) + (\zeta_m^i)_r)}{\sum_{m'=1}^{M} \exp(\sum_k \beta_{m'}^k x_{m'}^{ik} + \alpha(1 - 2Pr_{m'}^i) + (\zeta_{m'}^i)_r)} \right) \tag{10}$$

Here, R is the total number of draws from the distribution of ζ_m^i.

For drawing ζ_m^i from its multivariate normal distribution, Halton draws are used in this paper (*8*). However, this does not eliminate the problem of using a standard estimation technique for parameter estimation of HPL because of the presence of an endogenous variable. Since, the HPL probability function is a function of its own probability, a unique fixed point exists and iterations can lead to a convergence of parameters (*15*). In this research, the following iterative algorithm is used, which is programmed in GAUSS by repeatedly applying its maximum likelihood estimation routine, MAXLIK (*16*).

Algorithm

Iteration 0:

Consider $\alpha = 0$ in Equation 10 and estimate the parameter set of the model by using the maximum simulated likelihood estimation procedure and denote $(P^i_m)_0$ as the starting value for the next iteration.

Iteration 1:

Consider estimated parameters values in *iteration* 0 as the starting value and $P^i_m = (P^i_m)_1$ on right hand side of Equation 10 and re-estimate the parameter sets including α which has to be negative (an unconstrained estimation procedure is used in this case, but one can also use constrained optimization to enforce the negativity of α parameter).

Iteration 2:

Consider estimated parameter values in *iteration* 1 as the starting value and $P^i_m = (P^i_m)_1$ on right hand side of Equation 10 and re-estimate the parameter sets including α which has to be negative.

....

Iteration J:

Consider estimates from the previous iteration ($J-1$) as the starting values and $P^i_m = (P^i_m)_{J-1}$ on right hand side of Equation 10 and re-estimate the parameter sets $[\alpha^k, \gamma^k_m, \beta^k_m]_1$ and $P^i_m = (P^i_m)_J$.

For a well-specified utility function, the model converges within a limited number of iterations. It should be noted that a greater amount of estimation is required for the simulation estimation procedure than successive iterations required for the PL model estimation. Also, this algorithm is similar to that of an Expectation Maximization (EM) algorithm, where the value of P^i_m from the previous iteration is used to maximize the parameter values of the following step of iteration.

The proposed HPL model is an advanced version of the discrete choice model with social interaction as it accommodates preference heterogeneity along with social interactions (*7, 17, 18*). The social interaction effect is captured by the polarization effect as it explains the choice probability of an alternative as a function of the same probability function. Considering that the choice probability is a measurement, if not a perfect equivalent, of the market share of the same alternative, the model contains endogenous social interaction effects. However, unlike other social interaction effect models, such as Brock and Durlauf (*17, 18*), the proposed HPL contains a more heterogeneous measure of social interactions.

Study Area and Data for Empirical Study

The National Capital Region of Canada is used as the study area for empirical investigation of this paper. Data from the latest (2011) household travel survey of the NCR are used. Since the study is concerned with the competition of active modes (bicycle and walking) in a competitive modal arrangement, the trip records for which at least one active mode is feasible are considered in the investigation.

A total of 40,411 trip records are found to have all necessary information and are used for empirical modeling in this paper. The dataset has personal and household attributes along with trip origin, destination, and mode information. Modeling mode choice requires alternative level-of-service attributes of all feasible modes. A calibrated traffic assignment model (developed by the NCR planning agency and used for their regional planning studies) is used to impute level-of-service attributes (travel time, travel cost, etc.) into the dataset for modeling. In addition, land use and transportation network data are also compiled from secondary data sources to fuse with travel survey data for empirical modeling.

The final dataset has five feasible modes: car driving, car passenger, public transit, bicycle and walking and the maximum trip length of 16 km. Trips longer than 16 km make both active modes unfeasible and so are excluded from the subset. The sample modal share in the data for empirical modeling is 49% car driving, 18% car passenger, 12% public transit, 3% bicycle and 18% walking. It is clear that bicycle has the lowest modal share and is a minor mode in any competitive arrangement in the NCR. In this final dataset for investigation, the average trip length of observed bicycle trips is around 12 km with 8.7 km standard deviation. So, in fact, all 40,411 trips in the data set are bikeable trips even though bicycle has the lowest modal share.

The NCR planning authority has been putting a variety of efforts and plans in place to encourage cycling and eventually increase bicycle modal shares. So, in this investigation, bicycle networks in the study area are used to generate trip-specific bicycle network attributes in the dataset and these include whether the trip origin and/or destination has bicycle facilities (cycle lanes, pathways, etc.); length of bicycle facilities, grade difference between origin and destination location, and so forth. The average age of the individual in the dataset is 35 years and the dataset has an almost equal share of males and females.

Empirical Investigation

Empirical investigation is focused on developing a mode choice model for the NCR to evaluate factors that affect the choice of bicycle mode choices in the context of competition mode arrangements. That means for contexts/trips, the bicycle is one of the feasible modes. For such contexts in the NCR, the bicycle is a minor mode contributing around 3% of all bikeable trips. So, an accurate mode choice model is necessary for a clear understanding of the potential of promoting bicycle travel in the study area. For these, three types of models have been estimated: an MNL model, a PL model, and an HPL model. Final specifications of these three models are defined based on the statistical significance of the estimated parameters with expected signs. A 95% confidence limit (corresponding *t*-statistics of 1.64 for the parameters with expected sign) is considered for the statistical significance tests, but some with lower confidence are also retained if they provide behavioural insights. Each of these three final models is individually compared against a constant-only model (market share model) through the measure of adjusted rho-squared values. The final specifications of the MNL and PL models are then compared against the HPL model by using the Horowitz's non-nested test (*19*). This test gives the z-value (ϕ^{-1}) based on which the confidence level of the superiority of the HPL can be measured. For a value of z over 2.60, the confidence limit is 99%. The z value of Horowitz's tests is estimated as:

$$z = \Phi^{-1}\left(-\sqrt{-2(\bar{\rho}_u^2 - \bar{\rho}_l^2)L_0 + (B_u - B_l)}\right) \tag{11}$$

Here, $\bar{\rho}_u^2$ and $\bar{\rho}_l^2$ are adjusted Rho-squared value of the HPL model and the comparing model; L_0 is the Loglikelihood of the constant-only model; B_u and B_l are the number of estimated parameters of the HPL and the comparing model.

Summary statistics of the model comparisons and parameters of the final HPL model are presented in Table 1 (Part I and II). The goodness-of-fit of the presented HPL model is over 0.28, which indicates a very good fit to the observed data given the complexity in the model specification. The z value of Horowitz's test of the final model measured against the PL and MNL model are over 2.60. This indicates an over 99% confidence in the fact that the proposed Heteroscedastic Polarized Logit model is better than the Polarized Logit model proposed by Grange et al. (*7*) as well as an equivalent MNL model. Moreover, between the PL and the MNL model, the PL model stands out as superior to the MNL model with over 99% confidence. The following discussion is based on the HPL model of bikeable trip mode choice in the NCR.

Alternative Specific Constants (ASC) of the model capture the effects of systematic but unobservable factors influencing the choice of the modes. The empirical model considers the bicycle as the reference alternative and the estimated ASCs reveal that Drive Alone has negative unobserved factors (dis-benefit) even though it has the highest mode share in the bikeable trips. Perhaps, faster travel time stands out being the key winning factor for this mode even though there could be many systematic dis-benefits in choosing it. However, compared with the bicycle, car passenger, transit, and walking have more positive systematic unexplained benefits (utility) in the study area. Variables related to travel time, cost, distance, socioeconomic variables, topography (grades) and even presence/absence of bicycle infrastructure are used in the model, but still, there is considerable negative unexplained utility of choosing the bicycle compared with the car passenger, transit and walking modes. Perhaps, such negative factors are the systematic barriers that need to be overcome, which may require more than just investment in bicycle infrastructure.

The model correctly accommodates the negative coefficients of travel time and travel cost variables of driving and transit options, but the value of these variables is low. The resulting subjective Value of Travel Time (VOT) savings (the ratio time and cost coefficients) for driving and transit is as low as \$1.55 per hour. This explains the superiority of driving (driving alone and car passenger contributes 67% mode shares) over all other modes. Fast and relatively congestion free movements in the NCR for the bikeable trips systematically favor motorized modes over a bicycle. So, perhaps, increasing lane separation to give bicycles more free passage and at the same time constraining lane spaces for driving along with lowering speed limits for driving could be beneficial in increasing the relative attractiveness of cycling.

Public transit is the second minor mode having lower market share than that of walking and car passenger modes. Transit has additional negative utility because of the walking time required to access the transit station/stop, waiting time at the station/stop and time required to transfer between lines before reaching the destination. For the bikeable trips in the NCR, travellers are more (negatively) sensitive to walking access or transfer time than in-vehicle time while considering transit. Also they are more (negative) sensitive to waiting time at the station/stop time than walking access or transfer time.

Trip distance is considered to be the impedance variables for walking and bicycle mode choice. Instead of commonly used linear distance specification, a piece-wise non-linear specification is used. Nonlinearity of distance effects are captured by adding squared distance as variables in the utility functions of walking and cycling. The sign of the squared-distance is negative for walking, but

Table 1. Empirical Model

	Parameter	*t*-stat	
Alternative specific constant			
Drive alone	−5.57	−5.81	
Car passenger	6.02	7.70	
Transit	5.45	6.87	
Bicycle	0.00	—	
Walking	4.34	5.50	
In-vehicle travel time (minutes)			
Drive alone, car passenger, transit	−0.01	−2.38	
Total cost or fare			
Drive alone, car passenger, transit	−0.30	−16.35	
Walking access/egress time			
Transit	−0.02	−5.16	
Waiting time			
Transit	−0.04	−5.60	
Number of transfers			
Transit	−0.02	−0.76	
Distance for			
Bicycle			
<1 km	−0.66	−2.21	
1–3 km	−0.19	−2.31	
3–5 km	−0.24	−4.37	
5–6 km	−0.29	−5.41	
6–7 km	−0.32	−5.90	
7–8 km	−0.29	−5.27	
8–9 km	−0.29	−5.01	
9–10 km	−0.33	−5.42	
>10 km	−0.34	−4.62	
Squared distance	0.01	2.10	
Gender: Male			
Drive alone	−0.44	−5.42	
Car passenger	−1.12	−14.87	
Transit	−0.85	−10.96	
Bicycle	0.00	—	
Walking	−0.63	−8.08	
Number of private cars per household members			
Drive alone	3.47	18.99	
Car passenger	1.85	13.61	
Transit	0.53	3.53	
Bicycle	0.00	—	
Walking	1.35	9.40	
Trip purpose: Work and work related			
Drive alone	−0.68	−6.28	
Car passenger	−1.22	−13.05	
Transit	−0.53	−5.79	
Bicycle	0.00	—	
Walking	−0.30	−2.96	
Trip purpose: School			
Drive alone	−0.51	−2.82	
Car passenger	0.73	4.72	
Transit	0.85	5.21	
Bicycle	0.00	—	
Walking	0.51	2.99	
Error standard deviation			Nest Correlations
Bicycle and car driver	2.35	14.05	0.59
Bicycle and car passenger	2.04	−10.31	0.55
Bicycle and transit	0.17	−1.57	0.09
Bicycle and walking	0.86	−6.39	0.34
Polarization coefficients			

(continued)

Table 1. (continued)

	Parameter	t-stat
Trip purpose: Work	−0.84	−11.09
Trip purpose: School	−1.03	−11.53
Trip purpose: Shopping and others	−0.87	−11.97
Distance for		
Walking		
<1 km	−0.43	−2.37
1–1.5 km	−0.47	−4.46
1.5–2 km	−0.44	−4.96
2–2.5 km	−0.55	−6.10
2.5–3 km	−0.54	−5.71
>3 km	−0.49	−4.65
Squared distance	−0.08	−3.85
Origin-destination grade difference		
Bicycle	−0.01	−0.67
Origin-destination grade difference		
Walking	−0.06	−4.15
Bicycle infrastructure for bicycle mode choice		
Bicycle facility in origin, but not in destination	1.02	1.38
Bicycle facility in destination, but not in origin	1.26	1.71
Bicycle facility in origin and destination	1.25	1.73
No bicycle facility in origin and destination	0.00	—
Socio-economic variables		
Age in years (log transformed)		
Drive alone	2.26	12.14
Car passenger	−1.40	−17.04
Transit	−0.44	−4.79
Bicycle	0.00	—
Walking	−0.06	−0.71
Loglikelihood of final HPL model	−34276.89	
Loglikelihood of constant-only model	−47752.47	
# of parameters final HPL model	58	
# of parameters in constant-only model	4	
Adjusted Rho-squared of final HPL model	0.281066	
Loglikelihood of PL model	−34305.5	
# of parameters in PL model	54	
Adjusted Rho-squared of PL model	0.28055	
Loglikelihood of MNL model	−34382.12	
# of parameters in MNL model	51	
Adjusted Rho-squared of MNL model	0.278931	
Non-nested Horowitz hypothesis test of the final HPL model		
Φ^{-1} with respect to PL model	−7.295339	
Φ^{-1} with respect to MNL model	−14.52238	
Distance for walking		
<1 km	−0.43	−2.37
1–1.5 km	−0.47	−4.46
1.5–2 km	−0.44	−4.96
2–2.5 km	−0.55	−6.10
2.5–3 km	−0.54	−5.71
>3 km	−0.49	−4.65
Squared distance	−0.08	−3.85

positive for bicycle mode. This indicates that for walking mode, there is an increasing rate of negative effects with increasing trip distance, but the negative effect of trip distance has a diminishing rate of negative effects on the bicycle mode choice. Figure 1 presents the effects of trip distance on the systematic utilities of walking and cycling. Walking is not a feasible model for trip distances longer than 3 km, but the bicycle is found to be feasible for trips as long as 16 km in the NCR. It is also clear that trip distance has a more negative effect on choosing

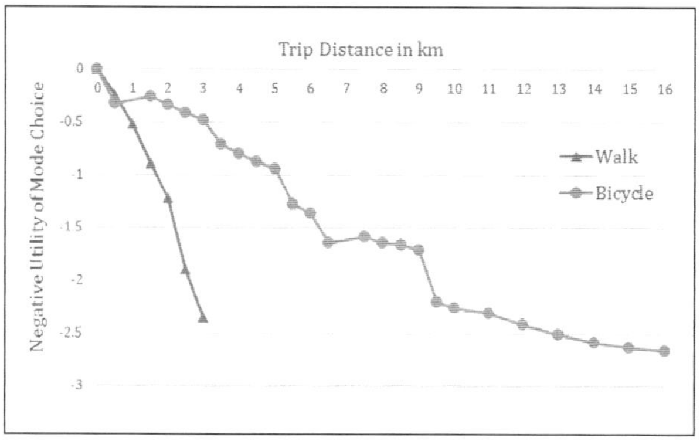

Figure 1. Negative utility of travel distance for walking and bicycle mode choice.

walking as the mode than on choosing the bicycle as the mode. Yet, walking mode has a six times bigger market share than that of the bicycle for the bikeable trips. This clearly demonstrates the fact that something needs to be done to make the people consider the bicycle as a travel mode in the first place.

To investigate the effects of topography on the choice of using a bicycle for bikeable trips, the grade differences between the origin and the destination locations are used as a variable in the bicycle mode choice utility function. The variable has an expected negative sign indicating the difficulty in using a bicycle on a steep slope, but it is not found statistically significant. With such a large dataset, an insignificant variable in the model indicates that it is not a serious issue in that study area. Even though this is perceived as a negative factor in choosing a bicycle, but there is no evidence that it has any real effect in the population of the study area. However, this is a significant variable affecting negatively the choice of bicycle as a travel mode. As expected, the presence of bicycle facilities (e.g., cycle lanes, cycle parking facility, etc.) has very strong and positive effect on the choice of the bicycle as a travel mode for the bikeable trips. Bicycle facilities at the destination of the trip have the largest positive effect on the choice of a bicycle for bikeable trips.

The empirical model also used socio-economic and trip context specific variables to develop a clear understanding of the issues related to not choosing the bicycle as a travel mode for the bikeable trips in the NCR. Driving is the most attractive mode for older travellers and, in fact, preference to drive alone increases twice as much as the age of the trip maker increases. Besides driving, the choice of the bicycle over all other modes seems to increase with the age. However, as in many other North American cities, the bicycle is not popular with women. The model reveals that women are, in general, less likely to choose the bicycle as a travel mode in the NCR.

Among all socio-economic variables, household car ownership is proven to be the most influential factor in increasing the probability of choosing driving over cycling even for the short-distance bikeable trips. Interestingly, a high level of car ownership in a household also negatively influences the choice of the bicycle over non-driving mode alternatives. This indicates the still-dominant car culture and its apparent negative influence on the bicycle as a travel mode in the NCR. Policies targeting the lowering of household car ownership levels seem to have a positive influence on the promotion of cycling in the NCR. According to the destination purposes of the trips, the bicycle is the most attractive mode for work and work-related trips, but it is not an attractive mode for school trips. This indicates that the policy initiatives are deemed necessary to promote cycling to school in the NCR.

The proposed HPL model is capable of capturing competitions and imperfection substitution of cycling with the other modes under investigation. The empirical model used the error-component approach to capture correlations of bicycle choice with other mode choices. Four error components are designed to investigate the other mode users who would be most likely to switch to the bicycle if bicycle choice utility can be increased. The estimated standard deviation of the error components gives an approximate estimation of such intra-alternative correlations within the nests induced by the error components (*8*).

The proposed error component specification induces a cross-nested correlation pattern in the HPL model and it reveals that bicycle can be nested with all other modes. This indicates that travelers are likely to switch from all other modes in different proportions. The error correlation of bicycle choice is the lowest with public transit and the highest with the drive-alone option. It indicates that bicycle and transit are not in strong competition in the NCR. This may be a good sign as both transit and bicycle are seen to be sustainable modes and low competition between these would reduce any necessity for a trade-off between investing in transit and bicycle infrastructure. Any initiatives to facilitate combining bicycle and transit, for example, bike-on-board, bicycle parking facilities at transit stations/stops will benefit both modes. The highest correlation between drive alone and bicycle reveal that, with proper policy initiatives, people who drive are most likely to switch to cycling in the NCR.

Conclusion and Policy Relevance

The main objective of the paper was to improve our empirical understandings of the factors that play critical roles in the choice of using a bicycle as a travel mode in the NCR of Canada. Data from the latest household travel survey of the NCR were used to identify the trips that are bikeable. Bikeable trips were defined as the trips with a trip distance less than 16 km, which was the longest bicycle trip distance observed in the dataset. A total of 40,411 trip records was separated for the investigation of which the bicycle share was around 3%. This raised the issue of the bicycle as the inferior/minor choice alternative in empirical investigation, which had been a problematic modelling context for using many classical or even advanced discrete model choice modeling approaches. The paper identified a recently proposed discrete choice model, the PL model, that could handle the issue of minor mode alternatives better than any other discrete choice models, with the caveat, however, that this model was not capable of handling correlations among the choice alternatives, which was also an important issue in the current investigation. So, the paper extended the PL model and proposed a new HPL model for investigating the competition of bicycle mode with all other feasible modes for the bikeable trips in the NCR. The model was estimated empirically and it proved that the proposed model is superior to the PL and the MNL models. This empirical model was then analyzed to gain evidence-based understanding of the factors influencing the choice of bicycle as a travel mode in the NCR.

Evidently, bicycle facilities (infrastructure) were found to be very strong in positively influencing the choice of using a bicycle as the travel mode for bikeable trips. However, it was also evident that there were other factors beyond such bicycle infrastructure that needed to be explored and addressed for a greater uptick in bicycle use. With relatively large and positive alternative specific constants in the utility functions of walking mode (in reference to that of a bicycle) along with the lower sensitivity of travel distance to bicycling than walking, walking had a three times larger market share than the bicycle. So, it became clear that people were not considering the bicycle as a mode for short trips (shorter than 3 km) in the first place as they were expected to. Similarly, the relatively low sensitivity of travel time, travel cost, and transit accessibility related factors revealed that the motorized modes had fast and congestion-free movement in the NCR for the bikeable trips reducing the competitive edge of bicycling as a travel mode.

Evidently, older people seemed to prefer the bicycle more than the younger people. Among the bicycle users in the NCR, cycling to work was preferable, but cycling to school was not. Also, as in many other North American cities, women in the NCR were less likely to choose the bicycle as a travel mode than men. So, there were more things to do than just investing in bicycle infrastructure to increase the use of the bicycle in the NCR. Empirical evidence from the HPL model indicated that systematic hindrances to motorized modes, perhaps by lowering speed limits, constraining available lane capacities, and so forth would be necessary to draw the bicycle out of its inferior position as an alternative travel mode for the bikeable trips. Any other efforts to promote consideration of cycling, for example, social marketing, bicycle promotion programs, education/publicity on the health benefits of cycling, and so forth, would be necessary to make cycling attractive in the first place. Programs, policies, or other initiatives to promote cycling among students and woman, in general, seemed to be imperative for a greater uptick of bicycle mode shares in the NCR. Household car ownership seemed to be one of the biggest contributors to the attractiveness of the drive-alone option and any effort to reduce household car ownership levels seems to benefit bicycle mode choice more than any other mode choices.

The paper contributes to the choice modeling methodology by proposing the HPL model, but it also contributes to the literature of bicycle travel demand by presenting an empirical model of mode choices in the context of bikeable trips. However, the research can be further extended in some directions, for example, joint modeling of mode choice and bicycle ownership and the application of prospect theory. It would reveal greater understanding of how to promote the choice of the bicycle as a travel mode. However, this poses a methodological challenge and is, therefore, recommended for further research.

Acknowledgments

The study was funded by an NSERC Discovery grant. The author acknowledges the support of TRANS Committee of the NCR by giving access to the dataset and related information used in this study. The author is grateful to Ahmad Subhani for facilitating such access and to Mohammed Mahmoud Salah for preparing bicycle network data used for the investigation. However, the author takes the sole responsibility for any comments and discussions presented in this paper.

Author Contributions

The author confirms sole responsibility for the following: study conception and design, data collection, analysis and interpretation of results, and manuscript preparation.

References

1. Wilson, A. G. *Entropy in Urban and Regional Modeling*. Pion, London, UK, 1970.
2. Ben-Akiva, M., and S. Lerman. Discrete Choice Analysis: Theory and Application to Travel Demand. MIT Press, Cambridge, MA, 1985.
3. Roorda, M. J., D. Passmore, and E. J. Miller. 2009. Including Minor Modes of Transport in a Tour-Based Mode Choice Model with Household Interactions. *ASCE*

Journal of Transportation Engineering, Vol. 135, No. 2, 2009, pp. 935–945.

4. Huang, Z., H. Zhao, and D. Zhu. Two New Prediction-Driven Approach to Discrete Choice Prediction. *ACM Transactions on Management Information Systems*, Vol. 3, No. 2, 2012, Article 9.
5. Anas, A. Discrete Choice Theory, Information Theory, and the Multinomial Logit and Gravity Model. *Transportation Research Part B: Methodology*, Vol. 17, No. 1, 1983, pp. 17–23.
6. Donoso, P., and L. De Grange. A Microeconomic Interpretation of the Maximum Entropy Estimator of Multinomial Logit Models and Its Equivalence to the Maximum Likelihood Estimator. *Entropy*, Vol. 12, 2010, pp. 2077–2084.
7. de Grange, L., F. González, I. Vergas, and J. C. Muñoz. A Polarized Logit Model. *Transportation Research Part A*, Vol. 53, 2013, pp. 1–9.
8. Train, K. *Discrete Choice Model with Simulation (2nd ed.)*. Cambridge University Press, Cambridge, MA, 2009.
9. Abou-Zied, M., and D. M. Scott. Current Issues in Mode Choice Modeling. *Transportation*, Vol. 38, 2011, pp. 581–585.
10. Idris, A. O., K. M. N. Habib, and A. Shalaby. An Investigation on the Performances of Mode Shift Models in Transit Ridership Forecasting. *Transportation Research Part A*, Vol. 78, 2015, pp. 551–565.
11. Mahmoud, M. S., A. Weiss, and K. M. N. Habib. Myopic Choice or Rational Decision Making? An Investigation into Mode Choice Preference Structures in Competitive Modal Arrangements in a Multimodal Urban Area, the City of Toronto. *Canadian Journal of Civil Engineering*, Vol. 43, 2016, pp. 423–428.
12. Gaudry, M., and M. Degenais. The Dogit Model. *Transportation Research*, Vol. 13B, 1979, pp. 105–111.
13. Habib, K. M. N., Y. Tian, and H. Zaman. Modelling Commuting Mode Choice with Explicit Consideration of Carpool in the Choice Set Formation. *Transportation*, Vol. 38, 2011, pp. 587–604.
14. Habib, K. M. N., and A. Weiss. Evolution of Latent Modal Captivity and Mode Choice Patterns for Commuting Trips: A Longitudinal Analysis using Repeated Crossectional Dataset. *Transportation Research Part A*, Vol. 66, 2014, pp. 39–51.
15. Fang, S., J. Rasajekera, and H. Tsao. *Entropy Optimization and Mathematical Programming (2nd ed.)*. Kluwer Academic Publisher, Dordrecht, the Netherlands, 1997.
16. Aptech. *GAUSS*, 2017. https://www.aptech.com/. Accessed July 2017.
17. Brock, W., and S. Durlauf. Discrete Choice with Social Interactions. *Review of Economic Studies*, Vol. 68, 2001, pp. 235–260.
18. Brock, W., and S. Durlauf. Multinomial Choice with Social Interactions. In *The Economy as an Evolving Complex System* (Blume, L., and S. Durlauf, eds.), Vol. 3. Oxford University Press, Oxford, UK, 2006.
19. Horowitz, J. H. Statistical Comparison of Non-nested Probabilistic Discrete Choice Models. *Transportation Science*, Vol. 17, 1983, pp. 319–350.

The Standing Committee on Transportation Demand Forecasting (ADB40) peer-reviewed this paper (18-03135).

Article

TRR
JOURNAL OF THE TRANSPORTATION RESEARCH BOARD

Transportation Research Record
2018, Vol. 2672(49) 69–80

DOI: 10.1177/0361198118794710
journals.sagepub.com/home/trr
SAGE

Time Allocation Behavior of Twentieth-Century American Generations: GI Generation, Silent Generation, Baby Boomers, Generation X, and Millennials

Annesha Enam[1] and Karthik C. Konduri[1]

Abstract

In recent years, time engagement behaviors of two generations, namely Baby Boomers and Millennials have sparked much interest because these generations constitute the bulk of the American population today and they also exhibit "atypical" activity–travel patterns compared with other generations. The objective of the current research is to conduct a systematic study of the time engagement behaviors of five American generations: the GI Generation (birth year: 1901–1924), the Silent Generation (birth year: 1925–1943), Baby Boomers (birth year: 1944–1964), Generation X (birth year: 1965–1981), and Millennials (birth year: 1982–2000). Particularly, the study aims at isolating heterogeneity in behaviors associated with structural changes in the society from those associated with inherent generational characteristics. Using data from four waves (1965, 1985, 2005, and 2012) of the American Heritage and Time Use Study, the analysis explores the time engagement behaviors while accounting for the age, period, and cohort effects in addition to different socioeconomic and demographic variables. The analysis reveals that Millennials have generally delayed participation in life-changing events such as marriage and workforce entry, and have exhibited prolonged student status compared with previous generations. Millennials show lower participation in work and higher participation in discretionary activities compared with individuals of the same age group from previous generations. On the other hand, Baby Boomers clearly exhibited increased travel engagement compared with the previous generations at different stages of their lives.

Researchers from different disciplines have approached the study of trends in time engagement behavior with different motivations and perspectives (*1*, *2*). Some examples include: study of gender convergence in time allocation behaviors by Fisher et al. (*3*); the trend in time spent with children studied by Fox et al. (*4*); the study of leisure time engagement trend by Aguiar and Hurst (*5*); and the study of cross-country time engagement trend by Gimenez-Nadal and Sevilla (*6*). Study of time engagement behavior has also garnered interest among transportation researchers in the recent past owing to the acknowledgment of the derived nature of travel.

There is a rich body of literature on time engagement choices of individuals in the field of transportation (*7*, *8*). More recently the out-of-home activity and travel engagement behavior of Baby Boomers and Millennials have sparked considerable interest among transportation researchers because of their "atypical" activity–travel patterns (*9*). For example, Millennials have been associated with declining rates of driving license holding, decreasing automobile dependency, decreasing employment participation, and use of virtual mobility (*10*, *11*). These trends in the activity–travel choices of Millennials have been attributed to their tendency to prolong years in education, delay workforce entry, and postpone marriage and consequently family formation (*12*). On the other side of the spectrum, the aging Boomers have been characterized to be more active than the previous generations, as demonstrated by increased out-of-home activity participation and late retirement tendency (*13*, *14*).

The overarching goal of the current research is to compare the activity–travel behaviors of these two generations, namely Baby Boomers (born 1944–1964) and Millennials (born 1982–2000) with that of previous

[1]Department of Civil and Environmental Engineering, University of Connecticut, Storrs, CT

Corresponding Author:
Address correspondence to Annesha Enam: annesha.enam@uconn.edu

generations while accounting for age, period, and cohort effects. The study focuses on understanding the similarities and differences in activity–travel pursuits of five American generations from the twentieth century: the GI Generation (born 1901–1924), the Silent Generation (born 1925–1943), Baby Boomers (born 1944–1964), Generation X (born 1965–1981) and Millennials (born 1982–2000). The analysis is conducted by combining data from four waves of the American Heritage Time Use Survey (AHTUS) from 1965 through 2012 to obtain a representative sample from different generations and across different stages of life. It must be noted that the definitions of the generational cohorts vary across the literature. The definitions in the current research are based on birth years and adopted from Strauss and Howe (*15*). Also, the range of birth years used to define different generational cohorts vary from 16 years (i.e., birth year between 1965 and 1981 for Generation X) to 23 years (i.e., birth year between 1901 and 1924 for the GI Generation).

The contribution of the current study is in the exploration of generational effect (also referred to as cohort effects) in relation to the two other types of dynamic effects: age and period effects. These three types of effects are defined below:

1. *Cohort effect* refers to activity–travel behavior differences associated with being born at a specific time in history. The cohort effect represents the attitudes, perceptions, beliefs, and traits that characterize individuals of same generation—these are formed by common epochal events that the individuals of the same generation experience (*15*). For example, Kostyniuk and Kitamura (*16*) note that a person growing up during an era of motorization would continue to be inclined toward private vehicle use for their mobility regardless of other factors such as living arrangements.
2. *Age effect* refers to differences associated with an individual's age. This component essentially captures the effects of life cycle stages.
3. *Period effect* refers to differences associated with a specific period that the individual experiences. This period is also generally experienced by more than one birth cohort. For example, during recessions, people belonging to all birth cohorts and age groups may curtail their discretionary spending and thus discretionary time allocation because of the financial and economic uncertainty they experience.

Despite the growing recognition of the study of cohort effects in understanding and forecasting travel behavior, there has been limited literature exploring the influence of cohort on activity participation (what types of activities are pursued?) and time allocation choices (how much time is spent in those activities?). Further, studies have not attempted to separate the cohort effect from the two other types of effects related to age and period. The research presented in this paper attempts to fill this gap by explicitly accounting for the cohort, age, and period when exploring the heterogeneity in activity participation and time allocation choices of Americans. The study also controls for other demographic and situational differences in exploring the heterogeneity. To the best of the authors' knowledge, the current study is the first of its kind that sheds light on the trends in in-home and out-of-home activity participation and time allocation behaviors of all American generations from the twentieth century.

The rest of the paper is organized as follows. The next section describes the AHTUS data. The third section presents a descriptive analysis of the trends in activity participation and time engagement behaviors of different birth cohorts of same-age groups as well as depicts the demographic differences across cohorts. The fourth section presents a multivariate modeling framework to explore the similarities and differences in participation and time engagement behaviors across cohorts while accounting for other factors. The fifth section presents results from the multivariate modeling analysis. The sixth and the final section provides a summary of the research findings and offers suggestions for future research.

Data Composition

In this research, we used data from the AHTUS, a database of cross-sectional time-use datasets of US individuals collected over six decades. The current study utilizes four datasets from the AHTUS database: (1) 1965–1966 dataset; (2) 1985 dataset, and (3) two waves of the American Time Use Survey (ATUS) from 2005 and 2012. The four cross-sectional datasets include information on demographic characteristics and living arrangements of the respondents along with a log of activity and travel episodes pursued over a 24-hour period. The 24-hour activity–travel log includes information about the type, location, accompaniment, and mode of the reported activity and travel episodes.

The pooled sample of four datasets includes 23,315 observations: 1,168 observations from 1965–1966, 2,160 observations from 1985, 10,224 observations from 2005, and 9,763 observations from 2012. Following data cleaning, the episode-level activity–travel information was aggregated up to the day level by activity type. At this stage, the detailed activity–travel purposes reported by AHTUS were consolidated into six broad categories: maintenance, discretionary, work, education, travel, and other. Further, the maintenance and discretionary

Table 1. Sample Composition of Five Cohorts by Age Groups

Age groups	18–24 years	25–34 years	35–44 years	45–54 years	55–64 years	≥65 years	Total
Periods[a]			GI Generation (birth year: 1901–1924)				
1965	NA	NA	121	254	161	NA	536
1985	NA	NA	NA	NA	98	246	344
2005	NA	NA	NA	NA	NA	351	351
2012	NA	NA	NA	NA	NA	NA	NA
Total	NA	NA	121	254	259	597	1,231
Periods[a]			Silent Generation (birth year: 1925–1943)				
1965	99	294	161	NA	NA	NA	554
1985	NA	NA	109	318	171	NA	598
2005	NA	NA	NA	NA	381	1,335	1,716
2012	NA	NA	NA	NA	NA	1,489	1,489
Total	99	294	270	318	552	2,824	4,357
Periods[a]			Baby Boomers (birth year: 1944–1964)				
1965	64	NA	NA	NA	NA	NA	64
1985	203	562	345	NA	NA	NA	1,110
2005	NA	NA	1,031	2,035	1,048	NA	4,114
2012	NA	NA	NA	1274	1,655	524	3,453
Total	267	562	1,376	3,309	2,703	524	8,741
Periods[a]			Generation X (birth year: 1965–1981)				
1965	NA	NA	NA	NA	NA	NA	NA
1985	95	NA	NA	NA	NA	NA	95
2005	109	1,796	1,494	NA	NA	NA	3,399
2012	NA	783	2,042	558	NA	NA	3,383
Total	204	2,579	3,536	558	NA	NA	6,877
Periods[a]			Millennials (birth year: 1982–2000)				
1965	NA	NA	NA	NA	NA	NA	NA
1985	NA	NA	NA	NA	NA	NA	NA
2005	644	NA	NA	NA	NA	NA	644
2012	593	845	NA	NA	NA	NA	1,438
Total	1,237	845	NA	NA	NA	NA	2,082

Note: NA = not available.
[a]Period refers to the survey year.

activities were disaggregated into in-home and out-of-home categories depending on the reported location. The above categorization scheme resulted in the following eight activity–travel groupings: maintenance in-home (MIH), maintenance out-of-home (MOH), discretionary in-home (DIH), discretionary out-of-home (DOH), work, education, travel, and other. Furthermore, the sample was restricted to adults (i.e., individuals of 18–85 years of age).

Table 1 presents the sample size of different cohorts by age group in the pooled dataset. The rows indicate the period to which the different cohorts belong. As can be seen from the table, it was only possible to observe the Silent Generation and Baby Boomers across different life cycle stages. For the rest of the generations it was only possible to capture individuals in either the older (e.g., the GI Generation) or the younger (e.g., Generation X and Millennials) life cycle stages.

Descriptive Analysis

In this subsection, a descriptive analysis is presented based on the subsample. Also, the results reported below are based on a weighted sample.

Trends in Activity Participation

Table 2 presents the percentage of people participating in different activities and travel across different cohorts of the same age group. Respondents used in the analysis participated in at least one episode of MIH; As a result, 100% participation in MIH is observed for all cohorts by all age groups. In general, participation in MOH is reducing across cohorts of different age groups except for individuals transitioning into adulthood (25–34 years) and elderly (above 65 years).

There is hardly any difference in DIH activity participation across cohorts in the 18–24 years age group. For the remaining age categories, the DIH activity participation decreases across different cohorts. However, the rate of decrease falls with increasing age. The decrease in DIH activity participation is generally accompanied by increases in DOH activity participation by different cohorts of all age groups. The increase in DOH participation is most prominent for the youngest age group.

Millennials in the younger age group exhibit a considerable decrease in work participation compared with the previous generations. Though the work participation of

Table 2. Participation Percentages across Cohorts by Age Groups

Age groups	Cohorts	MIH	MOH	DIH	DOH	Work	Education	Travel
18–24 years	Silent generation	100	79	89	48	57	1	92
	Baby boomers	100	79	91	59	50	5	94
	Generation X	100	76	91	65	45	10	91
	Millennials	100	75	90	71	39	13	92
25–34 years	Silent generation	100	77	96	52	48	0	93
	Baby boomers	100	76	94	49	56	0	94
	Generation X	100	81	90	63	57	3	93
	Millennials	100	80	89	62	54	3	93
35–44 years	GI Generation	100	81	91	45	62	NA	95
	Silent generation	100	76	96	51	53	NA	93
	Baby boomers	100	79	93	57	60	NA	92
	Generation X	100	80	91	60	58	NA	91
45–54 years	GI Generation	100	80	96	40	57	NA	90
	Silent generation	100	78	97	46	60	NA	92
	Baby boomers	100	75	95	61	61	NA	89
	Generation X	100	74	93	58	56	NA	87
55–64 years	GI Generation	100	78	97	44	48	NA	88
	Silent generation	100	70	98	58	38	NA	86
	Baby boomers	100	72	96	61	47	NA	86
≥65 years	GI Generation	100	57	99	49	9	NA	69
	Silent generation	100	59	98	59	11	NA	73
	Baby boomers	100	62	97	58	25	NA	78

Note: DIH = discretionary in-home; DOH = discretionary out-of-home; MIH = maintenance in-home; MOH = maintenance out-of-home; NA = not available.

Millennials increases as they transition into adulthood, it is still 3% less than that of Generation X. It can also be seen that Generation X continues to exhibit lower work activity participation than Baby Boomers even in the adult life stages. It is also interesting to note that Baby Boomers have higher participation in work activity compared with the Silent Generation for transition into older (55–64 years) and old-age (≥65 years) groups. On the other hand, young Millennials show higher participation in education compared with Generation X. Both Millennials and Generation X show higher participation in education compared with Baby Boomers and the Silent Generation for individuals in the 25–34 years age group.

For travel participation, small differences were observed between different cohorts at earlier life stages. For example, both Millennials and Generation X participate less in travel compared with Baby Boomers. This also brings about an important observation on widely held notions about Millennials' "atypical" travel patterns. It appears that, even before Millennials, the decline in travel participation was exhibited by Generation X at different life stages when compared with Baby Boomers. Also, the older Baby Boomers demonstrate higher participation (by 5%) in travel compared with the Silent Generation.

Trends in Time Allocation

Table 3 presents a comparison of average and median time allocation across cohorts belonging to each of the six age groups. It must be noted that the average and the median values of the time allocation were calculated considering only those individuals who reported participating in at least one episode of the said activity type. The inferences drawn from Table 3 are mostly based on the average time allocation values. In general, a comparison of the average and median time allocation reveals that the time allocation distribution for different activity types and travel are left skewed, resulting in a smaller median compared with the mean. The only exception is work activity with higher median than mean, indicative of a right-skewed distribution.

MIH time allocation has considerably reduced from the GI Generation to Millennials for all age groups. MOH time allocation has increased moderately from the GI Generation to Baby Boomers for those younger than 54 years old. For the young Millennials (18–24 years), daily MOH time allocation is lower compared with Generation X by 6 min.

DIH time spent is higher for later generations compared to preceding generations of the young age group. The difference is most significant from Baby Boomers to Generation X. Young adult Millennials (25–34 years) spend nearly 9 min more in DIH compared with Generation X and about 7 min less compared with Baby Boomers of the same age group. Generation X's DIH time allocation is lower than that of Baby Boomers of all age groups except for the young age group. On the other hand, the Baby Boomers' DIH time allocation has increased for all age groups compared with the Silent Generation except for the old-age group.

Table 3. Time Allocation Trend (Average and Median): Time Budget in Minutes across Cohorts by Age Groups[a]

Age groups	Cohorts	MIH	MOH	DIH	DOH	Work	Education	Travel
18–24 years	Silent generation	742 (685)	84 (60)	182 (180)	144 (128)	500 (485)	NA	98 (80)
	Baby boomers	694 (670)	101 (60)	199 (180)	195 (145)	448 (480)	223 (210)	100 (88)
	Generation X	726 (705)	116 (65)	232 (205)	148 (135)	409 (435)	271 (308)	85 (70)
	Millennials	683 (690)	110 (60)	239 (210)	195 (165)	435 (442)	233 (223)	87 (67)
25–34 years	Silent generation	789 (758)	80 (60)	201 (179)	150 (128)	468 (480)	NA	81 (65)
	Baby boomers	716 (690)	112 (70)	223 (190)	154 (125)	447 (480)	NA	96 (81)
	Generation X	719 (730)	101 (65)	207 (180)	139 (120)	459 (473)	189 (175)	88 (70)
	Millennials	713 (715)	106 (65)	216 (194)	152 (120)	465 (477)	155 (160)	84 (61)
35–44 years	GI Generation	755 (725)	83 (60)	183 (173)	155 (129)	460 (480)	NA	88 (79)
	Silent generation	760 (710)	99 (60)	204 (180)	143 (120)	452 (490)	NA	93 (73)
	Baby boomers	700 (694)	103 (62)	219 (195)	136 (105)	469 (480)	NA	89 (75)
	Generation X	723 (725)	98 (63)	214 (190)	137 (120)	464 (479)	NA	88 (70)
45–54 years	GI Generation	765 (714)	82 (60)	207 (180)	133 (125)	469 (480)	NA	83 (70)
	Silent generation	709 (675)	91 (60)	248 (221)	138 (110)	441 (480)	NA	91 (70)
	Baby boomers	678 (675)	95 (60)	256 (227)	128 (105)	465 (475)	NA	88 (65)
	Generation X	709 (694)	95 (65)	249 (223)	139 (120)	466 (480)	NA	85 (70)
55–64 years	GI Generation	750 (710)	92 (60)	257 (237)	149 (115)	462 (480)	NA	83 (72)
	Silent generation	714 (715)	116 (70)	316 (290)	146 (105)	442 (461)	NA	91 (70)
	Baby boomers	687 (679)	104 (60)	317 (290)	134 (120)	451 (470)	NA	82 (62)
≥65 years	GI Generation	801 (790)	94 (60)	426 (420)	143 (115)	405 (430)	NA	70 (55)
	Silent generation	745 (735)	110 (70)	437 (430)	164 (135)	360 (355)	NA	78 (60)
	Baby boomers	717 (705)	100 (60)	400 (370)	164 (123)	422 (415)	NA	82 (60)

Note: DIH = discretionary in-home; DOH = discretionary out-of-home; MIH = maintenance in-home; MOH = maintenance out-of-home; NA = not available.
[a]The values outside the parentheses represent the average duration calculated across people who reported at least one episode of the said activity type on the survey day. Similarly, the values inside the parentheses represent the median duration calculated across people who reported at least one episode of the said activity type on the survey day.

Like DIH time allocation, the DOH time allocation of Millennials seems to mimic that of Baby Boomers more closely than that of Generation X individuals. The young and transitioning into adult age Generation X individuals show a considerable reduction in the DOH time allocation compared with the Baby Boomers of the same age group. However, Generation X individuals converge to the Boomers' DOH time expenditure as they age.

The young age group (18–24 years) from both Generation X and Millennials show lower work time allocation compared with the Baby Boomers. However, when transitioning into the adult age group (25–34 years), Generation X and Millennials had work time allocation that was higher than that of Baby Boomers. The middle-age (35–54 years) Generation X individuals seem to be converging to the Baby Boomers.

Generation X and Millennials have significantly lower travel time allocation compared with Baby Boomers. The travel time expenditures of young Generation X individuals is significantly lower (15 min) than the Boomers' 100 min daily average. The young Millennials' travel time budget is not lower than that of the young Generation X individuals; rather, it is about 2 min higher. The 25–34-years-old Generation X individuals have an 8 min lower travel time expenditure compared with Baby Boomers, whereas the same-age Millennials have an additional 4 min reduction in travel time expenditure compared with Generation X individuals. The difference in travel time allocation decreases between Generation X and Baby Boomers for the 35–54-year age group. The old-age Baby Boomers continue to show an average 4 min higher travel time expenditure compared with the Silent Generation.

Demographic Differences across Cohorts

Table 4 presents the demographic profiles and living arrangements of different cohorts by age groups. In the interest of space, the presentation is limited to age groups 18–34 years and above 54 years. For living arrangements, more Generation X individuals live in urban areas compared with Baby Boomers and more Millennials live in urban areas compared with Generation X. For demographic characteristics, the percentage of student population among the 18–24 years Millennials is 45%, which is nearly 15% higher compared with Generation X individuals. The proportion of Millennials who are employed is lower than Generation X individuals for those in the 18–24 age category. The unemployed population is also nearly 8% higher for Millennials compared with Generation X in the 18–24 years group. The difference in proportions of married individuals is more pronounced between Generation X and Millennials compared with between Boomers and Generation X.

Table 4. Demographic Shifts across Cohorts by Age Groups

	Age group: 18–24 years		
	Baby boomers	Generation X	Millennials
Percentage living in urban areas	80	79	82
Student percentage	14	30	45
Worker percentage	72	65	64
Unemployed percentage	5	6	14
Married percentage	43	32	18
	Age group: 25–34 years		
Percentage living in urban areas	78	82	84
Student percentage	2	9	16
Worker percentage	79	83	78
Unemployed percentage	3	4	8
Married percentage	64	64	50
	Age group: 55–64 years		
	GI generation	Silent generation	Baby boomers
Percentage living in urban areas	80	74	80
Worker percentage	63	54	66
Retired percentage	18	28	14
	Age group: Above 65 years		
Percentage living in urban areas	72	77	79
Worker percentage	12	18	37
Retired percentage	81	74	48

Among the transitioning into older and old-age groups, a major shift in retirement age can be seen. There is a 14% drop in the retired population among the 55–64-year-old Baby Boomers compared with the Silent Generation. Similarly, there is a 24% decline in the retired population among the old-age Baby Boomers compared with the Silent Generation.

The discussion above clearly demonstrates the demographic differences in student status, (un)employment status, marital status, as well as living arrangements of the young age and transitioning into young age groups across generations—particularly for Millennials compared with Generation X and Baby Boomers. In addition, considerable differences are observed in retirement ages of Baby Boomers compared with the Silent Generation.

The next section presents a multivariate modeling analysis of the activity participation and time allocation choices while accounting for the age, cohort, and period effects, in addition to the contribution of different demographic characteristics.

Econometric Methodology

The multivariate modeling analysis utilizes the multiple discrete continuous extreme value (MDCEV) framework proposed by Bhat (*17*). The framework is capable of simultaneously modeling the activity participation and time allocation choices of different activity types subject to a resource constraint. The analysis combines four cross-sectional datasets from the years 1965, 1985, 2005, and 2012 to form a pooled dataset. This approach of combining multiple cross-sectional datasets to study the dynamics of choices has been used by other researchers in the past (*18–22*). The MDCEV model structure is briefly described below, along with a discussion of the necessary considerations for the pooled dataset.

MDCEV Model Framework

Following Bhat (*17*), the time engagement choices into different activities can be formulated as an allocation problem. According to the allocation problem, an individual allocates a certain available budget (i.e., 1,440 min available in a day) across multiple activities, such that the utility derived by engaging in the various activities is maximized.

The allocation problem, where an individual allocates $x = \{x_1, x_2, \ldots, x_k\}$ amounts of time to K activities, can be formulated as:

$$\max U(x) = \Psi_1 \ln(x_1) + \sum_{k=2}^{K} \gamma_k \Psi_k \ln\left(\frac{x_k}{\gamma_k} + 1\right) \quad (1)$$

subject to:

$$\sum_{k=1}^{K} x_k = T \quad (2)$$

where x is a ($K \times 1$) vector of time allocated to activities $1, 2, \ldots, K$. Individuals try to maximize the total utility U given by Equation 1 subject to the time budget constraint T given by Equation 2. $\Psi_k(>0)$ is known as the baseline marginal utility parameter and represents the gain in utility realized by allocating the first unit of time to activity k. On the other hand, $\gamma_k(>0)$ is known as the translation/satiation parameter and controls the amount of time allocation into different activity types. Ψ and γ are further parameterized as in Equations 3 and 4:

$$\Psi = \exp(\nu\beta + \varepsilon) \quad (3)$$

$$\gamma = \exp(w\alpha) \quad (4)$$

where Ψ and γ are ($K \times 1$) vectors of baseline marginal utility and satiation parameters respectively, ν and w are ($K \times D$) matrices of exogenous variables, β and α are ($D \times 1$) vectors of coefficients, and ε is a ($K \times 1$) vector of stochastic error term.

In the current study, ε is assumed to be independent and identically type I extreme value distributed across activities and individuals. However, the scale of the error can be different for different periods to account for the pooled nature of the dataset. More specifically, the scale of the error, σ, is parameterized as in Equation 5:

$$\sigma_i = \exp(\pi_i \delta_i) \quad (5)$$

where σ_i is the scale of the error for period i = 1965, 1985, 2005, and 2012, π_i is a period-specific parameter to be estimated along with α and β, and δ_i is an indicator variable which assumes the value 1 for period i and 0 otherwise. The exponentiation operator ensures that a non-negative scale is estimated for all periods.

Estimation Results

This section presents the estimation results of a joint model (MDCEV) of participation and time allocation into six activity types (1) MIH; (2) MOH; (3) DIH; (4) DOH; (5) work; and (6) education. Work (education) activity was only included in the choice set if the person was a worker (student). In addition to the demographic variables, age, period, and age-specific cohort indicators were included in the model to capture the age, period, and cohort effects.

Tables 5 and 6 present the coefficient estimates for the baseline marginal utility function and the translation/satiation function respectively. The t-statistics of the coefficient estimates are provided in parentheses next to each of the coefficient estimates. The final log-likelihood value of the estimated model is −117727.58 with 114 parameters compared with −119876.64 for a constant-only model. The three estimated scales of the errors are significantly different from 1, while the fourth scale of error (corresponding to the period 1965) is fixed to 1 for normalization. The scale of the error term for the recent years is less than 1, indicating a smaller variance of the error for recent years compared with the year 1965. Also, the scale of error for the years 2005 and 2012 are very similar in magnitude. This may be attributable to the proximity in time for the 2005 and 2012 datasets (i.e., 7 years) compared with the datasets from other years (i.e., 1965 and 1985) used in the analysis.

Heterogeneity across Demography

In the model, MOH was considered as the base alternative. Compared to MOH, all five other activities have lower propensity for participation, as can be seen from the negative constant values. MOH and DIH have the highest satiation (lowest time allocation) tendency and work and education have the lowest satiation (highest time allocation) tendencies. Females exhibit lower tendency to participate in all types of activities compared with males. However, they tend to spend more time in MOH and less time in DOH compared with males. Students tend to participate more time in MOH, DIH, and DOH compared with non-students. They also tend to spend less time in DIH activities. Like the students, workers tend to participate more into MOH, DIH, and DOH activities. However, they tend to spend less time in all three types of activities—this may be indicative of the time constraints experienced by workers owing to work commitments. Married people tend to participate less in all types of activities compared with MIH activities. All else being equal, retired individuals tend to spend more time in MOH, DIH, and DOH activities compared with non-retired individuals.

Trends across Age Groups

People in the middle-age group (35–64 years) tend to participate more in MOH activities compared with people in other age groups. Comparatively young people (younger than 55 years) participate less in DIH activities compared with people in the older age group. People belonging to the 35–54 years age group have the highest propensity to participate in work activity compared with people in the other age groups. People in the 25–44 years age group spend less time engaged in DIH and DOH activities compared with people in the other age groups.

Trend across Periods

Compared to 1965, the propensity to participate in MOH and DIH activities has decreased over the years. On the other hand, the propensity to participate in DOH activities has increased over the years compared with MIH

Table 5. Model Estimation Results for the Baseline Marginal Utility

	MOH	DIH	DOH	Work	Education
Constants	−6.036 (−80.18)	−4.913 (−51.2)	−7.42 (−88.33)	−6.007 (289.84)	−6.512 (−81.97)
Demographic variables					
Female indicator	−0.092 (−6.15)	−0.17 (−13.22)	−0.115 (−7.7)	−0.203 (−10.02)	−0.192 (−3.57)
Student indicator	0.143 (5.39)	0.051 (1.48)	0.124 (4.17)		
Worker indicator	0.473 (22.89)	0.21 (8.67)	0.349 (17.46)		
Unemployed indicator	0.104 (3.1)				
Married indicator	−0.062 (−4.45)	−0.109 (−8.16)	−0.122 (−8.14)	−0.068 (−4.39)	−0.127 (−1.8)
Retired indicator	0.073 (2.79)	0.096 (3.86)	0.096 (3.2)		
Age indicators (base: age 65 and above)					
Age 18–24 years		−0.227 (−7.23)			
Age 25–34 years		−0.212 (−8.05)	−0.068 (−1.73)		0.206 (2.31)
Age 35–44 years	0.078 (3.63)	−0.158 (−6.17)	−0.037 (−1.38)	0.143 (6.3)	
Age 45–54 years	0.064 (3.05)	−0.112 (−5.36)	−0.483 (−2.73)	0.032 (1.11)	
Age 55–64 years	0.089 (4.33)				
Period indicators (base: period 1 – 1965)					
Period 2 – 1985	−0.554 (−7.48)	−0.939 (−9.8)	0.711 (8.67)		
Period 3 – 2005	−0.543 (−7.4)	−0.956 (−10.22)	0.845 (10.31)		
Period 4 – 2012	−0.561 (−7.64)	−0.985 (−10.51)	0.826 (10.05)		
Cohort indicators specific to age group 18–24 years					
Generation X			0.114 (2.01)		0.276 (2.54)
Millennials			0.054 (1.43)	−0.151 (−3.79)	0.249 (3.11)
Cohort indicators specific to age group 25–34 years					
Generation X	0.091 (3.37)		0.038 (0.9)		
Millennials	0.038 (0.94)		0.055 (1.06)	−0.116 (−2.44)	−0.205 (−1.57)
Cohort indicators specific to age group 45–54 years					
Silent generation			0.55 (3.05)		
Baby boomers			0.528 (2.97)		
Generation X		−0.07 (−1.88)	0.448 (2.46)		
Cohort indicators specific to age group 55–64 years					
Silent generation			0.033 (0.78)		
Baby boomers			0.044 (1.59)		
Scale of the error					
Period 2			0.259 (95.06[a])		
Period 3			0.302 (126.85[a])		
Period 4			0.303 (122.37[a])		

Note: *t*-statistic with respect to 1.

activities. This trend of increased discretionary activity participation is also supported by previous literature on the topic (*2*, *5*). All the periods exhibit an increase in the amount of time allocation in all activities compared with the year 1965.

Trend across Cohorts

Major cohort-specific trends are observed in DOH, work, and education activity participation. For the work and education activities, the trends are observed for the people in the 18–34 years age group. For DOH activities, an increased tendency is observed across cohorts of all age groups. For work activity, Millennials show a tendency to participate less in work activity compared with the previous generations of the same age groups. The youngest Millennials also show a tendency to participate more in education compared with Generation X.

There is an increased tendency to allocate more time to work activities across different cohorts of different age groups. Generation X and Millennials belonging to the 25–34 years age group show a fall in MOH time allocation.

Relative Contribution of Demographic Shift and Cohort Attitude

To contrast the relative contribution of the demographic differences and the cohort-specific changes, the percentage change in the baseline marginal utility for different activity types were calculated using Equation 6 for different demographic variables and age-specific cohort indicators.

$$\frac{\Psi_{k,d,1}-\Psi_{k,d,0}}{\Psi_{k,d,0}} \times 100 = 100^{*}\left[\exp\left(\widehat{\beta_d}\right)-1\right] \quad (6)$$

Table 6. Model Estimation Results for the Translation/Satiation Parameter

	MOH	DIH	DOH	Work	Education
Constants	3.31 (35.33)	3.685 (33.8)	4.835 (41.73)	5.647 (66.01)	5.441 (6.07)
Demographic variables					
Female indicator	0.122 (3.1)		−0.169 (−3.55)	0.065 (1.07)	
Student indicator		−0.17 (−1.9)			
Worker indicator	−0.672 (−14.48)	−0.519 (−10.32)	−0.588 (−10.23)		
Age indicators (base: age 65 and above)					
Age 25–34 years		−0.078 (−1.44)	−0.089 (−1.29)		
Age 35–44 years	−0.074 (−1.55)	−0.125 (−2.44)	−0.066 (−0.98)		
Period indicators (base: period 1 – 1965)					
Period 2 – 1985	2.174 (20.79)	2.284 (18.55)	1.703 (12.75)	1.16 (10.51)	1.461 (1.54)
Period 3 – 2005	2.034 (22.04)	2.253 (20.26)	1.123 (9.8)	1.117 (11.15)	1.204 (1.32)
Period 4 – 2012	2 (21.43)	2.392 (21.42)	1.223 (10.62)	1.15 (11.36)	1.433 (1.55)
Cohort indicators specific to age group 18–24 years					
Generation X		0.316 (2.09)			
Millennials		0.205 (2.27)	0.408 (3.81)	0.611 (3.45)	
Cohort indicators specific to age group 25–34 years					
Baby boomers				0.216 (1.9)	
Generation X	−0.147 (−2.22)			0.209 (2.5)	
Millennials	−0.075 (−0.67)			0.479 (2.52)	
Cohort indicators specific to age group 35–44 years					
Baby boomers				0.298 (3.29)	
Generation X				0.284 (3.62)	
Cohort indicators specific to age group 45–54 years					
Baby boomers			−0.223 (−2.94)		
Generation X			−0.194 (−1.2)		
Cohort indicators specific to age group 55–64 years					
Silent generation				0.359 (2.03)	
Baby boomers				0.26 (2.47)	

[a]*t*-statistic with respect to 1.

where the left-hand side denotes the percentage change in the baseline marginal utility for the *k*th activity and the *d*th indicator variable and $\widehat{\beta_d}$ is the parameter estimate corresponding to the *d*th indicator variable.

Based on the above calculation, it was observed that being part of a demographic group of retired individuals was responsible for about 6.57% decrease in work activity participation propensity, whereas being a young Millennial contributed to a 14% decrease in work activity participation propensity. Similarly, a young Millennial exhibits a 28% increase in participation propensity in education activity compared with other segments of the population. On the other hand, a 25–34-year-old Millennial exhibits a 5.65% increase in DOH activity participation propensity and about 12% lower propensity to participate in work activity compared with other segments. Among the 55–64-year-old population, the Silent Generation shows 3.36% increase in DOH activity participation and Baby Boomers show 4.5% increase in DOH activity participation propensity.

Conclusion

This paper examines the activity–travel participation and time allocation trends of twentieth-century American generations using data from four waves (1965–1966, 1985, 2005, and 2012) of the AHTUS. The overarching goal of the research was to explore the activity–travel engagement choices of five generations—the GI Generation (born 1901–1924), the Silent Generation (born 1925–1943), Baby Boomers (born 1944–1964), Generation X (born 1965–1981), and Millennials (born 1982–2000). To the best of the authors' knowledge this is one of the very first attempts to compare the activity–travel pursuit of all five twentieth-century American generations while systematically isolating the age, period, and cohort effects in addition to the demographic differences.

Previous literature on the topic has defined the profile of Millennials based on their demographic characteristics as well as activity and travel pursuits. According to the literature, Millennials tend to live in dense urban places and exhibit low levels of car ownership (*23*). They also tend to delay marriage and family formation (*24*). On the travel behavior front, Millennials have been characterized by less driving, higher preference for alternative modes of transport, lower trip rates, and fewer vehicle miles (*25*). Additionally, Le Vine et al. (*26*) argued that Millennials have been trading off lower participation in out-of-home activities with higher participation in in-home virtual activities. The literature has also argued that the differences between Millennials and Generation X individuals are not significant enough to redefine the urban landscape of America (*27*). Additionally, according to some researchers, Millennials are just putting off entry into adulthood for a few additional years compared with the immediately previous generation—that is, Generation X (*28*).

The current study presented a detailed analysis of the trend in activity–travel pursuits as well as the demographic compositions of different generations at different stages of their lives. The trend analysis of the demographic composition revealed significant differences across various generations of similar age groups. Finally, the multivariate modeling analysis was conducted to quantitatively assess the contribution of the demographic- and cohort-specific differences on the heterogeneity of activity engagement. The key findings are summarized below:

1. The most prominent differences in the activity–travel pursuits between cohorts were observed at the early stages of individuals' lives (i.e., 18–34 years).
2. The differences across cohorts are lowest for the middle-age groups (i.e., 35–54 years).
3. Among the oldest age cohorts (i.e., above 65 years), Baby Boomers showed notable differences when compared with the Silent Generation for retirement age, work participation, and travel time allocation.
4. The most significant downward shift in the travel time allocation was noticed between Baby Boomers and Generation X, not between Generation X and Millennials. This is in contrast to what the literature has suggested (*11, 12*). However, the difference in travel time allocation tended to taper off between Generation X and Baby Boomers as they entered adulthood.
5. Many observations for the 18–34-year-old Millennials and Generation X individuals came from the 2012 AHTUS survey. This year was also characterized by low median household income ($53,331, in 2016 adjusted dollars, based on the Consumer Price Index for All Urban Consumers Research Series) compared with 2005 ($56,935). The median household income in 2012 was more comparable with 1985 ($52,068) in inflation-adjusted dollars than to 2005. Similarly, gasoline prices also saw a sharp rise in 2012 compared with 2005 (i.e., $3.66 per gallon [inflation adjusted] in 2012 compared with $2.71 per gallon in 2005). These manifestations of economic conditions can potentially explain time allocation behaviors of younger Millennials and Generation X individuals. In particular, the lower travel time expenditure of individuals in the 18–34 years age group of these two generations can partly be attributed to the economic climate in addition to their unique cohort characteristics.
6. Millennials' profile of discretionary time allocation matches most closely to that of Baby Boomers.
7. Fewer Millennials in the 18–34 years age group participate in work compared with Generation X. However, the work time allocation of the working Millennials has seen an increase compared with that of the working Generation X.
8. Being a Millennial accounts more for the decrease in workforce participation and increase in education participation than any of the demographic differences.

In general, the decreasing differences between the activity–travel pursuits of Baby Boomers and Generation X as they navigate adulthood (i.e., 35–54 years age group) suggest that the observed "atypical" activity–travel pursuits of Millennials—lower participation in the workforce, higher engagement in educational activities, less time allocated to travel, and more time allocated to discretionary activities—will fade out as they enter the adult stages of their lives. Another intriguing question is whether the comparatively active lifestyles of the old-age Baby Boomers will be replicated by the following generations—Generation X and Millennials—or if this is solely a characteristic of the Baby Boomer generation. The answers to this will have a significant impact on the need for transportation systems in the future.

It should be noted that the presented research did not make any explicit attempt to capture the effect of technological proliferation on the changing dynamics of recent generations. However, the influence of technology has been indirectly captured via period effects. For example, the study did not differentiate between in-home and out-of-home work activity time allocation. As a result, the study cannot directly provide information about the influence of technological innovations on the reduction of work commutes. Similarly, because of lack of mode information, the study could not comment on the impact of transportation alternatives facilitated by technological advances such as Uber and Lyft on the activity–travel behaviors. Such information can help comment on the increase in the use of alternative transportation services among the younger generations and, subsequently, characterize one of the many influences of technological proliferation on the changing mobility profile of the younger generations.

Information on some critical socioeconomic and demographic variables, namely income, auto ownership, and ethnicity, was missing from the AHTUS. Subsequently, the potential influences of these factors on the activity–travel behaviors of different generations could not be explored in this study. Furthermore, to draw inferences about the cohort, age, and period effects, the research sample was subdivided into subsamples. Some of the subsamples did not have enough observations to allow making reliable statistical inferences. Additional data can enhance the number of observations for the subsamples representing different combinations of cohort, period, and age. This would subsequently allow for a more reliable statistical analysis and better understanding of the trends across subsamples.

Author Contributions

The authors confirm contribution to the paper as follows: study conception and design: AE, KCK; data collection: not applicable; analysis and interpretation of results: AE, KCK; draft manuscript preparation: AE, KCK. All authors reviewed the results and approved the final version of the manuscript.

References

1. Gershuny, J., and J. P. Robinson. Historical Changes in the Household Division of Labor. *Demography*, Vol. 25, No. 4, 1988, pp. 537–552.
2. Aguiar, M., and E. Hurst. A Summary of Trends in American Time Allocation: 1965–2005. *Social Indicators Research*, Vol. 93, No. 1, 2009, pp. 57–64.
3. Fisher, K., M. Egerton, J. I. Gershuny, and J. P. Robinson. Gender Convergence in the American Heritage Time Use Study (AHTUS). *Social Indicators Research*, Vol. 82, No. 1, 2007, pp. 1–33.
4. Fox, L., W. J. Han, C. Ruhm, and J. Waldfogel. Time for Children: Trends in the Employment Patterns of Parents, 1967–2009. *Demography*, Vol. 50, No. 1, 2013, pp. 25–49.
5. Aguiar, M., and E. Hurst. Measuring Trends in Leisure: The Allocation of Time Over Five Decades. *The Quarterly Journal of Economics*, Vol. 122, No. 3, 2007, pp. 969–1006.
6. Gimenez-Nadal, J. I., and A. Sevilla. Trends in Time Allocation: A Cross-Country Analysis. *European Economic Review*, Vol. 56, No. 6, 2012, pp. 1338–1359.
7. Bhat, C. R., and Koppelman, F. S. A Retrospective and Prospective Survey of Time-Use Research. *Transportation*, Vol. 26, No. 2, 1999, pp. 119–139
8. Jara-Díaz, S., and J. Rosales-Salas. Beyond Transport Time: A Review of Time Use Modeling. *Transportation Research Part A: Policy and Practice*, Vol. 97, 2017, pp. 209–230.
9. Garikapati, V. M., R. M. Pendyala, E. A. Morris, P. L. Mokhtarian, and N. McDonald. Activity Patterns, Time Use, and Travel of Millennials: A Generation in Transition? *Transport Reviews*, Vol. 36, No. 5, 2016, pp. 558–584.
10. Delbosc, A., and G. Currie. Causes of Youth Licensing Decline: A Synthesis of Evidence. *Transport Reviews*, Vol. 33, No. 3, 2013, pp. 271–290.
11. McDonald, N. C. Are Millennials Really the "Go-Nowhere" Generation? *Journal of the American Planning Association*, Vol. 81, No. 2, 2015, pp. 90–103.
12. Polzin, S. E., X. Chu, and J. Godfrey. The Impact of Millennials' Travel Behavior on Future Personal Vehicle Travel. *Energy Strategy Reviews*, Vol. 5, 2014, pp. 59–65.
13. Goulias, K., L. Blain, N. Kilgren, T. Michalowski, and E. Murakami (2007). Catching the Next Big Wave: Do Observed Behavioral Dynamics of Baby Boomers Force Rethinking of Regional Travel Demand Models? *Transportation Research Record: Journal of the Transportation Research Board*, 2007. 2014: 67–75.
14. Miranda-Moreno, L. F., and M. Lee-Gosselin. A week in the life of baby boomers: how do they see the spatial–temporal organization of their activities and travel? *Transportation*, Vol. 35, No. 5, 2008, pp. 629–653.
15. Strauss, W., and N. Howe. Generations: The History of America's Future, 1584 to 2069. New York, New York: Quill, 1991.
16. Kostyniuk, L., and R. Kitamura. Effects of Aging and Motorization on Travel Behavior: An Exploration. *Transportation Research Record: Journal of the Transportation Research Board*, 1987, 1135: 31–36.
17. Bhat, C. R. The Multiple Discrete-Continuous Extreme Value (MDCEV) Model: Role of Utility Function Parameters, Identification Considerations, and Model Extensions. *Transportation Research Part B: Methodological*, Vol. 42, No. 3, 2008, pp. 274–303.
18. Glenn, N. D. Cohort Analysis. *Series: Quantitative Applications in the Social Sciences, 2nd Edition, 2005.*
19. Dargay, J.M. Determinants of Car Ownership in Rural and Urban Areas: A Pseudopanel Analysis, *Transportation Research E*, Vol. 2007, No. 38, 2002, pp. 351–366.
20. Huang, B. The Use of Pseudo Panel Data for Forecasting Car Ownership, *Dissertation*, University of London, London, 2007.
21. Hjorthol, R. J., L. Levin, and A. Siŕen. Mobility in Different Generations of Older Persons: The Development of Daily Travel in Different Cohorts in Denmark, Norway and Sweden. *Journal of Transport Geography*, Vol. 18, 2010, pp. 624–633.
22. Sobhani, A., N. Eluru, and A.R. Pinjari. Evolution Of Adults' Weekday Time Use Patterns From 1992 To 2010: A Canadian Perspective. Presented at 93rd Annual Meeting of the Transportation Research Board, Washington, D.C., 2014.
23. Nielsen. Millennials Prefer Cities to Suburbs, Subways to Driveways. Retrieved from http://www.nielsen.com/us/en/insights/news/2014/millennials-prefer-cities-to-suburbs-subways-to-driveways.html. Accessed August 15, 2017.
24. Lamberti, P. Millennials Just Aren't That Into Starting Families; Is That Such a Bad Thing? Retrieved from https://www.moneyunder30.com/millenials-waiting-to-start-a-family. Accessed August 15, 2017.
25. Lyons, G. Transport's Digital Age Transition. *Journal of Transport and Land Use, Vol.* 8, No. 2, 2015, pp. 1–19.
26. Le Vine, S., C. Latinopoulos, and J. Polak. Establishing the Links Between Online Activity and Car Use: Evidence From a Combined Travel Diary and Online-Activity

Pseudodiary Data Set. *Transportation Research Record: Journal of the Transportation Research Board*, 2014, 2405: 68–77.

27. Leanne, M. L., and D. L. Brett. Gen Y and Housing: What They Want and Where They Want It. Washington, DC: Urban Land Institute. Retrieved from https://americas.uli.org/report/gen-y-housing-want-want/. Accessed August 15, 2017.
28. Walker, A. Millennials Will Live in Cities Unlike Anything We've Ever Seen Before. Retrieved from http://gizmodo.com/millennials-will-live-in-cities-unlike-anything-weve-se-1716074100. Accessed August 15, 2017.

The Travel Analysis Methods Section (ADB00) peer-reviewed this paper (18-03136), and selected it to receive the 2018 Ryuichi Kitamura Award for best paper primarily authored by a student in collaboration with a professor or other mentor. The annual award honors the contributions of the late Ryuichi Kitamura to the field of activity–travel behavior analysis.

Article

TRR
JOURNAL OF THE TRANSPORTATION RESEARCH BOARD

Transportation Research Record
2018, Vol. 2672(49) 81–90

DOI: 10.1177/0361198118796731
journals.sagepub.com/home/trr
SAGE

Transferring Time-Series Discrete Choice to Link-Based Route Choice in Space: Estimating Vehicle Type Preference using Recursive Logit Model

Fabian Bastin[1], Yan Liu[2], Cinzia Cirillo[2], and Tien Mai[1]

Abstract

This paper considers a sequential discrete choice problem in a time domain, formulated and solved as a route choice problem in a space domain. Starting from a dynamic specification of time-series discrete choices, we show how it is transferrable to link-based route choices that can be formulated by a finite path choice multinomial logit model. This study establishes that modeling sequential choices over time and in space are equivalent as long as the utility of the choice sequence is additive over the decision steps, the link-specific attributes are deterministic, and the decision process is Markovian. We employ the recursive logit model proposed in the context of route choice in a network, and apply it to estimate time-series vehicle type choice based on Maryland Vehicle Stated Preference Survey data. The model has been efficiently estimated by a dynamic programming approach; the values of estimated coefficients provide important patterns on vehicle type preferences. Compared with a naive model based on sequential multinomial logit choices which are independent over time and a dynamic discrete choice model which considers agent's future expectations, the smaller root mean square error of recursive logit model indicates that it has a better performance in estimating sequential choices over time. We also compare the predictive powers and find that the proposed model outperforms the basic approach and the dynamic approach.

Discrete choice theory, introduced by McFadden (*1*), has proven to be an efficient framework to predict and explain individual choices. When individuals express a sequence of choices instead of a single one, the derivation and estimation of behavioral models become challenging. Despite an individual's sequence of choices being possible in a space domain or a time domain, in the literature of discrete choice analysis, most behavioral models capture time-series choices instead of link-based choices in space. Modeling the two types of decisions in sequence is in fact transferable as long as (a) the alternative-specific (link-specific) attributes are deterministic, (b) the perceived total utility is additive over the sequence of choices, and (c) the decision process is Markovian (*2*).

To estimate individuals' time-series choices, dynamic discrete choice models of consumer behaviors were first used in economics and social science (*3*). The earliest generation of studies includes Wolpin (*4*) on fertility and child mortality, Miller (*5*) on job matching and occupational choice, Pakes (*6*) on patent renewal, and Rust (*7*) on engine replacement. Rust (*7*) was the first to propose a dynamic logit model and to formulate the time-series discrete choices of bus engine replacement as an optimal stopping problem. His model is conceived for a single agent, a homogeneous product, and infinite time horizon. The utilities of the dynamic model are nonlinear, composed of information on current alternatives, expectations about future alternatives, and independent and identically Gumbel-distributed random components. Melnikov (*8*) extended Rust's formulation to consider binary decisions, whether to buy or postpone the purchase of a printer, based on the expected evolution of printer quality and price. His dynamic model considers heterogeneous products and homogeneous consumers, and assumes a consumer can only make one purchase over the time horizon. Lorincz (*9*) further extended the dynamic structure by considering the so-called persistent

[1]Department of Computer Science and Operations Research, University of Montreal and CIRRELT, Montréal, QC, Canada
[2]Department of Civil and Environmental Engineering, University of Maryland, College Park, MD

Corresponding Author:
Address correspondence to Cinzia Cirillo: ccirillo@umd.edu

effect, which allows consumers who already had a product to upgrade it instead of replacing it.

Dynamic discrete choice models for car ownership estimation have been developed in recent decades, including models accounting for inertia effect with lagged variables (*10*, *11*) and models capturing forward-looking behaviors with Bellman equations (*12–14*). Ben-Akiva and Abou-Zeid (*10*) proposed a dynamic formulation with the integration of Hidden Markov Chain to model sequence of discrete choices and evolution of latent variables, which was applied for driving behavior analysis. Realizing the importance of a consumer's future expectation in dynamic problems, Schiraldi (*12*) first introduced a dynamic structural approach with optimal stopping formulation to study car replacement decisions in a second-hand vehicle market. His model accounts for consumer's heterogeneity, future expectation, and price endogeneity in an infinite time horizon. However, it ignores the market evolution. To overcome the limitation, Cirillo et al. (*13*) further proposed a regenerative optimal stopping formulation with a stochastic diffusion process in order to capture not only the optimal car purchase time but also consumer's decisions on vehicle type in a dynamically changing vehicle market in a finite horizon. Taking advantage of approximate dynamic programming, the model has been estimated by the maximum likelihood technique. In the literature, a significant number of interesting applications aim at solving the empirical issues in economics, such as unobserved heterogeneity, initial conditions, state dependency, measurement error, endogeneity, and identification (*15*). However, the computational complexity of model estimation is considered to impede the development of these dynamic structures.

On the other hand, the idea of using a sequential link choice model to describe path choice becomes popular in the context of traffic assignment (*16–18*). More recently, Fosgerau et al. (*2*) developed the recursive logit (RL) model and applied it to the route choice problem by formulating each path choice as a sequence of link choices. At each node, individuals are assumed to choose the outgoing link with the maximum utility including an instantaneous cost and an expected downstream utility (i.e., value function) identified by the Bellman equation. This study builds a bridge between the sequential link-based route choice model and the finite path choice multinomial logit (MNL) model; it provides an interpretation of the route choice problem in a dynamic discrete choice formulation.

Based on the RL model, this paper proposes the transferability between the dynamic discrete choice problem and the route choice problem with an application on vehicle type preferences. To the best of our knowledge, this study is the first to establish that sequential choices in a time domain can be formulated as a route choice problem and solved as an individual-specific shortest path problem. Furthermore, the model estimation benefits from the computational advantages of the RL model. In particular, the value function can be quickly computed by solving systems of linear equations, and the overall path likelihood function has a closed form and can be efficiently maximized by a nonlinear optimization algorithm.

The remainder of this paper is organized as follows. First, the methodology is described and the dynamic discrete choice problem formulated as a route choice problem. Then, Maryland Vehicle Stated Preference Survey Data is introduced for model estimation; while the next section demonstrates an application on vehicle type choice using the RL model and performs an out-of-sample validation. The final section offers concluding remarks and avenues for future research.

Methodology

We consider the situation where an individual i, within a set $\mathrm{I} = \{1, ..., I\}$, has to make choices at time stages $t = 1, ..., T_i$, where T_i is the time horizon for the individual i, within finite choice sets A_t that can vary over time. At each stage, each alternative j presents some utility U_{ij} for individual i, and the total utility gained by an individual is supposed to be additive over the sequence of choices:

$$U_i = \sum_{t=1}^{T_i-1} U_{ij_t}, \tag{1}$$

where j_t is the alternative chosen at stage t. We assume that the individual aims to maximize his/her total utility, or in other words, to select the alternatives j_t such that U_i is maximized. Moreover, at each stage t, we assume that the individual situation can be represented by a state x_{it} and that the set of possible states X_{it} at each stage is finite. The state evolution over time is governed by a transition function, such that:

$$x_{i,t+1} = f(x_{it}, j_t) \tag{2}$$

The perceived utility is a function of the state, and therefore, the choice sequence is a solution to the problem:

$$\max_{j_t, t=1,...,T_i-1} \sum_{t=1}^{T-1} U_{ij_t}(x_{it}) \tag{3}$$

or equivalently:

$$\min_{j_t, t=1,...,T_i-1} - \sum_{t=1}^{T-1} U_{ij_t}(x_{it}) \tag{4}$$

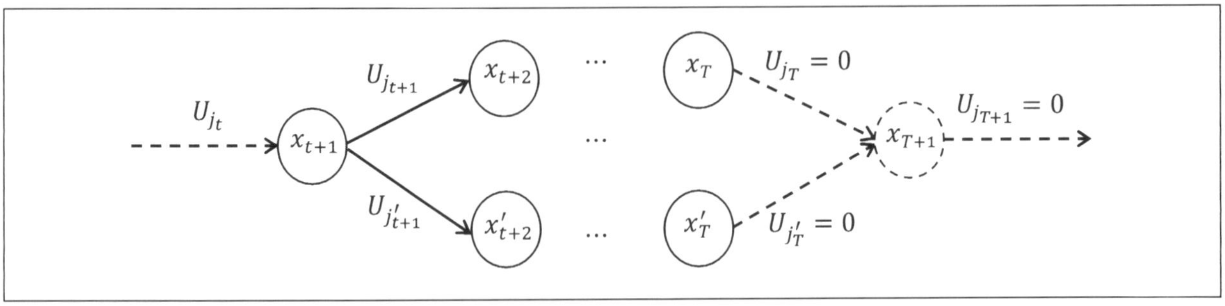

Figure 1. An example network with arbitrary root and artificial destination.

We can represent the problem with a graph where each node represents a possible state and each link corresponds to the transition from a state x_{it} to another state $x_{i,t+1}$ after the choice of some alternative j_t and has the weight $U_{ij_t}(x_{it})$. The graph can be rooted at $x_{i,t+1}$ and we add an artificial terminal node corresponding to an arbitrary state $x_{i,T+1}$ and artificial links connecting each possible state $x_{i,T}$ to $x_{i,T+1}$ with weight $U_{ij_T}(x_{iT}) := 0$. The network in Figure 1 shows a detailed illustration for the graph, with the dash line representing artificial node and links (individual index i is omitted in Figure 1 for simplification purpose).

The choice problem can be treated as a deterministic dynamic program, or equivalently as the computation of the shortest path from $x_{i,1}$ to $x_{i,T+1}$ where in addition the nodes can be numbered in topological order (*19*).

Unfortunately, the alternative utilities cannot be observed exactly, and as in classical discrete choice theory, we assume that the utility $U_{ij_t}(x_{it})$ can be decomposed as the sum of an observable utility $V_{ij_t}(x_{it})$ and a random term $\epsilon_{ij_t}(x_{it})$. The problem can then be viewed as a route choice problem with known origin and destination, and we will assume that the random terms are independent and identically distributed (i.i.d.) and Gumbel distributed. The sequence of choices can be viewed as the choice of a path in an oriented graph, where the origin is the individual state at the beginning of the choice process, and the destination is the artificial terminal node. In this way, a dynamic choice problem can be transferred to a route choice problem.

The estimation of route choices is usually viewed as a difficult problem, as the set of possible paths exponentially increases with the size of the network, so that this equivalence could seem useless in practice. However, recent developments in route choice theory have made this problem tractable. Especially, we can use the estimation techniques proposed by Fosgereau et al. (*2*) to estimate the model, e.g., the expected downstream utilities as well as their gradients can be computed by solving systems of linear equations, and the choice probability for each path is computed by the choice probabilities of the links on that path. We note that such model can be consistently estimated and quickly used for prediction without sampling of choice sets. We here rely on the MATLAB implementation from Mai (*20*) to perform the RL model estimation on time-series vehicle type choices.

Description of Maryland Vehicle Stated Preference Survey

The data used for the empirical analysis was collected from a stated preference survey, which was designed to analyze household vehicle preferences in a dynamic environment (we refer the reader to reference *21* for details in survey design). The survey was conducted under a self-interview, web-based format. Table 1 describes the survey methodology employed.

The stated choice experiment places respondents in a hypothetical nine-year future period starting from 2014. Each year includes two scenarios with a total of 18 scenarios possible. In each scenario, respondents are shown current prices of gasoline and electricity as well as characteristics of four vehicles—the current vehicle and three new vehicles: a gasoline, a hybrid electric, and a battery electric car. Respondents then choose whether to keep their current vehicle or to purchase a new one. If the respondents keep their current vehicle, they then go to the next scenario with a new set of vehicles. Otherwise, their chosen vehicle becomes their current primary vehicle and the respondents are accelerated three years into the future. After this acceleration, the respondents are returned to the scenario progression with the first scenario for the third year after purchase. Figure 2 shows the progression of vehicle type choice.

The stated preference panel data contains 3598 observations of vehicle type choices from 456 Maryland residents (households) over the hypothetical 18 scenarios. For each observation, the choice set contains four alternatives including residents' current vehicle and three new vehicles. New vehicle characteristics such as purchasing price, fuel economy, size, and recharging range are changing over time in the dynamic marketplace. The data is quite representative for Maryland residents, regarding the distributions of households'/respondents'

Table 1. Survey Design

Characteristic	Details
Time frame	May–June 2014
Target population	Maryland households
Sampling frame	Households with Internet access in the state of Maryland
Sample design	Recruitment panel
Use of interviewer	Self-administered
Mode of administration	Self-administered via Internet
Computer assistance	Computer-assisted, web-based self-interview
Reporting unit	One person aged 18 or older per household reports for the entire household
Time dimension	Cross-sectional survey with hypothetical longitudinal stated-choice experiment
Frequency	One 2-week phase of collecting responses
Levels of observations	Household, vehicle, person

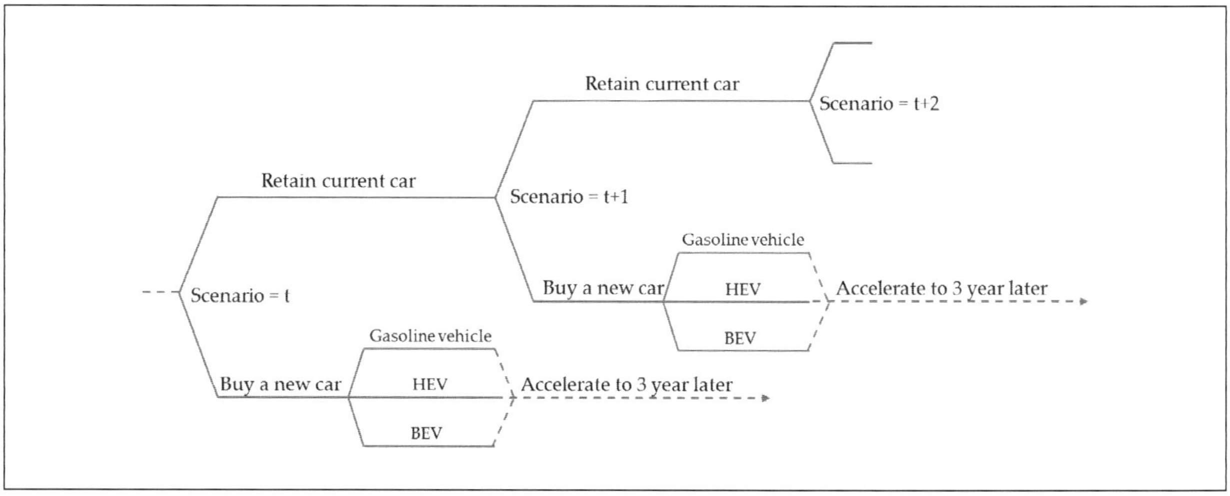

Figure 2. Decision progression of vehicle type choice.

characteristics and their current vehicle characteristics (reference *21* provides summary tables of descriptive data statistics).

Besides, the sequence of households' vehicle purchasing decisions over time provides essential evidence to capture households' preference on switching to greener vehicle types. According to the survey design, at each scenario households can be either in the market or out-of-market. If a household is in the market, he/she has an opportunity to decide either to retain his/her current car or to purchase a new car. Figure 3 shows the shares of households in the market, retaining current vehicle, buying a new gasoline vehicle (GV), a new hybrid electric vehicle (HEV), and a new battery electric vehicle (BEV) over the 18 scenarios.

Application on Vehicle Type Choice

In this section, we construct a link-based network to describe households' vehicle type choices over time and employ an RL model to predict future vehicle preferences on the time-series data in Maryland.

Link-Based Network to Represent Vehicle Type Choice

To transfer this dynamic vehicle type choice problem to a route choice problem, we focus on the sequence of households' vehicle type choices and vehicle characteristics over time. In particular, a link-based network is constructed to represent the decision progression of vehicle type choices over the 18 scenarios, with link attributes representing vehicle characteristics. Therefore, the sequence of choices made by a household over time is represented by a path in the network; with a total of 456 path observations available. The entire network contains 74 nodes and 129 links with 2971 possible paths from origin to destination. Figure 4 provides a partial network of vehicle type choices over four years (eight scenarios). If respondents retain their current vehicle at scenario (*j*-1), there will be four outgoing links from the black node

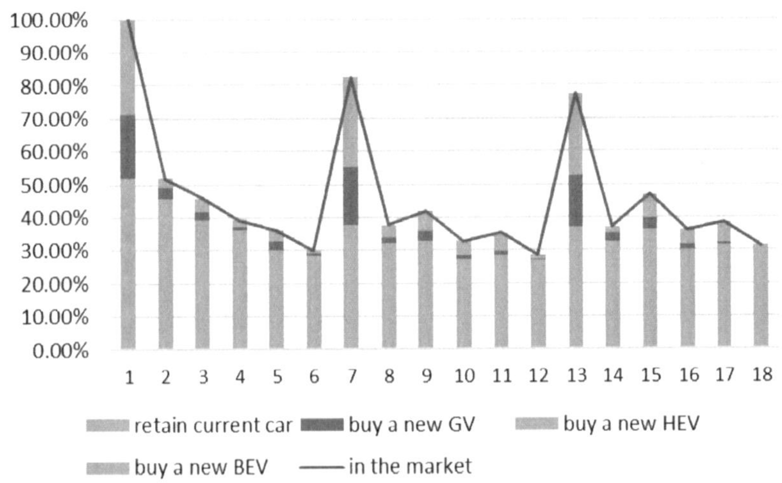

Figure 3. Shares of vehicle type choices for households in the market over time.

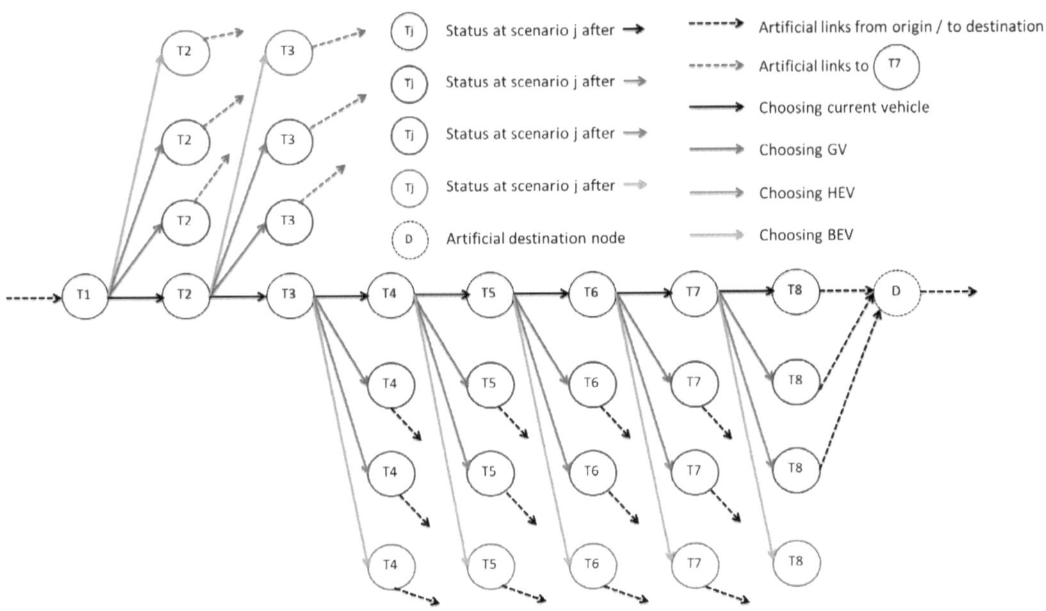

Figure 4. An example network representing vehicle type choices over eight scenarios.

at scenario (*j*-1): the four links represent keeping current vehicle (black arrow), buying a new GV (red arrow), buying a new HEV (blue arrow), and buying a new BEV (yellow arrow) at the next scenario *j*. Otherwise, respondents will jump directly to the first scenario in the third year from purchase (purple dashed arrow). This progression continues until respondents reach the artificial destination node. We assume all paths will eventually go to the unique destination. The main difference between the network considered in this paper with a typical transport network is that our network has only one origin and one destination.

Estimation Results of RL Model

We estimate the RL model on the path observations (sequence of vehicle type choices) of 456 Maryland residents. In the model specification, four outgoing links (alternatives) are considered at each scenario, and the link attributes include household social demographics and car characteristics such as number of workers within a household, respondents' age, education level and gender-related indicators, purchase price, fuel economy (i.e., miles per gallon), vehicle size, and recharging range. The instantaneous utility function (v) for each link is defined as:

$$\begin{aligned} v(g|k) &= \mathrm{ASC}_g + \beta_{\mathrm{VP},g}\mathrm{VP}_g + \beta_{\mathrm{FK},g}\mathrm{FK}_g \\ &\quad + \beta_{\mathrm{FU},g}\mathrm{FU}_g + \beta_{\mathrm{VS},g}\mathrm{VS}_g + \beta_{\mathrm{GP}}\mathrm{GP} + \varepsilon_g \end{aligned} \tag{5}$$

$$\begin{aligned} v(h|k) &= \mathrm{ASC}_h + \beta_{\mathrm{EF},h}\mathrm{EF}_h + \beta_{\mathrm{YH},h}\mathrm{YH}_h + \beta_{\mathrm{VP},h}\mathrm{VP}_h \\ &\quad + \beta_{\mathrm{FK},h}\mathrm{FK}_h + \beta_{\mathrm{FU},h}\mathrm{FU}_h + \beta_{\mathrm{VS},h}\mathrm{VS}_h + \varepsilon_h \\ v(e|k) &= \mathrm{ASC}_e + \beta_{\mathrm{EM},e}\mathrm{EM}_e + \beta_{YH,e}YH_e \\ &\quad + \beta_{\mathrm{VP},e}\mathrm{VP}_e + \beta_{\mathrm{FK},e}\mathrm{FK}_e \\ &\quad + \beta_{\mathrm{FU},e}\mathrm{FU}_e + \beta_{\mathrm{VS},e}\mathrm{VS}_e + \beta_{\mathrm{VR},e}\mathrm{VR}_e \\ &\quad + \beta_{\mathrm{EP}}\mathrm{EP} + \varepsilon_e \\ v(c|k) &= \beta_{\mathrm{VP},c}\mathrm{VP}_c + \beta_{\mathrm{NW},c}\mathrm{NW}_c + \varepsilon_c \end{aligned}$$

where

- k is the status retaining current vehicle
- g, g, h, and c are the outgoing links from status k that represent buying a new GV, HEV, BEV, and retaining current vehicle, respectively
- ASC_i is the alternative specific constant of link i, where $i \in \{g, h, e, c\}$
- EF_i, YG_i, EM_i, NW_i represent indicator of educated female (EF), indicator of young household head (YH), indicator of educated male (EM), and number of workers (NW)
- VP_i, FK_i, FU_i, VS_i, and VR_i represent vehicle price (VP), fuel economy for the group who knows the fuel economy of current car (FK), fuel economy for the group who does not know the fuel economy of current car (FU), vehicle size (VS), and vehicle recharging range (VR) of link i, respectively
- GP and EP are gas price (GP) and electricity price (EP)
- All β are alternative-specific coefficients to be estimated
- ϵ_i is the error component of link i following a Gumbel distribution

Table 2 compares the estimation results between an MNL model, an RL model, and a dynamic discrete choice model proposed by Liu and Cirillo (*22*). It should be noted that all attributes considered are observation specific.

In order to approve the capability of estimating vehicle type choice using the RL model, we compare the estimation results of the RL model not only with the static MNL model, but also with the dynamic discrete choice model.

In the specification of MNL model, choices made by the same household are treated as independent observations and panel effects are omitted. Their choices are assumed to be myopic and no forward-looking behavior is considered. We can observe that many estimation coefficients from the MNL model are not significant, such as fuel economy of GV, size of gasoline and hybrid vehicles, recharging range of electric vehicle, and fuel prices.

On the other hand, the RL model and the dynamic model account for the dependences in sequence of choices made by the same household. Further, they assume that households have expectations about future market. There are two main differences between the RL model and the dynamic model. First, the RL model assumes that households have perfect information of future market, while the dynamic model allows households to have limited forward-looking time. Second, the four alternatives in the vehicle type choice problem are treated equally by the RL model, while keeping the current vehicle and making a purchase are treated differently by the dynamic model. We refer readers to reference (*22*) for details in the formulation of the dynamic discrete choice model.

Most estimation coefficients of the RL model and the dynamic model are significant and have reasonable signs except gasoline price and electricity price. The values of "Rho-square" and "Run time" suggest that the RL model has even better performance in estimation than the dynamic model. The capability of estimating vehicle type choices using the RL model indicates that the estimation of dynamic discrete choice in a time domain can be efficiently solved by a route choice model in a space domain.

Some important patterns can be observed from the values of estimated coefficients in the RL model. The negative value of number of workers suggests that if there are more workers in a household, they tend to buy a new car. The indicators of young people are positive, which indicates households with a young household head have higher preference on hybrid and electric cars. The absolute values of vehicle price coefficients for GV (0.364), HEV (0.476), and BEV (0.817) increase, which indicates that households are more sensitive to the sale price of hybrid and electric cars, possibly because these vehicles have newer technology not fully known and they are more expensive. The coefficients for vehicle size of both gasoline-powered and electricity-powered vehicle have positive sign as households prefer larger cars. In addition, households care more about the size of BEV (0.884) than GV (0.459) and HEV (0.408), possibly because of the smaller size of electric cars currently in the market. The

Table 2. Comparison of Estimation Results between MNL, RL, and a Dynamic Model

Attributes [units]	Current	Gasoline	Hybrid	Electric	MNL estimate (standard error)	RL estimate (standard error)	Dynamic model estimate (standard error)
ASC_gas		X			−2.910 (0.47)	−5.838 (0.67)	—
ASC_hyd			X		−2.583 (0.50)	−6.059 (0.61)	2.148 (0.47)
ASC_ele				X	−6.293 (1.14)	−10.547 (1.45)	−3.198 (0.64)
Number of vehicles	X				0.042 (0.05)[a]	—	0.185 (0.01)
Number of workers	X				−0.121 (0.04)	−0.153 (0.02)	−0.027 (−0.01)
D_young_hyd [1/0]			X		0.158 (0.11)[a]	0.550 (0.19)	0.489 (0.12)
D_educated_female_hyd [1/0]			X		0.195 (0.11)[a]	0.245 (0.18)[a]	0.434 (0.11)
D_young_ele [1/0]				X	1.038 (0.17)	1.440 (0.25)	1.651 (0.17)
D_educated_male_ele [1/0]				X	0.410 (0.19)	0.461 (0.26)[a]	0.739 (0.16)
Vehicle_price_gas [$10,000]		X			−0.345 (0.10)	−0.364 (0.11)	−0.492 (0.06)
Vehicle_price_hyd [$10,000]			X		−0.497 (0.10)	−0.476 (0.13)	−0.592 (0.09)
Vehicle_price_ele [$10,000]				X	−0.614 (0.20)	−0.817 (0.24)	−0.794 (0.16)
Vehicle_price_cur [$10,000]	X				−0.097 (0.03)	−0.156 (0.03)	—
mpg_known_gas [100 mpg]		X			1.435 (0.98)[a]	5.115 (1.27)	10.111 (0.71)
mpg_known_hyd [100 mpg]			X		1.676 (0.68)	5.365 (0.83)	8.569 (0.60)
mpg_known_ele [100 mpg]				X	2.523 (0.67)	3.860 (0.80)	4.838 (0.50)
mpg_unknown_gas [100 mpg]		X			1.057 (1.00)[a]	3.715 (1.28)	8.631 (0.66)
mpg_unknown_hyd [100 mpg]			X		0.860 (0.69)[a]	3.715 (0.87)	5.630 (0.56)
mpg_unknown_ele [100 mpg]				X	2.603 (0.67)	3.372 (0.79)	4.060 (0.48)
Vehicle_size_gas [small, medium, large]		X			0.221 (0.13)[a]	0.459 (0.16)	0.905 (0.08)
Vehicle_size_hyd [small, medium, large]			X		0.166 (0.11)[a]	0.408 (0.13)	0.706 (0.09)
Vehicle_size_ele [small, medium, large]				X	0.637 (0.22)	0.884 (0.27)	0.752 (0.16)
Recharging_range [100 miles]				X	0.684 (0.43)[a]	1.225 (0.45)	2.010 (0.36)
Gasoline_price [$1]		X			−0.015 (0.09)[a]	0.005 (0.12)[a, b]	0.547 (0.07)[b]
Electricity_price [$1]				X	−0.154 (0.18)[a]	0.078 (0.25)[a, b]	0.123 (0.058)[b]
Number of observations					3598	456	456
Number of individuals					456	456	456
Null log-likelihood LL(0)					−8201.659	−8201.659	−8201.659
Final log-likelihood LL(β)					−3532.370	−2318.937	−2779.839
Rho-square					0.569	0.717	0.661
Adjusted rho-square					0.566	0.702	0.644
Run time					1 sec	6 min 34 sec	7 min 52 sec

Note: — = the corresponding variable is not considered.
[a]The coefficient is not significant at significant level of 0.05.
[b]The sign of the coefficient is not as expected.

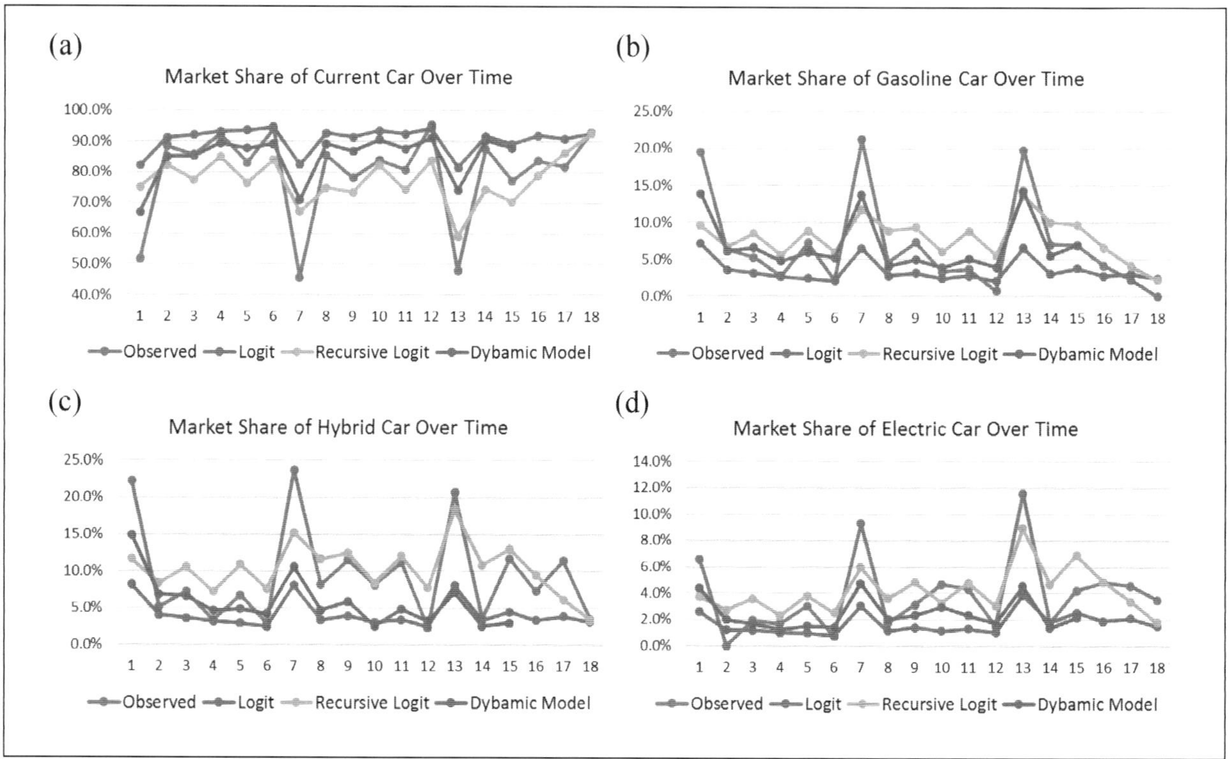

Figure 5. Comparison between observed and predicted market shares on testing dataset over time.

coefficients for recharging range of BEV are positive as expected since a larger range allows for longer trips.

To conclude, the RL model and the dynamic model give very reasonable estimation values which are consistent with general expectations. On the other hand, lower estimated values are obtained from the MNL model, suggesting that the MNL models more conservatively predict how households value fuel economy, vehicle size, and recharging range.

Out-of-Sample Model Validation

In order to validate the prediction performance of the RL model, the sample is randomly divided into two parts: a training dataset containing 80% of the sample, and a testing dataset containing the remaining 20% of the sample. We re-estimate the MNL, the RL, and the dynamic models on the training dataset and apply the estimated coefficients to the testing dataset for predicting the market share of different vehicle types. Figure 5 compares the observed and predicted market shares of the four alternatives along the 18 scenarios over the nine-year period.

From Figure 5, we observe that the MNL model predicts a more stable market share and it is incapable of capturinge fluctuations and peaks. Specifically, it predicts well only the upper bounds of the market share of retaining the current vehicle and the lower bounds of buying a new GV, HEV, and BEV. On the other hand, the RL model is able to recover the fluctuations and sudden changes in consumer demands. It has better performance in predicting the peaks and valleys of market share. The trend of market shares predicted by the dynamic model is in between that of the MNL and RL models. It is able to capture the fluctuations, peaks, and valleys of the actual market shares. However, the model can only predict market shares for 15 time periods, sacrificing the last three time periods to capture household forward-looking behavior.

To further measure the aggregate magnitude of errors in prediction over time, we compare the root mean square errors (RMSE) of market share for the MNL, the RL, and the dynamic models in Figure 6. It should be noted that the reported RMSE are the average values of five validation results. The values suggest that the MNL model has larger prediction errors, especially in reproducing the market share of the current vehicle; while in comparison the other two models perform well. Considering the average over four alternatives, the RL model produces the smallest error.

Concluding Remarks

This paper employed an RL model in the context of route choice in a network to capture sequential vehicle

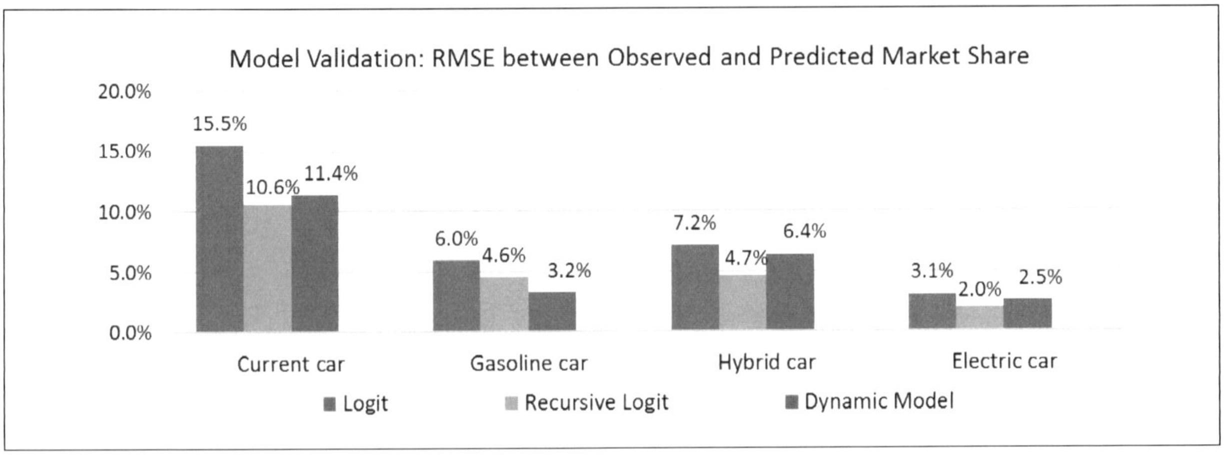

Figure 6. RMSE between observed and predicted vehicle market shares on testing data.

type choices over time and to forecast time-dependent vehicle type preferences. Starting from a dynamic specification of time-series discrete choices, we showed how it is equivalent to link-based route choices that can be formulated by a finite path choice MNL model. This study established that sequential choices over time and in space are transferrable for modeling, as long as the utility of sequential choices is additive, the link-specific attributes are deterministic, and the decision process is Markovian.

Based on the MATLAB implementation from Mai (*20*), we applied the RL model to estimate time-series vehicle type choices using Maryland Vehicle Stated Preference Survey data. The vehicle purchase decision progression of 456 Maryland residents over a nine-year period has been constructed as a link-based network; each node represents a possible state and each link corresponds to the transition between states after a decision is made. The model has been efficiently estimated by a dynamic programming approach; the values of estimated coefficients provide important patterns on vehicle type preferences. Compared with the MNL model and the dynamic model, the smaller RMSE of RL model indicates that it has a better performance in estimating sequential choices over time. The capability of estimating vehicle type choices using the RL model suggests that the estimation of dynamic discrete choice in a time domain can be efficiently solved by a link-based route choice model in a space domain.

There is a large literature on the estimation of time-series discrete choices in economics, social science, and engineering, i.e., patent renewal, welfare gains from industry innovation, machine replacement, car ownership and type choice. However, many dynamic applications of discrete choice models have mainly considered individuals' previous actions (i.e., inertia effect) and do not care much about future expectations. In addition, the computational complexity of model estimation is considered to impede the development of the dynamic structures. The methodology presented in this paper offers a novel and efficient way to model these time-series discrete choices econometrically.

The employed model however retains the well-known independence of irrelevant alternatives (IIA) property which is undesirable in a route choice setting (*23*), but also in the more general setting of sequence of decisions, as two sequences can share common states and decision. To relax the IIA property and to improve the prediction power, Mai, Forsgerau and Frejinger (*24*) extended the RL formulation to a nested RL model that allows path utilities to be correlated in a fashion similar to nested logit (*25*, *26*) and where links can have different scale parameters. Mai, Bastin and Frejinger (*27*) later applied the RL approach to estimate more complex route choice formulations, as mixed logit models, introducing a decomposition technique to alleviate the numerical costs. Such extensions should be investigated in future research. Another undesirable feature of the proposed approach is that individuals make choices with perfect information of the future. While this is necessary to develop the equivalence with a route choice model, this is clearly unrealistic. We could however combine the proposed approach to stochastic programming techniques (*13*) in a context of approximate dynamic programming (*28*) in future research.

Acknowledgments

This material is based upon work supported by the National Science Foundation under Grant No. 1131535 and Natural Sciences and Engineering Research Council of Canada (NSERC). The authors would like to acknowledge the reviewers for generously providing their valuable comments and suggestions.

Author Contributions

The authors confirm contribution to the paper as follows: study conception: Fabian Bastin; study design: Cinzia Cirillo; data collection: Cinzia Cirillo and Yan Liu; coding and model: Tien Mai; analysis and interpretation of results: Yan Liu; draft manuscript preparation: Yan Liu; Cinzia Cirillo, Fabian Bastin. All authors reviewed the results and approved the final version of the manuscript.

References

1. McFadden, D. Conditional Logit Analysis of Qualitative Choice Behaviour. In *Frontiers in Econometrics* (Zarembka, P., ed.), Academic Press, New York, NY, 1973, pp. 105–142.
2. Fosgerau, M., E. Frejinger, and A. Karlstrom. A Link Based Network Route Choice Model with Unrestricted Choice Set. *Transportation Research Part B: Methodological*, Vol. 56, 2013, pp. 70–80.
3. Liu, Y., and C. Cirillo. A Generalized Dynamic Discrete Choice Model for Green Vehicle Adoption. *Transportation Research Procedia*, Vol. 23, 2017, pp. 868–886.
4. Wolpin, K. I. An Estimable Dynamic Stochastic Model of Fertility and Child Mortality. *Journal of Political Economy*, Vol. 92, No. 5, 1984, pp. 852–874.
5. Miller, R. A. Job Matching and Occupational Choice. *Journal of Political Economy*, Vol. 92, No. 6, 1984, pp. 1086–1120.
6. Pakes, A. *Patents as Options: Some Estimates of the Value of Holding European Patent Stocks*. NBER Working Paper No. 1340. National Bureau of Economic Research, Cambridge, MA, April 1984. http://www.nber.org/papers/w1340.pdf.
7. Rust, J. Optimal Replacement of GMC Bus Engines: An Empirical Model of Harold Zurcher. *Econometrica: Journal of the Econometric Society*, Vol. 55, No. 5, 1987, pp. 999–1033.
8. Melnikov, O. Demand for Differentiated Durable Products: The Case of the US Computer Printer Market. *Economic Inquiry*, Vol. 51, No. 2, 2013, pp. 1277–1298.
9. Lorincz, S. *Persistence Effects in a Dynamic Discrete Choice Model-Application to Low-End Computer Servers*. Report No. MT-DP-2005/10. IEHAS Discussion Papers, 2005. http://hdl.handle.net/10419/108082.
10. Ben-Akiva, M., M. Abou-Zeid, and C. Choudhury. Hybrid Choice Models: From Static to Dynamic. In *Triennial Symposium on Transportation Analysis VI*, Phuket, Thailand, March 19, 2007.
11. Nolan, A. A Dynamic Analysis of Household Car Ownership. *Transportation Research Part A: Policy and Practice*, Vol. 44, No. 6, 2010, pp. 446–455.
12. Schiraldi, P. Automobile Replacement: A Dynamic Structural Approach. *The RAND Journal of Economics*, Vol. 42, No. 2, 2011, pp. 266–291.
13. Cirillo, C., R. Xu, and F. Bastin. A Dynamic Formulation for Car Ownership Modeling. *Transportation Science*, Vol. 50, No. 1, 2015, pp. 322–335.
14. Gillingham, K., F. Iskhakov, A. Munk-Nielsen, J. Rust, and B. Schjerning. *A Dynamic Model of Vehicle Ownership, Type Choice, and Usage*. Unpublished Working Paper, 2015. http://bschjerning.com/papers/iruc.pdf.
15. Aguirregabiria, V., and P. Mira. Dynamic Discrete Choice Structural Models: A Survey. *Journal of Econometrics*, Vol. 156, No. 1, 2010, pp. 38–67.
16. Bell, M. G. Alternatives to Dial's Logit Assignment Algorithm. *Transportation Research Part B: Methodological*, Vol. 29, No. 4, 1995, pp. 287–295.
17. Akamatsu, T. Cyclic Flows, Markov Process and Stochastic Traffic Assignment. *Transportation Research Part B: Methodological*, Vol. 30, No. 5, 1996, pp. 369–386.
18. Baillon, J. B., and R. Cominetti. Markovian Traffic Equilibrium. *Mathematical Programming*, Vol. 111, No. 1–2, 2008, pp. 33–56.
19. Bertsekas, D. P. *Dynamic Programming and Optimal Control (3rd ed.)* Athena Scientific, Belmont, MA, 2005.
20. Mai, T. *Dynamic Programming Approaches for Estimating and Applying Large-Scale Discrete Choice Models*. University of Montreal, Montreal, QC, Canada, 2016.
21. Cirillo, C., Y. Liu, and M. Maness. A Time-Dependent Stated Preference Approach to Measuring Vehicle Type Preferences and Market Elasticity of Conventional and Green Vehicles. *Transportation Research Part A: Policy and Practice*, Vol. 100, 2017, pp. 294–310.
22. Liu, Y., and C. Cirillo. A Generalized Dynamic Discrete Choice Model for Green Vehicle Adoption. *Transportation Research Procedia*, Vol. 23, 2017, pp. 868–886.
23. Mai, T., E. Frejinger, and F. Bastin. A Misspecification Test for Logit-Based Route Choice Models. *Economics of Transportation*, Vol. 4, No. 4, 2015, pp. 215–226.
24. Mai, T., M. Fosgerau, and E. Frejinger. A Nested Recursive Logit Model for Route Choice Analysis. *Transportation Research Part B*, Vol. 75, 2015, pp. 100–112.
25. Ben-Akiva, M. *The Structure of Travel Demand Models*. PhD thesis. Massachusetts Institute of Technology, Cambridge, MA, 1973.
26. McFadden, D. Modeling the Choice of Residential Location. *Transportation Research Record: Journal of the Transportation Research Board*, 1978. 673: 72–77.
27. Mai, T., F. Bastin, and E. Frejinger. A Decomposition Method for Estimating Recursive Logit-Based Route Choice Models. *EURO Journal on Transportation and Logistics*, Vol. 7, No. 3, 2018, pp. 253–275.
28. Powell, W. *Approximate Dynamic Programming: Solving the Curses of Dimensionality (2nd ed.)* Wiley, Hoboken, NJ, 2011.

The Standing Committee on Transportation Demand Forecasting (ADB40) peer-reviewed this paper (18-04149).

Any opinions, findings, and conclusions or recommendations expressed in this material are those of the authors and do not necessarily reflect the views of the National Science Foundation and NSERC.

TRR
JOURNAL OF THE TRANSPORTATION RESEARCH BOARD

Article

Transportation Research Record
2018, Vol. 2672(49) 91–100

DOI: 10.1177/0361198118796925
journals.sagepub.com/home/trr
SAGE

Dynamic Household Vehicle Decision Modeling Considering Plug-In Electric Vehicles

Fatemeh Nazari[1], Abolfazl (Kouros) Mohammadian[1], and Thomas Stephens[2]

Abstract

Plug-in electric vehicles (PEVs) offer alternatives to traditional vehicles that rely on petroleum-based fuels. While PEV customers can enjoy significant reductions in fuel costs, they incur a larger capital cost than users of traditional vehicles because PEV technology is still maturing. Therefore, consumers adopting PEVs face a trade-off between fuel cost savings with environmental benefits and extra capital cost. It is thus crucial for policy makers and PEV manufacturers to understand people's vehicle decision-making process while considering their socio-economic and travel pattern characteristics as well as the built-environment factors. This paper presents a connected, two-stage, dynamic model of PEV adoption and vehicle-transaction decision-making. Two connected nested logit (NL) models are estimated. The upper level is a two-level NL model to predict choice of vehicle type between four fuel types: gasoline, diesel, hybrid gasoline–electric, and PEV. The lower level is an NL model of vehicle-transaction choice which accommodates four transaction decisions of buy, trade, dispose, and do nothing, while accounting for the log-sum from the vehicle-type choice model, and is estimated using two waves of a panel data set. We find that households with higher levels of income and education are more likely to adopt a PEV. We also found that primarily decision makers take into account the accessibility to charging stations as a critical factor in choosing PEVs.

In the U.S., the transportation sector contributes to approximately one-third of greenhouse gasoline emissions that are primarily from fossil fuel consumption in vehicles (*1*, *2*). In the light of the consequent global warming issues and the concern of peak oil, vehicle manufacturers and technology developers have been cooperating to seek sustainable substitutes for the traditional fossil fuel vehicles. This has led to the development of plug-in electric vehicle (PEV) technologies in the past decade. In addition to the environmental and economic benefits of PEVs, individual users of PEVs are also expected to enjoy significant reductions in fuel and maintenance costs. However, because PEV technology is not mature enough at present, PEV customers may incur much larger capital cost than traditional vehicle (TV) users. In other words, consumers adopting PEVs face a trade-off between fuel cost savings with environmental benefits and extra capital cost. Despite the significant share of private car use in Americans' trips (87% of daily trips and 91% of commute trips in the U.S. are made by personal vehicles according to the U.S. Department of Transportation [*3*]), PEV market penetration is only less than 2% (*4*). However, the PEV stock in the U.S. was the second largest in the world in 2016 with an increasing trend (*5*). Hence, it is crucial for policy makers and PEV manufacturers to investigate people's vehicle decision-making process in the presence of PEVs and its interplays with their socio-economic and travel pattern characteristics as well as built-environment factors.

This paper aims to develop a behavioral PEV adoption model in conjunction with the longitudinal effects of historic vehicle decisions using a discrete choice approach. Therefore, the scope of this study is to explore vehicle decisions in terms of fuel type and transaction over time in a two-stage framework, among other dimensions involved in vehicle decision-making process (e.g., ownership, fleet size, body type, and vehicle usage). Indeed, information about these two decisions could suitably represent public reaction to a new vehicle

[1]Department of Civil and Materials Engineering, University of Illinois at Chicago, Chicago, IL
[2]Principal Transportation Systems Analyst, Argonne National Laboratory, Lemont, IL

Corresponding Author:
Address correspondence to Fatemeh Nazari: fnazar2@uic.edu

technology over time. First, a two-level nested logit (NL) model is estimated using *California Statewide Travel Survey* (*6*) to model vehicle-type choice behavior. We consider four vehicle choices in terms of fuel type including gasoline, diesel, hybrid gasoline–electric, and PEV. PEV includes both battery electric vehicle (BEV) and plug-in hybrid electric vehicle (PHEV). Second, an NL vehicle-transaction model is estimated using two waves of panel data set of the *Puget Sound Travel Survey* in the State of Washington (*7*). Specifically, we consider four alternatives in response to the availability of PEVs: (a) buy a new vehicle (add it to the current vehicle[s]); (b) trade an existing vehicle with another vehicle; (c) dispose one of the existing vehicles; and (d) do nothing. Furthermore, we determine the type of added vehicle for persons who decided to either add a vehicle or trade a vehicle with another one. In pursuing this aim, we consider the summation of the utilities of vehicle-type choice alternatives in logarithmic form known as inclusive value. In fact, the inclusive value is added to the utilities of persons who select either to add a vehicle or trade a vehicle with another one. The estimation results of the empirical analysis revealed that households with higher level of income and education attainment are more interested in adopting PEVs. We also find that people take into account the accessibility to charging stations, which is a representation of the built environment, as a critical factor in choosing PEVs.

The remainder of the paper is organized as follows. First, the previous studies related to vehicle decision models and PEV market penetration are reviewed. The modeling methodology is then presented. Next, the description of the data sets used for empirical analysis and the estimation results are discussed. The paper concludes in the final section with a summary of the research and key findings.

Background

Vehicle Decisions Models

Since the 1930s scholars have investigated vehicle decisions in terms of the number of cars and usage at aggregate national, regional, or zonal levels (*8*). The earlier models failed to capture the behavioral process of vehicle ownership decisions, which strongly impacts individuals' travel behavior. In other words, owning a car is one of the significant factors forming day-to-day travel behavior and activity pattern of individuals and households, including their frequency, location, duration, scheduling, travel mode, and time of day of out-of-home activities (*9*). This reason motivated scholars to shift from aggregate-level car ownership models to disaggregate household level. One of the earliest disaggregate vehicle-ownership models was developed by Lerman and Ben-Akiva (*10*). These models are mainly econometric-based and typically focus on behaviorally investigating one aspect of vehicle decisions. (Examples of econometric-based models can be found in [*11*, *12*].) For instance, discrete choice models are used to estimate the number of vehicles (*13*), vehicle type (*14*), and vehicle transaction (*15*) depending on the lifestyle, travel pattern, socio-economic, and built-environment factors.

One of the improvements to behavioral vehicle decision models is endogenously modeling several aspects of vehicle decisions in one framework. For instance, Paleti et al. (*16*) modeled a vehicle-transaction model taking into consideration fleet composition and vehicle utilization. Most of the above-mentioned models were estimated using static data, discounting the fact that people's lifestyle shapes their vehicle choice decisions over time. Consideration of the dynamic vehicle decision process requires panel data, known as longitudinal data, consisting of cross-sectional data collected at different points in time. In addition, using panel data for analyzing the travel behavior of households solves the "self-selection" problem to some extent (*17*). In light of this fact, a new generation of studies investigated the evolution of vehicle decisions during time (*18*, *19*). As an example, Eluru et al. (*20*) developed an endogenous dynamic vehicle decision modeling. (Interested readers are referred to a review study of vehicle decision models based on two factors of data availability and methodological framework by Anowar et al. [*21*].)

Market Penetration of PEVs

Several aggregate models of PEV market penetration have been proposed. In a recent study, the total cost of PEV ownership and models of PEV consumer adoption were reviewed by Al-Alawi and Bradley (*22*). They found out that the lower net cost of PEV ownership than conventional vehicles results in a shorter payback period and higher consumer preference. Given the uncertain share of PEVs in the future market (*23*) and its impacts on society and private car users, it is essential to investigate the market penetration of PEVs.

From a market standpoint, introducing a new technology is associated with the process of individuals' learning, adapting, and adopting. This process is shaped based on their socio-economic attributes, affordability, and contextual variables. On top of that, wide variety of policy and macroeconomic scenarios can influence consumers' acceptance of a new technology. In a review study by Al Alawi and Bradley (*24*) on hybrid, PHEV, and BEV market penetration models, three main modeling approaches were identified, including; agent-based, diffusion rate and time series, and discrete choice-based models.

The first group of market penetration models are developed based on examining the interactions of agents with specific characteristics with each other, as well as with those of the system (*25, 26*). These studies have defined different operating agents in the modeling environment including consumers, automakers, policymakers, and fuel suppliers. Furthermore, several examples of the second group of models (i.e., diffusion rate and time series) can be found in the literature (*23, 27*). The last type of model was originally developed based on the theory of economic choice behavior (*28*). Santini and Vyas (*29*) developed a multinomial logit to model choice by two groups of early adopters and majority buyers. This approach was developed further by Lin and Green (*30*). To the best knowledge of the authors, currently there is only one work in the literature that has utilized a revealed preference data set to estimate a binary logit model. The work is accomplished by Javid and Nejat (*31*). They also estimated a backward regression model using cross-sectional data set to predict the market penetration of PEVs.

Methodology

This paper focuses on modeling the process of vehicle ownership decision considering dynamic longitudinal effects and the availability of new vehicle technology. For this purpose, vehicle transaction and vehicle type are modeled in a connected two-stage framework. In order to estimate this framework, a bottom-up approach is used. First, the vehicle type is estimated using an NL model. An appropriate representative of vehicle type which reflects the new vehicle technology is fuel type of the vehicle. As depicted in Figure 1, the choice of fuel type is modeled in two levels. At the upper level, the decision makers decided on choosing between TV and PEV. Decision makers who select TV at the upper level, choose one of the three fuel alternatives of gasoline, diesel, and hybrid vehicles at the lower level. This nesting structure underlies that there is a trade-off between TV and PEV at the upper level as the result of the different purchasing price and fuel/maintenance cost during the vehicle's lifetime. In the next step, the decision maker selects a vehicle with a specific fuel type within the selected group at the upper step (Figure 1).

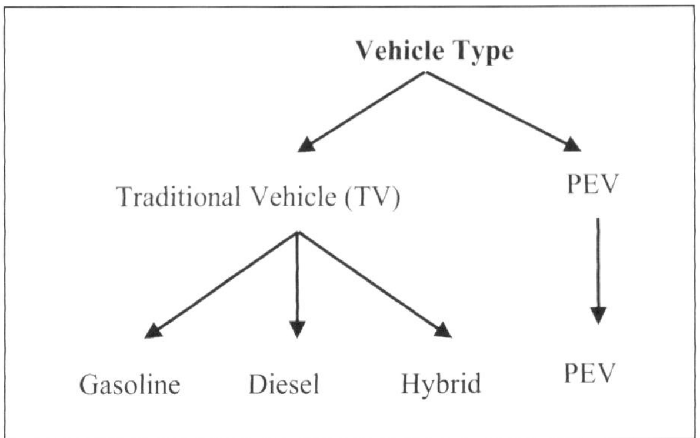

Figure 1. Vehicle-type choice model (NL).

The next dimension of vehicle decision is the choice of vehicle transaction with a sequential NL model borrowed from Mohammadian and Miller (*15*) that is shown in Figure 2. In the sequential estimation of the NL model, the standard errors of the parameters in the upper-level model are biased downward. However, the simultaneous estimation of the models using full information maximum likelihood (FIML) is not practical. As discussed by Mohammadian and Miller (*15*), the decision tree is too "deeply nested" to estimate in one step using FIML and in turn enlarges the alternative set drastically. Furthermore, it causes an extreme difficulty of convergence. Therefore, we use a sequential NL approach to estimation vehicle-transaction model.

The first transaction alternative is engaging in no transaction which means that the decision maker continues utilizing their current vehicle fleet. Decision makers who choose to trade an existing vehicle with another one, first decide on disposal of the existing vehicle. Their next decision is determining the type of the next vehicle to be added to their fleet. The third group of the decision makers choose to buy a vehicle and thus determine the type of added vehicle. The last alternative of the vehicle-transaction model is disposal of an existing vehicle. In order to consider the type of added vehicle, the results of the vehicle-type choice model (i.e., lower level) are considered in the utilities of decision makers who choose adding or trading alternatives. These inputs are in the form of the logarithm of the summation of the utilities of the vehicle type model (known as inclusive value or logsum). Furthermore, type of the disposed vehicle for persons who select trade or dispose in the vehicle-transaction model, is dependent on the explanatory variable of the age of the disposed vehicle.

Data Sets

The models of vehicle type and vehicle transaction in the proposed two-stage framework are estimated using the following data sets.

Vehicle-Type Data Set

The vehicle-type model is estimated using the latest released databases of *California Statewide Travel Survey* (*6*). This database was collected in 2012 and 2013 containing a sample of 42,426 households who own a vehicle, 109,113 individuals, and 75,624 vehicles. Four fuel types are observed in this sample: gasoline, diesel, hybrid, and PEV. PEV includes both BEV and PHEV. The fuel

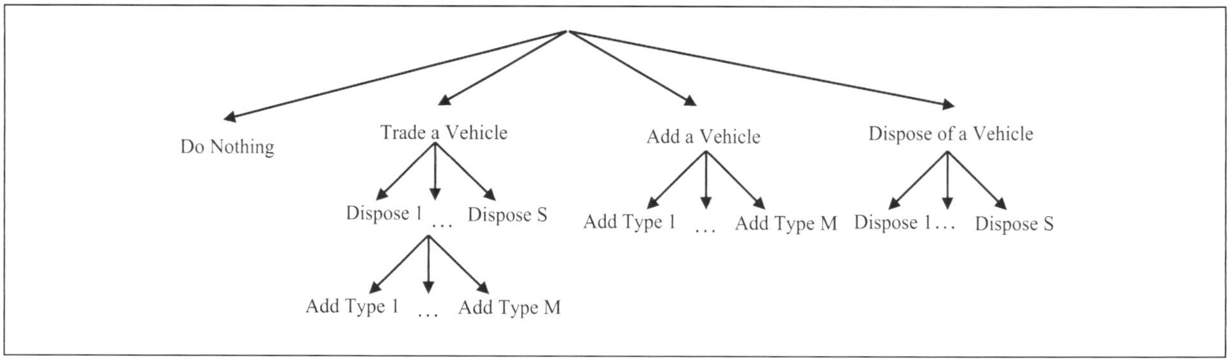

Figure 2. Vehicle transaction and type choice model (*15*).

Table 1. Sample Characteristics of Vehicle-Type Data Set

Sample characteristics	Range	All households 75,624	100.0%
Socio-economic characteristics of household			
Household size	1	8,907	11.8%
	2	29,502	39.0%
	3	14,603	19.3%
	4 +	22,612	29.9%
Number of adult females	>=# adult males	17,531	23.2%
	=# adult males	41,724	55.2%
	<# adult males	16,369	21.6%
Presence of a senior individual (age $\geq$ 66)	Yes	20,042	26.5%
	No	55,582	73.5%
Income	<75K	38,431	50.8%
	75K $\leqslant$ <100K	11,794	15.6%
	100K $\leqslant$ <150K	13,674	18.1%
	150K $\leqslant$ <200K	6,050	8.0%
	$\geq$200K	5,675	7.5%
Number of persons with bachelor's degree	[0–5]	-	-
Number of persons with graduate degree or higher	[0–5]	-	-
Built-environment characteristics			
Residential type	Single-family detached house	62,828	83.1%
	Single-family attached house	4,294	5.7%
	Apartment	6,582	8.7%
	Other	1,920	2.5%
Accessibility to PEV charging station (# of stations in census tract)	[0–65]	-	-
Travel attitude of household			
Number of household's daily trips	[0–32]	-	-
Presence of transit user	Yes	13,208	17.5%
	No	62,416	82.5%
Presence of car-sharing member	Yes	13,208	17.5%
	No	62,416	82.5%
Vehicle (fuel) type alternative	Gasoline	69,857	92.4%
	Diesel	2,225	2.9%
	Hybrid	3,189	4.2%
	PEV	353	0.5%

types, as vehicle types, and the corresponding sample shares are shown in Table 1.

The vehicle-type choice of a household is modeled by three categories of explanatory variables including socio-economic characteristics of the household, built-environment characteristics, and travel attitude of the household. The independent variables of each category and the related sample share are presented in Table 1.

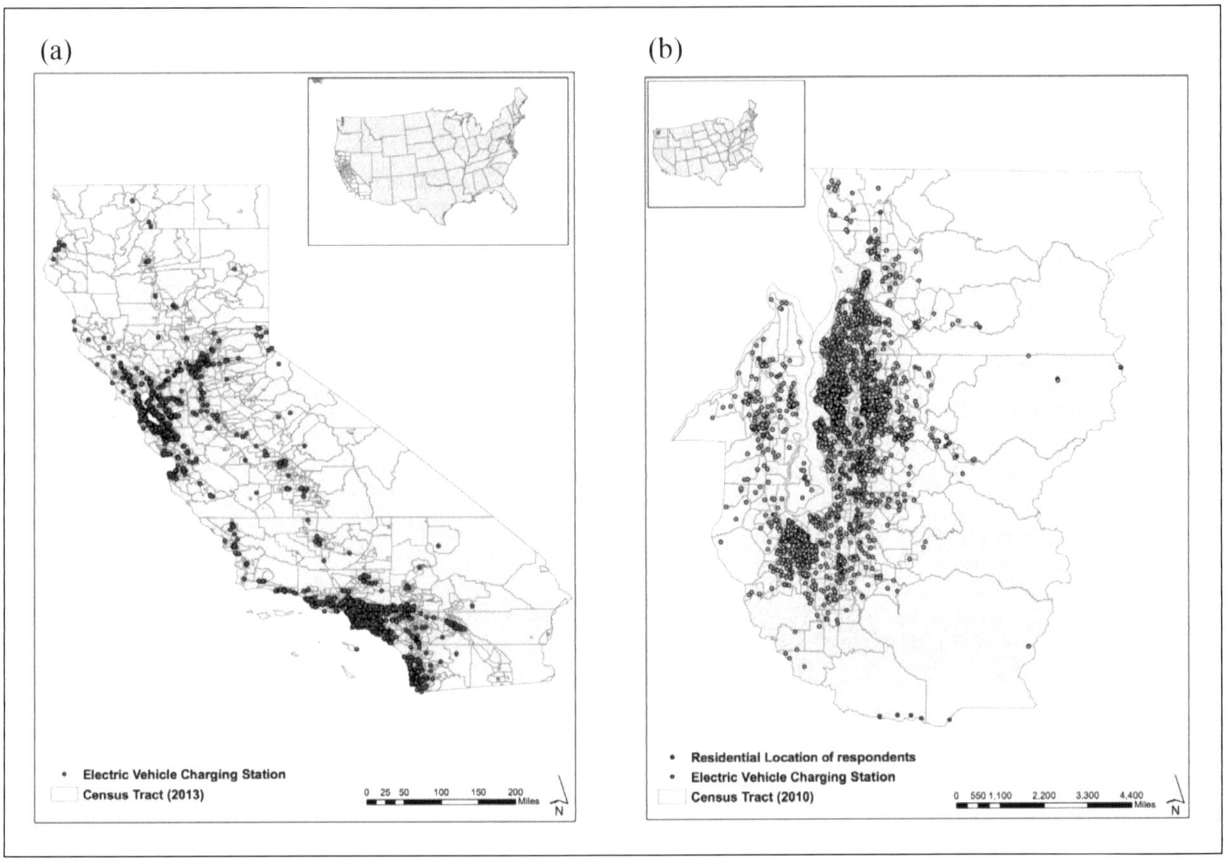

Figure 3. Electric vehicle charging stations: (*a*) state of California (*32, 33*); (*b*) Puget Sound region (*32–34*).

One of the characteristics of built environment in this model is accessibility to public PEV charging stations which is measured using Geographic Information System (GIS). This variable is defined as the total number of PEV charging stations in the census tract of residential location of each household. The dispersion of PEV charging stations with respect to the census block of the state of California and Puget Sound region are depicted in Figure 3.

Vehicle-Transaction Data Set

The decision of vehicle transaction at the household level over time are influenced by change(s) in households' lifestyle and lifecycle. For instance, as a child is added to a family, they may decide on buying a vehicle or exchanging the current vehicle for a larger one. This behavior is better understood by collecting a time-varying database so that we can follow the households' changes over time (e.g., in a panel survey). To this end, the vehicle transaction at the household level is modeled using recent *Puget Sound Travel Survey* (*7*) which is a two-year panel data (2014–2015). The total number of participants in this survey is 1,600 households representing 2,399 vehicle transactions over two years.

Four alternatives of vehicle-transaction choices are considered: do nothing, add a vehicle, trade an existing vehicle with another one, and dispose of a vehicle, with the corresponding sample shares of 88.2%, 2.8%, 4.0%, and 4.9%, respectively. The vehicle-transaction behavior of households is modeled using three categories of explanatory variables: the socio-economic characteristics of the households in wave 2 (i.e., year 2), observed changes in characteristics of the household from wave 1 to wave 2, and their vehicle fleet attributes. The independent variables of each category and related sample shares are provided in Table 2.

Results

Due to data scarcity, the vehicle-type model is estimated using *California Statewide Travel Survey* (*6*) and then is spatially transferred to Puget Sound region. In order to test the transferability of the estimated coefficients, the likelihood ratio test is conducted. Based on the results, the transferability of the estimated coefficients cannot be rejected at significance level of 5%. The estimated results of vehicle-type and vehicle-transaction models are discussed below.

Table 2. Sample Characteristics of Vehicle-Transaction Data Set

Sample characteristics	Range	All households 2399	100.0%
Socio-economic characteristics of household			
Income	<100K	1385	57.7%
	≥100K	878	36.6%
	NA	136	5.7%
Number of vehicles in wave 2	0	318	13.3%
	1	680	28.3%
	2	944	39.3%
	3 +	457	19.0%
Number of driver licenses	<# vehicles	429	17.9%
	=# vehicles	1444	60.2%
	>**#** vehicles	526	21.9%
Residential location	Suburb	1573	65.6%
	Urban	612	25.5%
	Central business district	211	8.8%
	Other	3	0.1%
Change in household characteristics			
Change in number of people in the household from the previous year	Decreased	145	6.0%
	Same	2085	86.9%
	Increased	169	7.0%
Change in number of children in the household from the previous year	Decreased	63	2.6%
	Same	2252	94.0%
	Increased	84	3.5%
Change in number of workers in the household from the previous year	Decreased	255	10.6%
	Same	1945	81.1%
	Increased	199	8.3%
Fleet attributes			
Age of pervious vehicle	[0–34]	-	-
Status of added vehicle	0 (new)	1067	44.5%
	>0 (used)	1014	42.3%
Inclusive value	[−1.4–2.5]	-	-
Vehicle-transaction alternative			
	Do nothing	2,116	88.2%
	Buy	68	2.8%
	Trade	97	4.0%
	Dispose	118	4.9%

Vehicle-Type Model

As depicted in Figure 1, vehicle-type choice is modeled in a two-level NL by three groups of independent variables (Table 3). Almost all of the estimated coefficients are significant at the level of 5%.

The first explanatory variable is household size. The first choice of households with large number of members is gasoline vehicle. Also, households with larger size are less likely to choose hybrid vehicles or PEVs. Considering the fact that median income decreases by increasing household size, the results suggest that the value of PEV cost per income increases for households with larger size.

The vehicle decision of a household usually is made by the adult members (age > 16) who are the potential license holders. The gender of the adults is significant in vehicle-type choice as a dummy variable: whether the number of females is more than males in a household or not. Households with more female adults are likely to be interested in hybrid vehicle, gasoline vehicle, and PEV in a decreasing order.

From the age perspective, presence of a senior person (age ≥ 66) in a household can affect the choice of vehicle type. The estimation results revealed that households with at least one senior person prefer gasoline vehicles. Their interest in hybrid vehicle is also positive but they are more interested in gasoline vehicles. Also, the negative coefficient of this variable in the utility of PEV shows unwillingness of the elderly with respect to PEVs. As expected, we can conclude that younger people are more interested in PEVs than older people.

From the users' perspective, the main difference between PEV and TV is the initial and fuel cost. As expected, households with income less than $75K/year (~ average income of Americans) tend to choose a

Table 3. The Results of Vehicle-Type Model

	Gasoline	Diesel	Hybrid	PEV
Constant	—	–3.057 (–41.525)	–2.546 (–35.590)	–5.807 (34.042)
Socio-economic characteristics of household				
Household size	0.156 (7.579)	—	–0.098 (–3.393)	–0.246 (–4.118)
# adult females > # adult males	0.582 (9.213)	—	0.712 (9.332)	0.230 (1.347)
Presence of a senior person (age ≥ 66)	0.184 (3.544)	—	0.086 (1.279)	–0.380 (–2.440)
Income < 75K	0.303 (8.219)	—	—	—
100K < income < 150K	—	—	0.240 (4.483)	0.272 (1.737)
150K < income < 200K	—	—	0.401 (6.205)	0.675 (3.902)
Income > 200K	—	—	0.559 (8.816)	1.141 (7.295)
# persons with bachelor's degree	0.225 (7.120)	—	0.652 (15.777)	0.538 (6.361)
# persons with graduate degree or more	0.376 (9.875)	—	1.064 (23.052)	0.901 (9.291)
Number of vehicles	–0.307 (–17.011)	—	–0.659 (–21.411)	—
Built-environment characteristics				
Residential type (single-family attached house)	0.844 (6.210)	—	0.915 (5.956)	—
Residential type (apartment)	1.363 (9.376)	—	1.133 (7.079)	—
Number of charging stations in CT	—	—	—	0.035 (4.263)
Travel attitude of household				
Number of household's daily trips	0.019 (1.800)	—	0.089 (6.776)	0.044 (1.656)
Presence of transit user	0.213 (3.153)	—	0.250 (3.081)	0.248 (1.625)
Presence of car-sharing system user	–0.516 (–3.797)	—	—	1.784 (7.120)
Inclusive value				1.000 (fixed)
Inclusive value of TV nest = 0.757 (6.221)				
Log-likelihood at zero = −104,837.120				
Log-likelihood with constants = −25,377.840				
Log-likelihood at convergence= −23784.980				

Note: — = not significant and/or not intuitive; *t*-Statistics in parentheses.

gasoline vehicle. Furthermore, we expect to observe a greater willingness to choose PEVs and hybrid vehicles in households with a high level of income. The results show that households with more than \$100K/year (considered here as high-income households) have a positive inclination toward PEVs and hybrid vehicles. From the estimated coefficients on the income variables, it is seen that (other things being equal) higher-income groups tend to favor hybrids over TVs and PEVs over hybrids, and this preference increases with increasing income.

As indicated in Table 3, vehicle choice of households depends on the educational attainment of household members. Note that diesel is the base alternative in vehicle-type model. As indicated by the results, households with the holders of bachelor's degree or higher are more likely to be interested in owning hybrid vehicles, PEVs, and gasoline vehicles, respectively.

The last significant socio-economic characteristic is the number of vehicles owned by a household. Our analysis revealed that owning more cars would decrease the probability of buying both hybrid and gasoline vehicles. However, this negative tendency is almost twice as large for hybrid vehicles than gasoline vehicles.

Taking into consideration the inevitable connection between transportation and land use, we use built-environment characteristics as explanatory variable. The characteristics of a neighborhood (e.g., residential density) could be summarized in the residential type of a household. As indicated by the estimated results of Table 3, residents of a single-family attached house, or an apartment prefer gasoline and hybrid vehicles. It implies that families who live in a single-family detached house with lower residential density neighborhood (e.g., suburbs) tend to be more interested in PEVs. Probably, due to the fact that single-family detached houses provide the option of charging PEV in their private garage rather than at public charging stations. The other characteristic of a built environment is accessibility to PEV charging stations, which was defined as the number of PEV charging stations in the census block of the household's residence. This variable intuitively has a positive sign in the utility of PEV, since it is expected that neighborhoods with larger number of PEV charging stations encourage the residents to buy PEVs.

Travel behavior of a household is represented by the number of daily trips, transit use, and car-sharing use. On average, as the number of household trips increases, their vehicle miles travel (VMT) also increases. The coefficients of the model show that the households with higher number of trips may realize slightly more utility from hybrid vehicles and PEVs. It implies that fuel cost may be considered more important as VMT of households increases. Furthermore, choosing a hybrid vehicle over a PEV may be influenced by the range limitation of

PEVs because of the charging capacity of the batteries. This limitation could previously be inferred from positive sign of the variable of number of charging station in the CT in PEV utility. As a policy recommendation, increasing the number of charging stations of PEVs by infrastructure investments and informing consumers about infrastructure availability will encourage more people to buy PEVs.

The presence of transit users in a household has positive sign in the utility of PEV, hybrid, and gasoline vehicles with slightly different coefficients. This shows a higher desire of transit users to own PEVs or hybrid cars over gasoline cars. Finally, households with car-sharing members show a greater tendency to select PEVs and are less likely to choose a gasoline vehicle over hybrid and diesel vehicles. It may be that use of car sharing is a technology-savviness proxy which shows the willingness of people to accept new technologies in transportation, which may influence adoption of a new vehicle technology such as PEVs.

Vehicle-Transaction Model

As discussed in the methodology section, a sequential NL is estimated to model the vehicle transaction. The estimated results are provided in Table 4. The behavior of vehicle transaction is explained by three groups of independent variables that are shown in the rows of Table 4, including current socio-economic characteristics of household, changes in characteristics of households, and fleet attributes. The value of *t*-statistics presented in the parentheses show that almost all of the estimated coefficients are significant at the level of 5%. Furthermore, the summation of the utility of four vehicle types in logarithmic form, namely the inclusive value (or log-sum) are considered in the utility of buy and trade transaction choices. The estimated parameters of inclusive value are significant at the level of 5% with acceptable values in range of 0–1 which is consistent with NL model formulation.

As a socio-economic characteristic of households, income threshold of $100K/year divides respondents into two categories of low and high incomes with different vehicle-transaction behavior. The coefficient for the "buy" alternative is almost twice that of the "trade" alternative for the higher-income category.

The next significant characteristic of a household is the current number of vehicles owned. The results show that households with higher number of vehicles are more interested in adding another vehicle rather than choosing the other alternatives. In addition, the number of vehicles in a household is significant with respect to the potential users of vehicles in the household. Households with equal number of vehicles and driver-license holders are interested in trading the current vehicle or engaging in no transaction.

Taking into consideration the role of built environment in shaping the travel behavior and important travel behavior–vehicle interactions, the next explanatory variable is residential location of respondents. Suburban residents show a preference for trade or do nothing with a 70% higher coefficient for the trade option.

Table 4. The Results of Vehicle-Transaction Model

	Do nothing	Buy	Trade	Dispose
Constant	3.311 (13.907)	−1.768 (−3.259)	−1.489 (−3.271)	—
Socio-economic characteristics of household				
Income > 100K	—	0.586 (2.217)	0.338 (1.663)	—
Number of vehicles	—	0.571 (4.229)	—	—
# of driver license holders = # of vehicles	1.517 (7.956)	—	1.944 (6.920)	—
Residential location (suburb)	0.159 (0.857)	—	0.552 (1.912)	—
Change in household characteristics				
Household size +	—	1.335 (4.024)	—	—
Household size −	—	—	—	1.134 (3.701)
Number of children +	—	—	0.925 (2.342)	—
Number of children −	—	—	1.141 (2.698)	—
Number of workers +	—	0.818 (2.368)	—	—
Number of workers −	—	—	—	0.766 (2.694)
Fleet attributes				
Age of previous vehicle	—	—	0.058 (4.664)	0.064 (5.586)
Status of added vehicle (new = yes)	—	—	0.503 (2.443)	—
Inclusive value		0.889 (3.340)	0.372 (2.155)	
Log-likelihood at zero = −3,325.720				
Log-likelihood with constants = −1,329.240				
Log-likelihood at convergence = −1,052.878				

Note: — = not significant and/or not intuitive; *t*-statistics in parentheses.

One of the most important changes in the characteristics of a household is increasing and decreasing the number of members regardless of age. Increasing the size of a household leads to adding a vehicle while households with lessened size are more interested in disposal of one of the existing vehicles. In addition to the household size, changes in the number of workers in a household is significant in this model. This variable, which is correlated with the household income, is found to be significant in the model, where adding more workers to a household means increasing the household income and in turn increasing the utility of buying a vehicle. Reversing the same pattern leads to disposal of a vehicle when a household loses one of its workers.

When a couple adds a child to their family, they may also become interested in trading their current car. Households with children often need a larger vehicle. On the other hand, as a child reaches the age of sixteen, he/she is counted as an adult in a household. A decrease in the number of children has a positive sign in the utility of the trade alternative, which is consistent with a household no longer needing the larger vehicle and intending to trade in for a smaller one.

In the vehicle-transaction model, type of disposed vehicle is determined by vehicle age according to the methodology Section. As shown in the estimation results (Table 4), as the age of a vehicle increases the probability of its disposal and trade tend to increase. The fact of increasing vehicle-maintenance cost with vehicle age should justify this finding. Furthermore, age of the added vehicle in wave 2 is significant as a dummy variable of new or used car. This variable has positive coefficient in the utility of trade transaction, which is consistent with a high correlation between trading a vehicle in and purchasing a new vehicle vs. a used one. However, the average age of vehicles in the U.S. is increasing since 1995 (*35*) which shows that households keep their vehicles for more than 10 years on average. This fact is an important limitation on the rate of penetration of new vehicle technologies (e.g., PEVs).

Conclusions

Two linked, disaggregate models of vehicle type and transaction choices are estimated using a two-stage framework and sequential NL structures, to determine the market share of PEVs. As demonstrated by the empirical analysis, the adoption of PEVs is correlated with income, education level, and number of car-sharing members of a household. Moreover, the findings of this paper revealed that neighborhoods with more accessibility to PEV charging stations may encourage the residents to adopt PEVs. From the perspective of travel patterns, PEVs are more likely to be adopted by households with medium number of daily trips compared to those with low or high number of daily trips. Furthermore, change in the structure of a household, income, and the features of their fleets are significant in the decision to add a vehicle or trade an existing vehicle with another one. The highlighted findings of vehicle-type and vehicle-transaction models can showcase the capability of these models in a sequential two-stage framework to model the dynamic PEV adoption process.

Acknowledgments

This work is funded by the Vehicle Technologies Office of the U.S. Department of Energy's Office of Efficiency and Renewable Energy and performed under a collaborative effort of the University of Illinois at Chicago and Argonne National Laboratory (Argonne). Argonne is a U.S. DOE laboratory managed by UChicago Argonne, LLC under contact DE-AC02-06CH11357. The authors express their gratitude to California Department of Transportation (*6*) and Puget Sound Regional Council (*7*) for providing data used in this study.

Author Contributions

All authors contributed to the study conception and design, data collection, analysis and interpretation of results, and draft manuscript preparation. All authors reviewed the results and approved the final version of the manuscript.

References

1. U.S. Environmental Protection Agency. *Inventory of US Greenhouse Gas Emissions and Sinks: 1990-2015*. EPA 430-17-001. U.S. Environmental Protection Agency, Washington, D.C., 2017.
2. U.S. Energy Information Administration. *Annual Energy Outlook 2017 with Projections to 2050*. AEO2017. U.S. Energy Information Administration, Washington, D.C., 2017.
3. U.S. Department of Transportation, Bureau of Transportation Statistics. *National Household Travel Survey Daily Travel Quick Facts*, 2018. https://www.rita.dot.gov/bts/sites/rita.dot.gov.bts/files/subject_areas/national_household_travel_survey/daily_travel.html. Accessed February 26, 2018.
4. Zhou, Y. *Light Duty Electric Drive Vehicles Monthly Sales Updates*. Transportation R&D Center, Argonne National Lab., US Department of Energy, 2016. http://Www.Transportation.Anl.Gov/Technology_analysis/Edrive_vehicle_monthly_sales.Html. Accessed December 20, 2016.
5. International Energy Agency. *Global EV Outlook 2017, Two Million and Counting*. OECD/IEA. International Energy Agency, Paris, France, 2017.
6. California Department of Transportation. *California Household Travel Survey*, California Department of Transportation, Sacramento, CA, 2017. http://www.dot.ca.gov/hq/tpp/offices/omsp/statewide_travel_analysis/chts.html, Accessed March 10, 2017.
7. Puget Sound Regional Council. *Household Travel Survey Program*, 2017. https://www.psrc.org/household-travel-survey-program. Accessed March 10, 2017.

8. de Wolff, P. The Demand for Passenger Cars in the United States. *Econometrica: Journal of the Econometric Society*, Vol. 6, No. 2, 1938, pp. 113–129.
9. Roorda, M. J., J. A. Carrasco, and E. J. Miller. An Integrated Model of Vehicle Transactions, Activity Scheduling and Mode Choice. *Transportation Research Part B: Methodological*, Vol. 43, No. 2, 2009, pp. 217–229.
10. Lerman, S. R., and M. Ben-Akiva. Disaggregate Behavior Model of Automobile Ownership. *Transportation Research Record: Journal of the Transportation Research Board*, 1976. 569: 34–55.
11. Mamdoohi, A., F. Nazari, and M. H. Noruzoliaee. An Investigation of Human Factors Impact on Tehran Urban Crash Severity. *Journal of Traffic and Logistics Engineering*, Vol. 2, No. 2, 2014.
12. Nazari, F., S. Seyedabrishami, and A. R. Mamdoohi. A Direct Demand Model of Departure Time and Mode for Intercity Passenger Trips. *International Journal of Transportation Engineering*, Vol. 3, No. 2, 2015, pp. 125–141.
13. Bhat, C. R., and V. Pulugurta. A Comparison of Two Alternative Behavioral Choice Mechanisms for Household Auto Ownership Decisions. *Transportation Research Part B: Methodological*, Vol. 32, No. 1, 1998, pp. 61–75.
14. Mohammadian, A., and E. Miller. Empirical Investigation of Household Vehicle Type Choice Decisions. *Transportation Research Record: Journal of the Transportation Research Board*, 2003. 1854: 99–106.
15. Dynamic Modeling of Household Automobile Transactions. *Transportation Research Record: Journal of the Transportation Research Board*, 2003. 1831: 98–105.
16. de Abreu e Silva, J., and K. G. Goulias. *Using Structural Equations Modelling to Unravel the Influence of Land use Patterns on Travel Behavior of Urban Adult Workers of Puget Sound Region*. University of California Transportation Center, Berkeley, CA, 2008.
17. Krizek, K. J. Residential Relocation and Changes in Urban Travel: Does Neighborhood-Scale Urban Form Matter? *Journal of the American Planning Association*, Vol. 69, No. 3, 2003, pp. 265–281.
18. Golob, T. F. The Dynamics of Household Travel Time Expenditures and Car Ownership Decisions. *Transportation Research Part A: General*, Vol. 24, No. 6, 1990, pp. 443–463.
19. Mohammadian, A., and T. Rashidi. Modeling Household Vehicle Transaction Behavior, Competing Risk Duration Approach. *Transportation Research Record: Journal of the Transportation Research Board*, 2007. 2014: 9–16.
20. Eluru, N., C. R. Bhat, R. M. Pendyala, and K. C. Konduri. A Joint Flexible Econometric Model System of Household Residential Location and Vehicle Fleet Composition/Usage Choices. *Transportation*, Vol. 37, No. 4, 2010, pp. 603–626.
21. Anowar, S., N. Eluru, and L. F. Miranda-Moreno. Alternative Modeling Approaches used for Examining Automobile Ownership: A Comprehensive Review. *Transport Reviews*, Vol. 34, No. 4, 2014, pp. 441–473.
22. Al-Alawi, B. M., and T. H. Bradley. Total Cost of Ownership, Payback, and Consumer Preference Modeling of Plug-in Hybrid Electric Vehicles. *Applied Energy*, Vol. 103, 2013, pp. 488–506.
23. McManus, W., and R. Senter, Jr. *Market Models for Predicting PHEV Adoption and Diffusion*. University of Michigan Transportation Research Institute, Ann Arbor, MI, 2009.
24. Al-Alawi, B. M., and T. H. Bradley. Review of Hybrid, Plug-in Hybrid, and Electric Vehicle Market Modeling Studies. *Renewable and Sustainable Energy Reviews*, Vol. 21, 2013, pp. 190–203.
25. Cui, X., C. Liu, H. K. Kim, S. Kao, M. A. Tuttle, and B. L. Bhaduri. *A Multi Agent-Based Framework for Simulating Household PHEV Distribution and Electric Distribution Network Impact*. TRB Committee on Transportation Energy (ADC70), 2010.
26. Eppstein, M. J., D. K. Grover, J. S. Marshall, and D. M. Rizzo. An Agent-Based Model to Study Market Penetration of Plug-in Hybrid Electric Vehicles. *Energy Policy*, Vol. 39, No. 6, 2011, pp. 3789–3802.
27. Lamberson, P. *The Diffusion of Hybrid Electric Vehicles. Future Research Directions in Sustainable Mobility and Accessibility*, 2008. http://www.umsmart.org/project_research/future_directions.pdf.
28. McFadden, D. *Econometric Models of Probabilistic Choice*. In C. F. Manski and D. L. McFadden (eds.), *Structural Analysis of Discrete Data with Econometric Applications*, The MIT Press, Cambridge, MA, 1981.
29. Santini, D. J., and A. D. Vyas. *Suggestions for a New Vehicle Choice Model Simulating Advanced Vehicles Introduction Decisions (AVID): Structure and Coefficients*. ANL/ESD/05-1. Center for Transportation Analysis, Argonne National Laboratory, Lemont, IL, 2005.
30. Lin, Z., and D. L. Greene. *A Plug-in Hybrid Consumer Choice Model with Detailed Market Segmentation*. Presented at 89th Annual Meeting of the Transportation Research Board, Washington, D.C., 2010.
31. Javid, R. J., and A. Nejat. A Comprehensive Model of Regional Electric Vehicle Adoption and Penetration. *Transport Policy*, Vol. 54, 2017, pp. 30–42.
32. Alternative Fuels Data Center. *Alternative Fueling Station Locator*. U.S. Department of Energy Office of Energy Efficiency & Renewable Energy, 2018. https://www.afdc.energy.gov/locator/stations/. Accessed February 28, 2018.
33. U.S. Census Bureau. *Geography, TIGER Line, Shapefiles and TIGER Line Files*, 2018. https://www.census.gov/geo/maps-data/data/tiger-line.html, Accessed February 26, 2018.
34. Puget Sound Regional Council. *Data and Sources*, 2018. https://www.psrc.org/data-and-resources/data-psrc. Accessed February 26, 2018.
35. U.S. Department of Transportation, Bureau of Transportation Statistics. *Average Age of Automobiles and Trucks in Operations in the United States*, 2017. https://www.rita.dot.gov/bts/sites/rita.dot.gov.bts/files/publications/national_transportation_statistics/html/table_01_26.html_mfd, Accessed November 1, 2017.

The Standing Committee on Transportation Demand Forecasting (ADB40) peer-reviewed this paper (18-04260).

The authors are solely responsible for the findings of this research.

TRR
JOURNAL OF THE TRANSPORTATION RESEARCH BOARD

Article

Transportation Research Record
2018, Vol. 2672(49) 101–112

DOI: 10.1177/0361198118796971
journals.sagepub.com/home/trr
SAGE

Comparison of Four Types of Artificial Neural Network and a Multinomial Logit Model for Travel Mode Choice Modeling

Dongwoo Lee[1], Sybil Derrible[1], and Francisco Camara Pereira[2]

Abstract

Discrete choice modeling is a fundamental part of travel demand forecasting. To date, this field has been dominated by parametric approaches (e.g., logit models), but non-parametric approaches such as artificial neural networks (ANNs) possess much potential since choice problems can be assimilated to pattern recognition problems. In particular, ANN models are easily applicable with their higher capability to identify nonlinear relationships between inputs and designated outputs to predict choice behaviors. This article investigates the capability of four types of ANN model and compares their prediction performance with a conventional multinomial logit model (MNL) for mode choice problems. The four ANNs are: backpropagation neural networks (BPNNs), radial basis function networks (RBFNs), probabilistic neural networks (PNNs), and clustered probabilistic neural networks (CPNNs). To compare the modeling techniques, we present the algorithmic differences of each ANN technique, and we assess their prediction accuracy with a 10-fold cross-validation method. Furthermore, we assess the contribution of explanatory variables by conducting sensitivity analyses on significant variables. The results show that ANN models outperform MNL, with prediction accuracies around 80% compared with 70% for MNL. Moreover, PNN performs best out of all ANNs, especially to predict underrepresented modes.

In travel mode choice modeling, and based on a set of given attributes, the objective is to model the choice processes and behaviors of travelers faced with several travel mode alternatives. Being able to model mode choice and predict future trends is critical to transportation planners and policy makers. Travel behaviors are typically examined by using statistical survey techniques such as reveal preference (RP) and stated preference (SP), in which respondents are asked to present their choices and preferable scenarios, respectively (*1*). The surveys yield discrete choice data that can then be used to calibrate travel mode choice models that are fundamental components of disaggregate travel demand modeling approaches, for example, activity-based modeling (ABM).

As the main parametric modeling approach, random utility modeling has been the dominant technique used in the literature since the 1980s to investigate parametric relationships between mode choice and its possible determinants. In particular, logit models—one of the random utility models—gained popularity in the travel mode choice realm because they are based on simple mathematical formulations while accounting for unobserved variables (i.e., stochasticity). Logit models (e.g., multinomial logit [MNL]), however, assume that each choice is independent and identically distributed (IID), which can lead to biased estimations and misleading predictions. Moreover, interactions between the explanatory variables (e.g., the nonlinearity of attributes) are often neglected since it is difficult to represent them in the utility function (e.g., conditional and threshold effects). One way to take this issue into account in the linear utility function is by introducing additional variables (e.g., polynomial and interaction terms) either through considering all possible combinations of explanatory variables or by measuring empirical relationships (*2*). Nonetheless, it is practically impossible to identify all the interactions between all the variables as well as to identify the necessary variables. MNL tries to overcome this issue by using domain knowledge as much as possible. Although the flexible random utility model (RUM) methods such as

[1]Department of Civil and Materials Engineering, University of Illinois at Chicago, Chicago, IL
[2]DTU Management Engineering, Technical University of Denmark, Kongens Lyngby, Denmark

Corresponding Author:
Address correspondence to Dongwoo Lee: dlee226@uic.edu

mixed logit (*3–5*) can have better modeling performance than MNL by relaxing IID assumptions, predetermined structures and linear characteristics of underlying functions still make it difficult to capture or infer high degrees of nonlinearity in a dataset (*3*, *6*, *7*).

In contrast, algorithmic non-parametric approaches have proven to be incredibly powerful to study urban systems (*8–15*). Specifically, artificial neural networks (ANNs) present higher adaptability in identifying nonlinear interactions (*16*, *17*). Contrary to the logit model that deals with the nonlinearity issue by reducing the "complexity" of the dataset, ANNs can capture nonlinear properties more easily through additional units (e.g., hidden layers) without assuming predetermined functional forms (*1*, *18–20*). Nonetheless, a commonly acknowledged disadvantage of ANNs is their lack of interpretability, and they are often associated with black boxes. Another disadvantage is their relative inability to use previously acquired knowledge (i.e., domain knowledge). Despite these challenges, ANNs tend to outperform logit models (from the multinomial to nested logit models [*20–22*]). Among different types of ANN models, several studies have used backpropagation neural networks (BPNNs) to model mode choice (*19*, *21*, *23*). BPNNs tend to achieve high prediction accuracy and are easy to apply. Nonetheless, many different types of ANN models exist, often superior to BPNNs, combining both a strong statistical background with machine learning features. Probabilistic neural networks (PNNs), for example, are derived by incorporating statistical features (e.g., Bayesian decision rule and kernel density estimation [KDE]) into the structure of the neural network. In addition, additional treatments on the dataset or network structure are applied to obtain more efficient and reliable models.

In this article, we investigate the feasibility and capability of four ANNs to model discrete mode choice behaviors and compare their prediction performance with a MNL. Specifically, there have been a limited number of articles that conduct comprehensive analyses of different types of ANN algorithms to model discrete choice in the field of transportation. In this study, we focus on four types of ANNs: BPNNs, radial basis function networks (RBFN), PNNs, and clustered probabilistic neural networks (CPNNs). As a typical travel survey, the Chicago Metropolitan Area for Planning (CMAP) Travel Tracker Survey dataset collected from 2007 to 2008 is used. This study consists of six sections. After this introduction, we present the five methods for discrete choice modeling. We then explain how the data were prepared. Thereafter we provide the model specifications for MNL and ANN models and follow this with the results and a discussion, before the conclusion.

Methods for Discrete Choice Modeling

Random Utility Model: Logit Models

The logit model is the most popular type of random utility model derived from consumer economics theory, and it was initially developed by McFadden (*3*, *24*). In utility maximization behavior, an individual i makes a decision to select one choice among discrete alternatives, by evaluating their associated attributes $\boldsymbol{X}$. The individual i chooses the alternative m that provides the largest utility:

$$U_{im} > U_{ik} \quad \forall m \neq k \tag{1}$$

In reality, researchers do not observe the complete utility of the individual. Thus, the utility can be classified into two parts: an observed utility V_{im} and an unobserved utility ε_{im}. The observed utility generally contains two sets of attributes: 1) covariates associated with both the individual and the alternative X_{im}; and 2) decision-maker characteristics, S_i (*3*). The observed (stated) utility (V) is a value determined from a linear combination of the attributes used, which captures the attractiveness of an alternative bounded to the given model specification as follows:

$$V_{im} = V(X_{im}, S_i) \tag{2}$$

In contrast, the unobserved utility ε_{im} cannot be observed by researchers. This unobserved part mainly results from the specification of the observed utility V_{im}. In practice, it is impossible for statistical approaches to include all possible attributes. Therefore, researchers treat the unobserved terms as a stochastic element. Specifically, the logit model is derived by assuming that each unobserved term, ε_{lm}, is IID extreme values, i.e., Gumbel and type 1 extreme values. By combining two utilities, we can get the probability of individual i choosing alternative j by solving the mathematical formulation:

$$P_{ij} = \frac{e^{\beta' X_{im}}}{\sum_{k=1}^{M} e^{\beta' X_{ik}}} \tag{3}$$

where $\boldsymbol{X}_{ik}$ is the vector of observed explanatory variables to choose a given alternative, and β' is parameters for the observed utility. For more technical details about logit models, see (*3*).

ANNs

ANNs essentially emulate decision-making processes occurring in the brain. As opposed to logit models that assume linearity in estimation, ANNs can algorithmically construct models that are based on nonlinear relationships between determinants. Figure 1*b* shows a

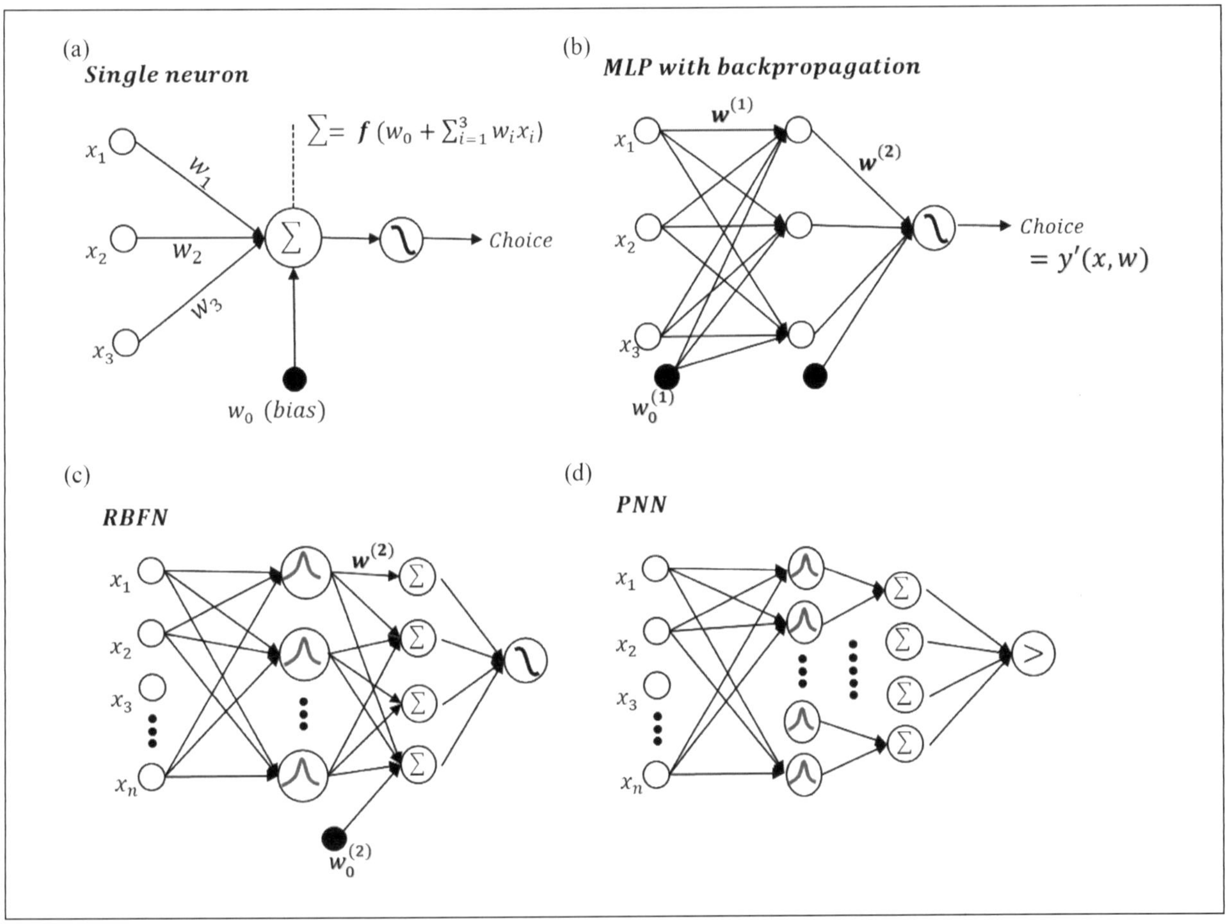

Figure 1. Structure of an ANN: (*a*) single neuron; (*b*) MLP with backpropagation; (*c*) RBFNN; (*d*) PNN.

multi-layer perceptron MLP that has been widely adapted in most neural network models.

In Figure 1*a*, the process of predicting output using a single neuron (i.e., a basis function) is similar to the logit model, since it is based on a linear function. ANNs, however, are constructed by combining multiple neurons to identify nonlinear relationships between a choice and its associated explanatory variables.

Several types of neural networks exist, and the differences between them mainly come from the nature of the basis function for each model. In the following subsections, we detail four different types of neural networks and their structural and mathematical differences. Table 1 contains the main characteristics, advantages, and disadvantages of the four ANN models.

BPNN. BPNN is the most popular and simplest algorithm of MLP. The backpropagation process usually minimizes the error function (Equation 4) of the MLP by changing the value of the weights using a gradient descent method (GDM):

$$E = \sum_{i=1}^{N} \|y_i' - y_i\|^2 \tag{4}$$

where E is the error, i is the number of training samples, y' is the predicted values through the training, and y is the actual values from the training dataset. Each neuron follows the same activation process as shown for a single neuron. Typical activation functions include the step, sigmoid, tanh, rectified linear unit (ReLU), and the identity functions (*25*).

The hidden neurons are "activated" by receiving the sum products of input vectors $\boldsymbol{x}$ and their associated weights $\boldsymbol{w}$. The output layer also obtains information from the hidden layer in the same manner. The final output, y' is:

$$y'(\boldsymbol{x}, \boldsymbol{w}) = \sigma(\sum_{j=1}^{3} w_j^{(2)} \cdot h(\sum_{i=1}^{3} x_i w_i^{(1)} + w_0^{(1)}) + w_0^{(2)}) \tag{5}$$

where $\boldsymbol{x}$ is input vector, $\boldsymbol{w}$ is the vector of associated weights, $w^{(1)}$ are the weights between the input layer and the hidden layer, $w^{(2)}$ is the weights between the hidden

Table 1. Comparison of Four Types of Neural Networks

	BPNN	RBFN	PNN	CPNN
Application scope	Classification/regression	Classification/ regression	Classification	Classification
Classification process	Minimize sum squared errors by updating weights	RBF and BPNN process	Parzen PDF classifier (KDE and Bayesian decision rule)	Use clustering methods and PNN process
Advantages	• Simple application to predict the patterns • Does not require any statistical features in the learning process • Easy to identify the magnitude of attributes based on weights (relative importance) • A variety of applications are available	• Simpler format of Gaussian function enables faster learning process than other Gaussian models • Radial basis function nodes can be substituted with different functional forms • Performs relatively well in both smaller and larger dataset	• Simple architecture (no backpropagation) • More ways to manage the algorithm by determining the shape of bell curve, specifically width (σ) • (more specific than RBFN) • Relatively good accuracy in classification problems • Insensitive to the noise points	• Smaller network size than ordinary PNN • Can avoid saturation of Parzen window that leads to misclassification • May be more applicable because it provides knowledge of relative importance between explanatory variables • Faster training time
Disadvantages	• Easily get stuck in local minima, resulting in suboptimal solution • Like a black box (not sure how to estimate the model) • Need sufficient observations • Prone to overfitting	• Difficult to determine the σ values • Constructing networks architecture is complicated. • Long training time	• More computationally expensive than BPNN (pre-stored pattern neurons) • Saturated Gaussian function can lead some misclassification	• May not provide higher prediction accuracy than PNN for discrete choice data • Varied by number of clusters determined in K-means clustering

layer and the output layer, and the $h()$ and $\sigma()$ functions are the activation functions (see Figure 1*b*).

The drawbacks of the backpropagation process notably come from the GDM that does not guarantee a global minimum is reached when local minima exist. Thus, the results usually vary each time the BPNN is trained. Second, no "optimal" rule exists to set the structural parameters of the BPNN such as the number of hidden layers and their associated number of neurons; although it is generally believed that a larger number of neurons per hidden layer increases the accuracy of the model estimation up to a point (*25*). Nonetheless, in general, one or a few hidden layers are sufficient. Despite these limitations, it is worth mentioning that the BPNN is still the most dominant type of ANN used because it is simple to apply and it ensures relatively high accuracy.

RBFNs. A RBFN is also a feedforward network with three layers: an input, a radial basis function (RBF) hidden, and an output layer (see Figure 1*c*). In contrast to the BPNN, the RBFN is formulated by a linear combination of Gaussian basis functions $g(x)$ as the outputs of the hidden layer:

$$Output\ Score = \sum wh(g(x)) \tag{6}$$

where $h()$ is an activation function, w is a weight between a hidden neuron and an output neuron, and *Output score* is an output for each given class. The basis function $g(x)$ is conceptually obtained by calculating the distance between two vectors based on the Gaussian function whose outputs are inversely proportional to the distance from the mean:

$$g(x) = \frac{1}{\sigma(2\pi)^{\frac{1}{2}}}\exp\left(-\frac{\|\boldsymbol{x}-\boldsymbol{c}_i\|^2}{2\sigma^2}\right) = \exp\left(-\beta\|\boldsymbol{x}-\boldsymbol{c}_i\|^2\right) \tag{7}$$

where $\boldsymbol{x}$ is an input vector with n variables (vector to be classified) and $\boldsymbol{c}$ is a "prototype" vector. To be specific, this equation represents the function of the Euclidean distance between the new input vector and the prototype vector. Essentially, the term β determines the width of the Gaussian curve, which controls the decision boundaries. The prototype vector, $\boldsymbol{c}_i$, is pre-stored in each hidden neuron, and it is a preselected data point from the training set, whether randomly or not (discussed later).

Each hidden neuron compares the input vector to its prototype vector to measure how "similar" they are.

The neurons in the output layer are the sum (linear combination) of the weighted hidden layer neurons (i.e., output score), and the final output, y' is determined by activation functions such as sigmoid function:

$$y' = f\left(\sum_{i=1}^{N} w_i \cdot \exp(-\beta\|x - c_i\|^2)\right) \quad (8)$$

To calibrate the weights, the GDM is also used. By using the outputs (i.e., activations) from each RBF hidden neuron as inputs, the gradient descent process is separately conducted for each output class. Although it tends to be superior to the BPNN network structure, similar to the BPNN, there is no guarantee that a global minimum in found when optimizing the weights in a RBFN.

The performance of an RBFN depends on the number of neurons in the RBFN hidden layer and their prototype vectors. The simplest way to determine the k prototypes is to select them randomly. While it can improve the accuracy of the model, using more prototypes increases the computational costs of the algorithm, and it may generate overfitting issues as well. Another approach to select the prototypes is first to use a clustering algorithm (discussed later).

PNNs. PNNs add a Bayesian decision rule into the framework of RBFNs (see Figure 1*d*). Technically, a PNN is derived from Bayes–Parzen classifiers (*26*), which include Bayesian classification and classical estimators for probabilistic density function (PDF) (*27*, *28*). Essentially, Parzen's method (*29*) is first used to estimate the PDF of each class (i.e., travel modes) from the provided training samples based on a certain type of kernel. These estimated PDF can then go through a Bayesian decision rule to find a final decision.

For any classification problem, a sample vector x is taken from a collection of samples belonging to several classes. If we assume that the prior probability that a sample belongs to class i is h_i, the loss associated with misclassifying that sample is c_i, and the PDF for each of the populations $f_k(x)$ is known. The Bayesian decision rule for classifying an unknown sample into the ith class can be applied as:

$$h_i c_i f_i(x) > h_j c_j f_j(x) \quad \forall j \neq i \quad (9)$$

The PDF for class i, $f_i(x)$, represents the concentration of class i samples around the unknown sample. That is, the Bayesian decision rule favors a class that has high density adjacent to the unknown sample, or if the cost of misclassification or prior probability is high. However, in the Bayesian decision, the PDF for each class is usually known. Therefore, the Parzen's method (*29*) is usually used to estimate the PDF. The univariate KDE was developed by Parzen and it was then extended to the multivariate case by Cacoullos (*30*). The multivariate KDE for n random (independent) variables can be defined as:

$$\varnothing_i(x_n) = \frac{1}{N\lambda}\sum_{k=1}^{N_i} W\left(\frac{x - x_{ik}}{\lambda}\right) \quad (10)$$

where x_n is the vector of independent variables with $n$numbers, and x_k is kth training vector choosing class i; N_i is the total number of observations in the training dataset for class i; λ represents all kernel width for number of variables (n), $\lambda = [\sigma_1, \sigma_2, \cdots, \sigma_n]$; σ is a smoothing parameter representing kernel width (standard deviation; σ is important to obtain an optimal PNN); W is the weight function, and the most popular type of weight function is Gaussian Kernel, which is mainly derived from the concept of Euclidean distance as:

$$W\left(x - x_{ik}/_{\lambda}\right) = e^{\left(-\|x - x_{ik}\|^2/2\sigma^2\right)} \quad (11)$$

By employing a Gaussian weight function, the PDF for choosing class i in the given multivariate input vector x can be expressed as follows:

$$\varnothing_i(1x) = \frac{1}{N_i(2\pi)^{\frac{n}{2}}\sigma^n}\sum_{k=1}^{N_i}\exp\left(-\|x - x_{ik}\|^2/2\sigma^2\right) \quad (12)$$

The final classification decision for the input vector x can be obtained by applying Bayes decision rule as follows:

$$C(x) = arg\max_{i=}(\varnothing_i(x)) \quad (13)$$

CPNN. An ANN using Gaussian KDE such as the PNN may encounter issues related to the size of the network. Specifically, without any pretreatment of the input dataset, the number of pattern neurons containing the prototype vectors is determined either by the number of training samples or by random selection during the training. Since each prototype vector plays a role in KDE by locating the center of the Gaussian distribution for measuring Euclidean distance, the number of pattern neurons can affect the model performance. As the number of pattern neurons increases, the structure of the PNN will become more complicated and more computationally expensive. In previous studies, two feasible methods have been applied to the conventional PNN: 1) K-means clustering to determine the number of pattern neurons; and 2) self-adaptive learning for smoothing parameter (σ) (*31–33*). Here, this study employs the K-means clustering method. Although determining σ is directly related to the decision boundaries, this study simply assigns a certain value according to the knowledge of the dataset and values selected in other studies.

As an unsupervised learning technique, *K*-means clustering has been widely used to classify the data set into the number of clusters (*34*). Then, the centroid of a cluster (i.e., the mean value of the observations within the cluster) is defined by minimizing the distance between the observations belonging to a cluster and the center of the cluster (for details about *K*-means, see [*34*]). The centroids represent the centers of clusters in the Euclidean space, and they become new prototype vectors to measure a Euclidean distance in the pattern neurons. The *K*-means clustering process is separately applied to each choice alternatives.

Data

The data used to test the four models comes from the CMAP Travel Tracker Survey that was collected in 2007 and 2008. The RP survey includes travel diary information as well as detailed individual and household socio-demographic information, for 10,500 households. The original database consists of four interconnected datasets containing household, person, place, and vehicle information. The first three datasets are consolidated to make the dataset.

To merge the datasets, the ESRI ArcGIS package is used to identify interrelated geographical and travel information such as the location of the origins (O) and the destinations (D), walk accessibility, and the actual distance between Os and Ds. Since our focus is on the comparison of the five models, we only look at home-based trips. Other trips, such as the access and egress trips, transfer trips, and other nonhome-based trips are excluded from the dataset used for this study.

In addition, the alternatives for the home-based trip are classified into four classes: walk, bike, transit (CTA bus and train), and auto. The original database contains only 2–3% of transit observations, which is an issue since the proportion of transit observations in the training and testing datasets can be significantly different (e.g., one of the two could have 0 transit observations). To be able to fairly compare the model accuracies, we select to instead use a sub-sample of the original dataset to include more transit observations. Nonetheless, to assess the performance of each model to select poorly represented modes, we left the bike observations intact, which only account for about 4% of all observations. In the end, the dataset used includes 4,764 observations. It includes two sets of attributes as independent variables: (1) individual/household socio-demographic attributes; and (2) travel attributes (e.g., travel time, cost, accessibility) (see Table 2).

Model Specification

Normalization

Discrete choice data contain a variety set of variables with varying scales and ranges. The different numerical

Table 2. Description of Statistics

Variable	Description	Mean	Std
HHSIZ	Number of household members	2.80	1.43
HHWK	Number of workers in household	1.49	0.92
HHSTU	Number of students in household	0.87	1.14
FEMALE	1: if a traveler is female, 0: otherwise	0.52	0.50
EMPLY	Employment status for traveler	1.59	1.05
EDUCA[a]	Education level	3.77	1.91
STUDE	Student grade	2.64	0.74
HHVEH[a]	Number of vehicles in household	1.66	1.06
BIKES[a]	Number of bikes in household	1.48	1.76
CAPINCOME[a]	Capita income of household	0.75	0.37
WalkTT	Travel time for walk mode (hour)	4.76	3.92
BikeTT	Travel time for bike mode (hour)	1.32	1.53
AutoTT	Travel time for auto mode (hour)	0.88	0.91
TransitTT	Travel time for CTA mode (hour)	0.90	1.30
AUTO_COST	Total trip cost for auto (gas, toll, parking / $)	0.90	1.11
Transit_COST	Total trip cost for CTA ($)	0.96	1.12
AGE[a]	Traveler's age	44.02	21.20
ACT_DUR	Actual activity duration (hours)	3.48	3.58
WALK_ACC.	1: if the walking accessibility is within 0.30 miles, 0: otherwise	0.08	0.28

Note: Std. = Standard deviation.

[a]Employed with either the type of dummy or nominal or categorical.

Table 3. Multinomial Logit Model Estimation for Mode Choice

Variable	Coefficient	t-stat
Alternative specific constant (auto is base)		
Walk	2.442	3.48
Bike	−2.033	−11.41
Transit	−1.097	−3.85
Travel time × auto	−0.92	−2.06
Travel time × transit	−1.51	−3.71
Travel time × bike	−0.05	13.87
Auto operating cost × auto	−1.25	−12.931
Transit fare × transit	−2.229	−16.71
Walk accessibility × transit	−0.755	−8.12
Walk accessibility × bike	0.424	2.74
Bikes in HH × bike	1.794	13.18
Number of vehicles in HH × walk	−0.941	−2.08
Number of vehicles in HH × transit	−1.131	−1.98
Age over 75 × walk	−0.358	−4.91
Age over 75 × bike	−1.224	−11.94
Loglikelihood at constant: −4822.92		
Loglikelihood at final convergence: −4121.61		

properties among variables may result in estimation biases. Specifically, the PDF for the Gaussian kernel-based ANN (i.e., RBFN and PNN) cannot be estimated without any pretreatment of the input dataset. Moreover, it is preferable to normalize the input data when the sigmoid function is used in BPNN. Therefore, we normalize all values of attributes in the input dataset before training (estimating) the neural networks specifically; the non-normalized data was used for the MNL as is common in practice. The max-min normalization allows all attributes to be located ranging from 0– 1:

$$x' = \frac{x - \min A}{\max A - \min A} \tag{14}$$

where x' is the new value of the attribute, x is the original value of the attribute; A is the set of all values for a variable from entire data set (training and testing), and $\min A$ and $\max A$ represent the minimum and maximum value of a variable respectively.

Cross-Validation

Cross-validation is a technique used to evaluate model accuracy to prevent overfitting. To evaluate the performance, this study used both k-fold cross-validation method and hold-out method. The hold-out method consists in simply separating the dataset into a training (60%) and a testing (40%) set. Although the holdout method has been widely used for evaluating model performance, by its nature, it may lead to overfitting since only one dataset is used for training. To overcome this issue, the 10-fold cross-validation is also employed, which ensures reliable model performance while minimizing overfitting problems common in machine learning.

The 10-fold cross-validation process functions as follows. The data set is divided into 10 subsets, and the ANN is trained 10 times. Each time, one of the 10 subsets is used as the test set, and the 9 other subsets are put together to form a training set. The average accuracy across all 10 trials is computed. In this study, the 10-fold cross-validation is applied to the four ANN, and the MNL used to measure the overall model accuracy.

Logit Model

The same training dataset was used to calibrate the MNL, adding two extra dummy variables. The MNL is calibrated with SPSS and Biogeme (*35*). The results suggest that household size, age, walk availability, vehicle fleets, travel time, and travel costs (i.e., gas price and transit fare) are statistically significant (see Table 3).

ANN Model

The input dataset for the ANN consists of 14 input values (i.e., explanatory variables) and four target values (i.e., mode alternatives). The process of data transmission in the neural networks differs by ANN type. For the BPNN model, we adopt the sigmoid activation function. We also use one single hidden layer (using multiple hidden layers did not improve the model performance). The number of hidden neurons was determined through the training process, and we used 21 neurons in the hidden layer.

Table 4. Confusion Matrix for Model Accuracy

	Predicted choice					
Test dataset (BPNN)	Walk (114)	Bike (44)	Auto (958)	CTA (473)	Mode accuracy	Overall accuracy
Observed choice						
Walk (187)	68	13	73	33	36.4%	79.4%
Bike (64)	22	17	11	14	26.6%	
Auto (824)	7	3	781	33	94.8%	
Transit (514)	17	11	93	393	76.5%	
Test dataset (RBFN)	Predicted choice					
	Walk (93)	Bike (27)	Auto (859)	CTA (610)	Mode accuracy	Overall accuracy
Observed choice						
Walk (187)	64	13	67	43	34.2%	78.4%
Bike (64)	2	10	23	29	15.6%	
Auto (824)	17	3	719	85	87.3%	
Transit (514)	10	1	50	453	88.1%	
Test dataset (PNN)	Predicted choice					
	Walk (115)	Bike (45)	Auto (804)	CTA (625)	Mode accuracy	Overall accuracy
Observed choice						
Walk (187)	94	7	36	50	50.3%	82.9%
Bike (64)	6	27	23	8	42.2%	
Auto (824)	2	5	723	94	87.7%	
Transit (514)	13	6	22	473	92.0%	
Test dataset (CPNN)	Predicted choice					
	Walk (140)	Bike (48)	Auto (819)	CTA (582)	Mode accuracy	Overall accuracy
Observed choice						
Walk (187)	97	11	45	34	51.9%	83.3%
Bike (64)	9	31	10	14	48.4%	
Auto (824)	16	4	733	71	89.0%	
Transit (514)	18	2	31	463	90.1%	
Test dataset (MNL)	Predicted choice					
	Walk (108)	Bike (114)	Auto (990)	CTA (377)	Mode accuracy	Overall accuracy
Observed choice						
Walk (187)	55	41	67	24	29.4%	70.5%
Bike (64)	2	21	33	8	32.8%	
Auto (824)	28	13	729	54	88.5%	
Transit (514)	11	27	160	316	61.5%	

To run the kernel-based ANNs (i.e., RBFNs and PNNs), the number of hidden neurons and activation functions have to be determined. For the RBFN, the hidden neurons are activated by the Gaussian function, and the sigmoid function is applied to the output neurons. The number of hidden neurons is usually determined based on the size of data or in a random manner during the training process. In this study, the number of hidden neurons is optimally selected by the algorithm. We set the width of Gaussian function β to 0.2 based on the knowledge of the data and on previous values selected in other studies. For the PNN models, most modeling parameters are similar to RBFNs, except for the activation function of the output neurons, which is not required in PNNs because it is determined by the Bayesian decision rule. For the CPNN, 10 clusters for each alternative are partitioned with *K*-means clustering.

The ANNs are trained using the following four Python open source packages: Scikit-learn (*36*) for BPNN and 10-fold cross-validation, a combination of

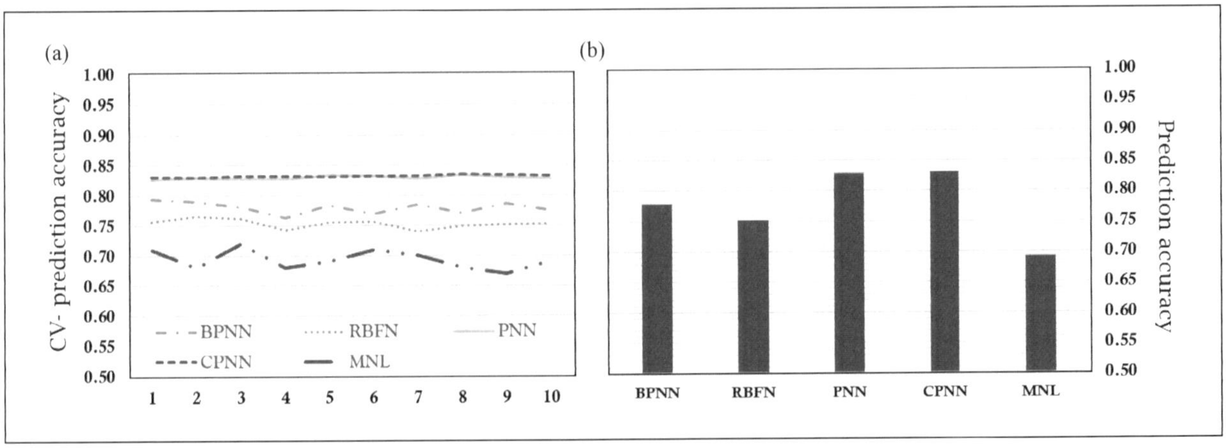

Figure 2. (*a*) Ten-fold cross-validation result for the models; (*b*) comparison of accuracy between the models.

Theano (*37*) and NeuPy (*38*) for PNN and RBFN, and a combination of Scikit-learn (*36*) and NeuPy (*38*) for CPNN. In addition, all ANN models were run on a laptop computer with an average hardware specification (i.e., 6th generation inter processor with 8 GB of RAM). Specifically, computational runtimes for ANN models were similar except for PNN that took 15% longer because of the KDE (specifically for the pattern layer). However, this extra time can be reduced with better hardware specification. In addition, it can be cut down by selecting a limited number of neurons in the pattern layer.

Results

The results of the 10-fold cross-validation for the ANN model are presented in Figure 2*a*. The results indicate that the CPNN model is relatively less sensible over each validation iteration than the BPNN and RBFN. This is because the BPNN and RBFN models are trained with GDM, which may find slightly different minimum values during cross-validation. Furthermore, we compare the average model accuracy between all five models in Figure 2*b*. The results show that the four ANN models achieve better prediction accuracies than the MNL model.

Table 4 presents the confusion matrix, in which each row and each column indicates the observed and predicted the number of travelers for each mode respectively. The overall model accuracy of the four ANNs is around 80%, thus higher than the accuracy of the MNL with 70.5%. While the four ANNs present similar accuracies, the prediction accuracies of each individual mode differ by ANN type. The PNN and CPNN notably show better prediction performance for poorly represented modes. Specifically, the CPNN has slightly better matching rates for the walk and bike modes, in part thanks to the preprocessing with *K*-means clustering.

In addition to the overall accuracy, this study identifies the sensitivity of mode choice decisions from the two most important explanatory attributes: transit costs and auto costs. Unlike traditional regression models, ANN models cannot estimate the impact of explanatory variables on an outcome variable. Thus, sensitivity analysis can be exploited as a sensible option for examining the impact of explanatory variables. Figure 3, *a* and *b* present the result of sensitivity analysis and compare the results by the models for two different variables: auto cost and transit cost. In particular, we can see that BPNN and MNL models are relatively more sensitive to the variations in two variables. As expected, transit users are more likely to change their mode to auto when the price of gas decreases. In contrast, auto users are relatively insensitive to increases in auto costs of 15% or lower.

Conclusion

This article has investigated the feasibility, capability, and performance of four ANN methods in dealing with travel mode choice modeling. The prediction performance of ANN and MNL models has been evaluated by conducting cross-validation tests with the 10-fold cross-validation method. In addition, a confusion matrix has been used to evaluate the matching accuracy between the models. The cross-validation results revealed that the four ANNs achieve better prediction accuracies (around 80%) than the MNL (around 70%). In particular, the CPNN showed the highest performance among the ANN models in part thanks to the *K*-means clustering procedure used. The confusion matrix also showed that the matching rates for bike and walk modes were relatively poor because they were poorly represented in the dataset, although the PNN and CPNN performed best. We suggest that this phenomenon is linked to the fact that ANNs are better able to identify nonlinear relationships in a multivariate survey dataset. Nonetheless,

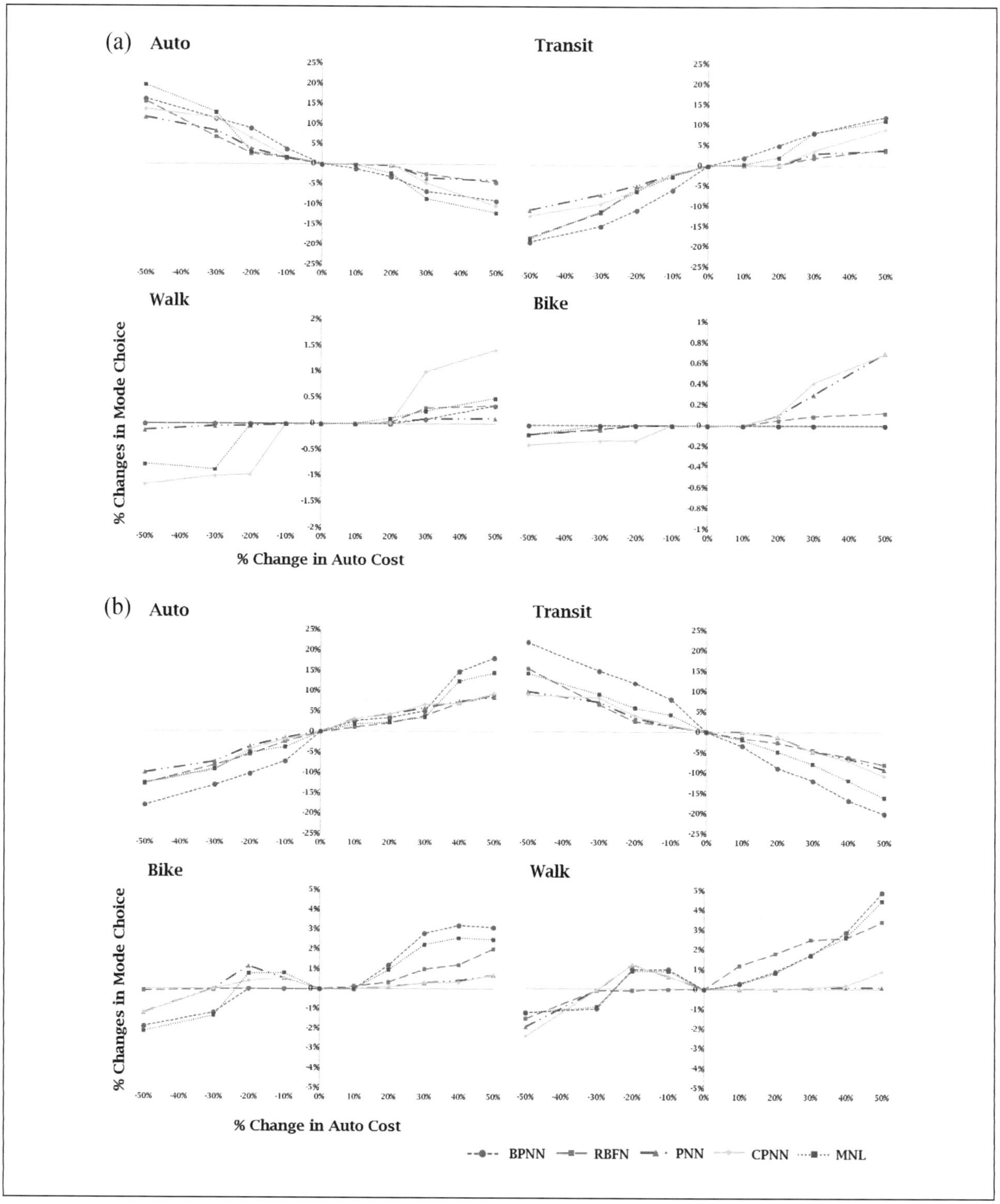

Figure 3. The result of sensitivity analysis—comparison between models based on two scenarios: (*a*) percent changes in auto cost (*b*) percent changes in transit cost.

although the model accuracy of the ANN outperformed the MNL in mode choice modeling, we must acknowledge that ANNs suffer from a lack of interpretability. As a partial strategy to tackle this issue, we conducted a sensitivity analysis for two significant attributes (i.e., transit costs and auto costs) when choosing a mode.

Both the MNL and ANN have advantages and disadvantages. As a parametric approach, logit models are generally used for extracting parameters based on given behavioral patterns. However, when the behaviors being modeled are complex, it becomes difficult to design an accurate logit model. For instance, the MNL is susceptible to unobserved biases, and it is often impractical for identifying the relationship and configuration of explanatory variables. In contrast, ANNs are able to capture nonlinearity and biases in the data by using additional information processing units (e.g., hidden layers). Although ANN is often assimilated to a "black box," its relatively easy applicability and its ability to capture complex patterns make it particularly powerful and promising for the future of mode choice modeling.

For future work, novel non-parametric models have emerged recently (*16*). For instance, deep learning (DL) generally requires more complex algorithmic features, and they are computationally more expensive than ANNs, but in theory, they should be even better able to capture complex and nonlinear relationships in a dataset. Advanced data mining methods such as DL may, therefore, possess a bright future in mode choice modeling in particular, and in travel demand forecasting in general.

Acknowledgments

This research is partly supported by the National Science Foundation (NSF) CAREER award #155173 and by the NSF Cyber-Physical Systems (CPS) award 1646395.

Author Contributions

The authors confirm contribution to the paper as follows: study conception and design: DL, SD; data collection: DL; analysis and interpretation of results: DL, SD, FP; draft manuscript preparation: DL, SD, FP. All authors reviewed the results and approved the final version of the manuscript.

References

1. De Carvalho, M., M. Dougherty, A. Fowkes, and M. Wardman. Forecasting Travel Demand: A Comparison of Logit and Artificial Neural Network Methods. *Journal of the Operational Research Society*, 1998, pp. 717–722.
2. Bentz, Y., and D. Merunka. Neural Networks and the Multinomial Logit for Brand Choice Modelling: A Hybrid Approach. *Journal of Forecasting*, Vol. 19, No. 3, 2000, pp. 177–200.
3. Train, K. E. *Discrete Choice Methods with Simulation*. Cambridge University Press, Cambridge, UK, 2009.
4. Bhat, C. R. Quasi-random Maximum Simulated Likelihood Estimation of the Mixed Multinomial Logit Model. *Transportation Research Part B: Methodological*, Vol. 35, No. 7, 2001, pp. 677–693.
5. Shen, J. Latent Class Model or Mixed Logit Model? A Comparison by Transport Mode Choice Data. *Applied Economics*, Vol. 41, No. 22, 2009, pp. 2915–2924.
6. Kuhn, M., and K. Johnson. *Applied Predictive Modeling*. Springer New York, New York, NY, 2013.
7. James, G., D. Witten, T. Hastie, and R. Tibshirani. *An Introduction to Statistical Learning*. Springer, New York, NY, 2013.
8. Derrible, S. Urban Infrastructure is not a Tree: Integrating and Decentralizing Urban Infrastructure Systems. *Environment and Planning B: Planning and Design*, 2016.
9. Ahmad, N., S. Derrible, T. Eason, and H. Cabezas. Using Fisher Information to Track Stability in Multivariate Systems. *Royal Society Open Science*, Vol. 3, No. 11, 2016, p. 160582.
10. Ahmad, N., S. Derrible, and H. Cabezas. Using Fisher Information to Assess Stability in the Performance of Public Transportation Systems. *Royal Society Open Science*, Vol. 4, No. 4, 2017, p. 160920.
11. Ahmad, N., and S. Derrible. Evolution of Public Supply Water Withdrawal in the USA: A Network Approach. *Journal of Industrial Ecology*, Vol. 19, No. 2, 2015, pp. 321–330.
12. Derrible, S., and N. Ahmad. Network-Based and Binless Frequency Analyses. *PloS One*, Vol. 10, No. 11, 2015, p. e0142108.
13. Derrible, S. Complexity in Future Cities: The Rise of Networked Infrastructure. *International Journal of Urban Sciences*, Vol. 21, 2017, pp. 68–86.
14. Cottrill, C. D., and S. Derrible. Leveraging Big Data for the Development of Transport Sustainability Indicators. *Journal of Urban Technology*, Vol. 22, No. 1, 2015, pp. 45–64.
15. Karduni, A., A. Kermanshah, and S. Derrible. A Protocol to Convert Spatial Polyline Data to Network Formats and Applications to World Urban Road Networks. *Scientific Data*, Vol. 3, 2016, p. 160046.
16. Wong, M., B. Farooq, and G.-A. Bilodeau. Latent Behaviour Modelling Using Discriminative Restricted Boltzmann Machines. *Proc., 5th International Choice Modeling Conference*, 2017, pp. 1–18.
17. Bengio, Y., and S. Bengio. Modeling High-Dimensional Discrete Data with Multi-Layer Neural Networks. *Proc., Advances in Neural Information Processing Systems*, 2000.
18. Dougherty, M. A Review of Neural Networks Applied to Transport. *Transportation Research Part C: Emerging Technologies*, Vol. 3, No. 4, 1995, pp. 247–260.

19. Sayed, T., and A. Razavi. Comparison of Neural and Conventional Approaches to Mode Choice Analysis. Vol. 14, No. 1, 2000, p. 23.
20. Xie, C., J. Lu, and E. Parkany. Work Travel Mode Choice Modeling with Data Mining: Decision Trees and Neural Networks. *Transportation Research Record: Journal of the Transportation Research Board*, 2003. 1854: 50–61.
21. Hensher, D. A., and T. T. Ton. A Comparison of the Predictive Potential of Artificial Neural Networks and Nested Logit Models for Commuter Mode Choice. *Transportation Research Part E: Logistics and Transportation Review*, Vol. 36, No. 3, 2000, pp. 155–172.
22. Mohammadian, A., and E. Miller. Nested Logit Models and Artificial Neural Networks for Predicting Household Automobile Choices: Comparison of Performance. *Transportation Research Record: Journal of the Transportation Research Board*, 2002. 1807: 92–100.
23. Golshani, N., R. Shabanpour, S. M. Mahmoudifard, S. Derrible, and A. Mohammadian. Comparison of Artificial Neural Networks and Statistical Copula-Based Joint Models. Travel Behaviour and Society, Vol. 10, 2017, pp. 21–32. doi:10.1016/j.tbs.2017.09.003.
24. Domencich, T. A., and D. McFadden. *Urban Travel Demand: A Behavioral Analysis*. Elsevier, Amsterdam, the Netherlands, 1975.
25. Bishop, C. M. *Pattern Recognition and Machine Learning*. Springer, New York, NY, 2006.
26. Masters, T. *Advanced Algorithms for Neural Networks: A C + + Sourcebook*. John Wiley & Sons, Inc., Hoboken, NJ, 1995.
27. Specht, D. F. Probabilistic Neural Networks. *Neural Networks*, Vol. 3, No. 1, 1990, pp. 109–118.
28. Specht, D. F. Probabilistic Neural Networks for Classification, Mapping, or Associative Memory. *IEEE International Conference on Neural Networks*, San Diego, CA, 1988, pp. 525–532.
29. Parzen, E. On Estimation of a Probability Density Function and Mode. *The Annals of Mathematical Statistics*, Vol. 33, No. 3, 1962, pp. 1065–1076.
30. Cacoullos, T. Estimation of a Multivariate Density. *Annals of the Institute of Statistical Mathematics*, Vol. 18, No. 1, 1966, pp. 179–189.
31. Kim, D. K., D. H. Kim, S. K. Chang, and S. K. Chang. Modified Probabilistic Neural Network Considering Heterogeneous Probabilistic Density Functions in the Design of Breakwater. *KSCE Journal of Civil Engineering*, Vol. 11, No. 2, 2007, pp. 65–71.
32. Monfort, M., B. M. Lake, B. Ziebart, P. Lucey, and J. Tenenbaum. Softstar: Heuristic-Guided Probabilistic Inference. *Proc., Advances in Neural Information Processing Systems*, 2015.
33. Yi, J.-H., J. Wang, and G.-G. Wang. Improved Probabilistic Neural Networks with Self-Adaptive Strategies for Transformer Fault Diagnosis Problem. *Advances in Mechanical Engineering*, Vol. 8, No. 1, 2016, p. 1687814015624832.
34. Park, H.-S., and C.-H. Jun. A Simple and Fast Algorithm for K-Medoids Clustering. *Expert Systems with Applications*, Vol. 36, No. 2, 2009, pp. 3336–3341.
35. Bierlaire, M. BIOGEME: A Free Package for the Estimation of Discrete Choice Models. *Proc., Swiss Transport Research Conference*, Monte Verità, Ascona, Switzerland, 2003.
36. Pedregosa, F., G. Varoquaux, A. Gramfort, V. Michel, B. Thirion, O. Grisel, M. Blondel, P. Prettenhofer, R. Weiss, and V. Dubourg. Scikit-Learn: Machine Learning in Python. *Journal of Machine Learning Research*, Vol. 12, 2011, pp. 2825–2830.
37. Al-Rfou, R., G. Alain, A. Almahairi, C. Angermueller, D. Bahdanau, N. Ballas, F. Bastien, J. Bayer, A. Belikov, and A. Belopolsky. Theano: A Python Framework for Fast Computation of Mathematical Expressions. *arXiv preprint*, 2016.
38. Shevchuk, Y. *NeuPy: Neural Networks in Python*. http://neupy.com/pages/home.html.

The Standing Committee on Transportation Demand Forecasting (ADB40) peer-reviewed this paper (18-04636).

Article

Transportation Research Record
2018, Vol. 2672(49) 113–124

DOI: 10.1177/0361198118783113
journals.sagepub.com/home/trr
SAGE

Exploring the Impact of Different Cut-off Times on Isochrone Measurements of Accessibility

Yang (Luna) Xi[1], Eric J. Miller[1], and Shoshanna Saxe[1]

Abstract

Isochrone analysis and assessments of cumulative opportunities are a common way to quantify accessibility. However, different time cut-offs have been used by different researchers, with little investigation into what is the 'best' cut-off time. Outstanding questions remain concerning the most effective or predictive cut-off time and the potential implications of choosing one time limit over another. The primary objective of this paper is to explore how different cut-off times affect the calculation of isochrone-based accessibility measurements and their potential to predict travel-mode choice. Fifty dissemination areas (DAs) within the Greater Toronto and Hamilton Area (GTHA) are selected to test the impact of different isochrone cut-off times in 5-minute intervals for public transit, automobile, and walking accessibility. The relative predictive power of 30- and 45-minute isochrones in modeling mode choice is also examined. This paper finds that different cut-off times do impact the interpretability of accessibility measurements in the isochrone approach, but a defined cut-off time for general use cannot be determined based on the analysis.

Accessibility is widely used in the transportation field to interpret or predict travel behavior (*1–4*). Accessibility is often quantified through the calculation of travel time isochrones and the cumulative opportunities of population, jobs, or other activities captured within the isochrone. Different time cut-offs have been used by different researchers, with 30 and 45 minutes being common choices (*5–8*). Outstanding questions remain around the most effective or predictive time cut-off and the potential implications of choosing one time limit over another. The primary objective of this paper is to explore how different cut-off times affect the calculation of isochrone-based accessibility measurements and their potential to predict travel mode choice.

Fifty dissemination areas (DAs) from across the Greater Toronto Hamilton Area (GTHA) were selected to test the change in calculated accessibility at 5-minute intervals. Accessibility is calculated from 0~90 minutes for automobile and 0~60 minutes for walking and public transit. Additionally, accessibility was calculated for all DAs (9035) within the GTHA at 30 and 45 minutes. Bivariate linear regression is used to check the relative predictive power of accessibility to mode share at 30 and 45-minute isochrone cut-offs.

The structure of this paper is as follows. The next section presents a literature review of accessibility research. We then discuss the methods adopted and the selection of 50 sample DAs. We follow this with a presentation of the isochrone maps and accessibility results using graphical, ranking, Interquartile Range (IQR), and linear regression analyses. Finally, the paper summarizes the significance of the findings and recommends paths for future research.

Literature Review

Accessibility is defined as "the potential of opportunities for interaction" that measures the spatially distributed activities for a specific location considering the abilities and desires of travelers (*9*). The measurement of accessibility is complicated because of the variability of individual trips and activities (*10*). Many approaches for measuring accessibility have been developed. As summarized by Geurs & Van Wee (*11*), there are four types of accessibility measures, including infrastructure-, person-, utility-, and location-based measures. Isochrone approaches and cumulative opportunities assessments within the isochrone are a popular method for quantifying accessibility without defining a destination. The isochrone defines the achievable locations from a fixed starting point within the given cut-off time using a transportation mode, with the isochrones shape being defined by the network of the mode in question.

[1]Department of Civil Engineering, University of Toronto, Toronto, ON, Canada

Corresponding Author:
Address correspondence to Yang (Luna) Xi: luna.xi@mail.utoronto.ca

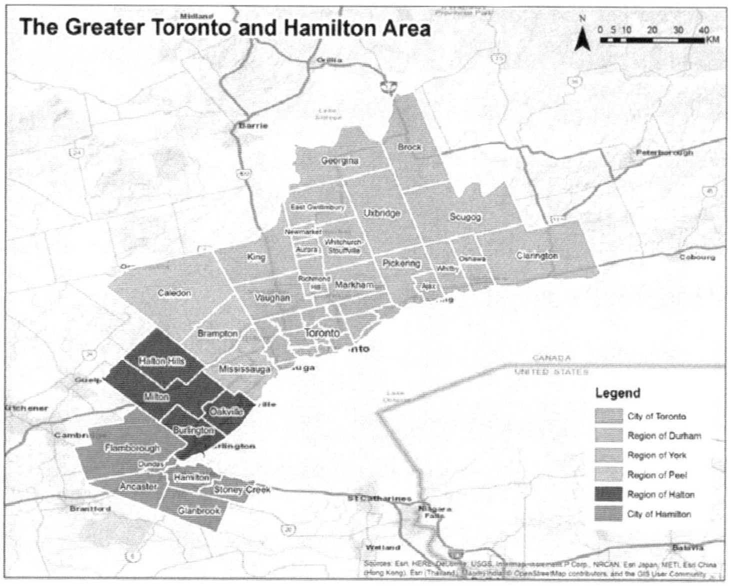

Figure 1. The Greater Toronto and Hamilton Area (GTHA) map.

The isochrone approach has been widely used for quantifying accessibility through the counting of the opportunities (cumulative opportunities) within the isochrone – commonly jobs and or population. To cite a few examples, it is used to evaluate: block-level accessibility measures with space-time constraints, the influence of accessibility on commuting mode share, spatial-temporal dynamics of transit supply, and the relationship between accessibility and land use (*1*, *3*, *6*, *12*).

Despite the widespread use of this approach, researchers vary in their choice of cut-off times. For example, Chen et al. calculate different accessibility indicators using 10, 20, and 50-minute cut-off times (*1*), while Owen and Levinson use cut-off times ranging from 10 to 60 minutes with a 10-minute interval to calculate accessibility to jobs (*3*). El-Geneidy et al. investigate the correlation between gravity-based accessibility measures and cumulative measures of accessibility with cut-off times varying from 20 to 45 minutes (*2*), while Martens analyzes the changes of accessibility deficiency between 20- and 30-minute cut-off times (*4*). A recent study uses a 30-minute cut-off time to investigate an accessibility-based regression model (*13*), while another recent study analyzes public transit accessibility with a 30-minute cut-off time and daily grocery trips with 10 or 20 minutes (*6*). A 30-minute cut-off time is sometimes selected based on the argument that the average commuting time in one direction is often found to be about 30 minutes (*5*). In U.S. federal transportation projects, the cut-off time is frequently set as 30 or 45 minutes (*7*, *8*). Considering various preferences for choosing cut-off times in research studies, this paper evaluates the impact of picking different cut-off times on accessibility measurements when using the isochrone approach.

Methodology

Study Area and Transportation Networks

The study area for this work is the Greater Toronto and Hamilton Area (GTHA) illustrated in Figure 1. The GTHA consists of two single-tier municipalities, the Cities of Toronto and Hamilton, and four regional municipalities, Durham, Halton, Peel, and York (*14*). The total population of GTHA was 6,954,000 in 2016, which is approximately 19.8% of Canada's population (*15*).

Automobile, public transit, and walking are the three most common travel modes in the GTHA (*16*) and are the modes examined in this paper. All networks were built using morning–peak-hour conditions.

The walking network was constructed based on the National Road Network (NRN) datasets with an assumed uniform walking speed of 4.8 km/h (3.0 mph) (*17*), excluding high-speed roads where walking is prohibited. Automobile travel times were simulated using the EMME network modeling software with the combination of the NRN datasets (*17*) and the 2011 GTAModel V4.0, developed by the University of Toronto's Travel Modelling Group (*18*). For the major roads, such as expressways, the travel speeds were defined based on the predicted morning–peak-hour speeds from GTAModel. For minor roads not included in the GTAModel, such as local streets, the travel speeds were based on the posted speed limit in NRN datasets. The public transit network was built by combining the walking network with the available transit routes and travel times, which were extracted from the GTAModel simulated through EMME software to include waiting and transferring times for

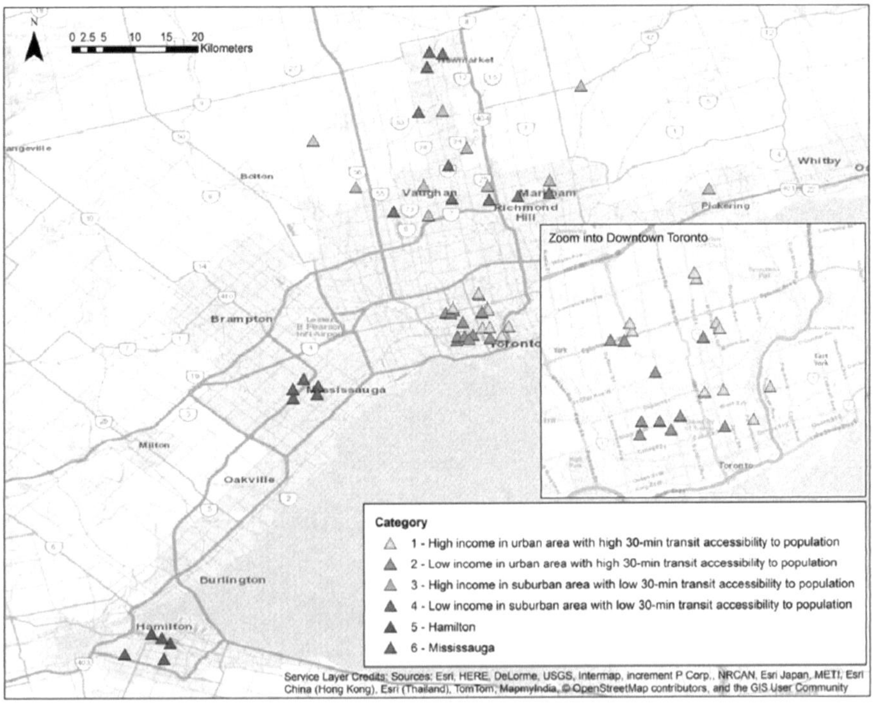

Figure 2. Geographical locations of 50 test cases.

point-to-point transit travel times between all transit nodes in the GTHA (*13*). The types of public transit available in the GTHA include TTC subways, streetcars, buses, GO trains, and local buses (*18*).

Selection of Sample DAs

Dissemination area (DA) is a spatial agglomeration for the Canadian Census defined by Statistics Canada; it is a small geographic unit consisting of 400~700 persons (*19*). Fifty DAs were selected as test cases for an exploratory analysis based on household characteristics including income level, location, and 30-minute transit accessibility to population.

As many studies have discussed, the choice of transportation mode for travel and associated accessibility are correlated with household income (*20–22*). Therefore, income level is one of three criteria used for selecting test cases. Based on the 2011 National Household Survey, the average household income in Ontario is \$95,842 before tax (*23*). In this paper, the high-income DAs are defined as where the DA's median household income is in the range of \$125,000 ~ \$260,000 and low-income DAs have a median household income of \$25,000 ~ \$50,000.

The second criterion is household location. DAs were selected to have some inside the historic City of Toronto and current downtown, some in areas with more suburban forms, and some in Mississauga and Hamilton, the second and third largest cities within the GTHA.

Finally, estimates of the 30-minute transit accessibility to population are used as the third criterion, to ensure selection of sample DAs with a variation in access to public transit. Previous research shows that the average 30-minute transit accessibility to population in the GTHA is approximately 79,000 (*13*). Therefore, in this paper, the DAs with high access are defined as those in the range of 150,000–700,000 and the DAs with low access have the results of 500–50,000.

Based on these criteria, 50 DAs within the GTHA were selected as test cases to evaluate the change of impact that different cut-off times have on accessibility measurements for various types of household and modes of travel. The 50 test cases were divided into 6 categories based on location and income as shown in Figure 2.

Calculation of Isochrones

Isochrones for each of the 50 test cases were calculated using the service area tool in ArcGIS (*24*). The cut-off time was set in 5-minute increments for walking, public transit, and automobile trips. Accordingly, the sets of isochrones are as follows: (i) every 5 minutes up to 60 minutes for walking; (ii) every 5 minutes up to 60 minutes for public transit; and (iii) every 5 minutes up to 90 minutes for automobile. The starting points of trips are the centroids of the test-case DAs; the routes or travel times are based on the transportation networks discussed below. The accessibility

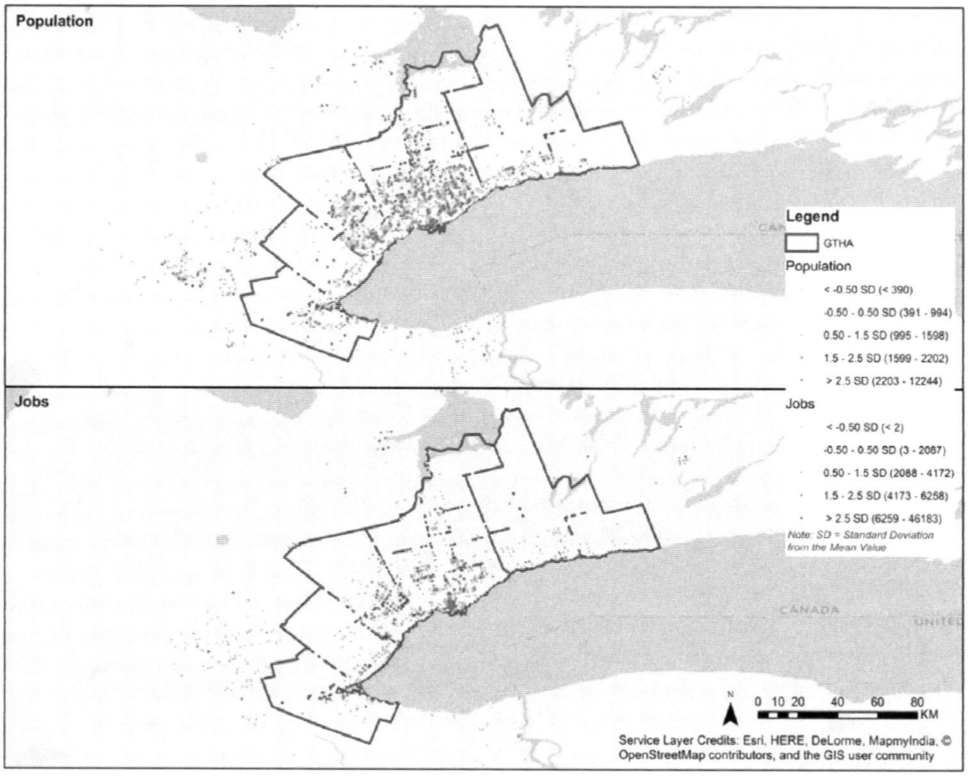

Figure 3. Distribution of population and jobs.

to population for each test case is calculated as the cumulative population within the isochrones, which varies with different cut-off times and transportation modes. The population data was extracted from the datasets of the 2011 Census of Population (*25*) and agglomerated to DAs. Similarly, the accessibility to jobs for each test case is computed as the cumulative opportunities of jobs within the isochrones. Job data was obtained from the 2011 Transportation Tomorrow Survey (TTS) and agglomerated to TAZ-level (*26*). TAZs are 4~7 times larger than DAs (*27*), resulting in reduced accuracy for job-based accessibility. The distributions of population and jobs are illustrated in Figure 3. Linear regression models investigated the correlations of travel mode shares with accessibility at different cut-off times (10, 15, 20, 30, 40, 45, and 50 minutes) for the 50-DA test case. The regression equation is:

$$(Mode\ Share) = Intercept + Coefficient * (Accessibility)$$

Where Mode Share is the percentage of total 24-hour weekday trips completed by the transportation mode (transit or automobile) originating in the DA, and Accessibility is the cumulative opportunities of population or jobs captured within isochrones defined by the cut-off time. Mode share calculations were based on TTS trip data.

Extended Examination

According to the literature review, 30 and 45 minutes are commonly used cut-off times in the isochrone approach (*5–8*). Therefore, an extended examination was conducted for 30- and 45-minute cut-off times for all 9,035 DAs within GTHA. In addition, bivariate linear regression models were built for each combination of accessibility and mode share (i.e., population/job vs. automobile/public transit) with similar steps described in Section 3.3. The regression models were also run excluding DAs with zero automobile and/or transit mode shares, all DAs with more than 10 trips, and all DAs with more than 30 trips to eliminate the impact of DAs with low trip numbers and the potential for associated artificially high mode shares.

Findings

Fifty Test Cases

Isochrones were drawn for the 50 test cases with different cut-off times and transportation modes. A sample isochrone map for the 30-minute cut-off time with six selected test cases (i.e., each represents a different category) is shown in Figure 4 to illustrate the procedure used. In addition, Figure 5 delineates sample public transit isochrones with different cut-off times.

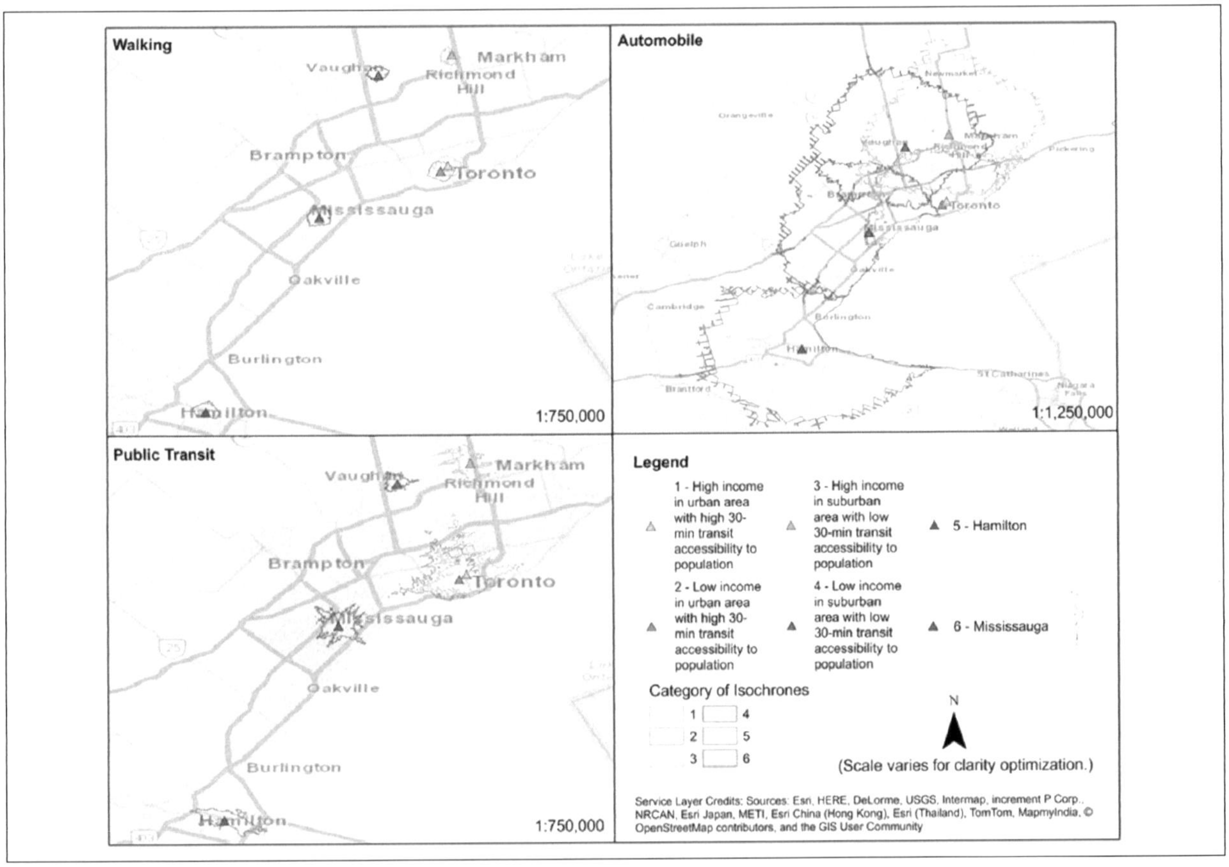

Figure 4. Sample isochrone map for 30-minute cut-off time with selected test cases.

The maps reveal stark changes in the size and shape of isochrones between different modes and in different geographic locations. As would be expected, the walking and automobile isochrones are more circular, while the transit isochrones reflect the location of the transit network. The walking isochrones have a similar size at any location reflecting relatively uniform walking speeds, while the size of automobile or transit isochrones change between urban city and other regions. A relatively steady expansion in size is observed as the cut-off times increase for all transportation modes.

The graphical analysis of cut-off times and associated accessibility to people and jobs shows more complicated relationships in Figure 6. Had the relationships been linear any cut-off time would accurately describe the line. However, the graphs indicate the relationship is not completely linear for any transportation mode, although some parts of the curves have linear trends (e.g., transit accessibility after a 40-minute cut-off time) or are approximately bilinear (e.g., green curves in automobile accessibility).

Unlike the walking access to population, which increases constantly with the cut-off time for all test cases, the job accessibility of walking rises quickly in the urban core, especially for cut-off times greater than 20 minutes. For the automobile, the rate of increase in population accessibility keeps changing as the cut-off time rises and the inflection points are different for each category. Such patterns are also observed in the job accessibility of travel by automobile, which suggests the possibility that a slightly different cut-off time used for measuring accessibility could lead to large differences in the results. For instance, over 800,000 more jobs become accessible for a suburban test DA as the cut-off time increases from 45 to 50 minutes. This 5-minute change facilitates accessibility to large job concentrations in the downtown core and airport employment. The population accessibility of public transit shows more distinctive patterns, where a divergence occurs between the city and other regions (i.e., suburban areas and Mississauga/Hamilton). Because of limited local transit services, no additional transit accessibility beyond walking is observed when the cut-off time is less than 20 minutes. As the cut-off time increases, transit accessibility increases sharply, showing an inflection point when the isochrones capture major shopping malls and transit terminals. The significant difference before and after the inflection point demonstrates the impact of picking cut-off times. A similar divergence appears in the job accessibility of public transit, and the rate of increase is much faster in

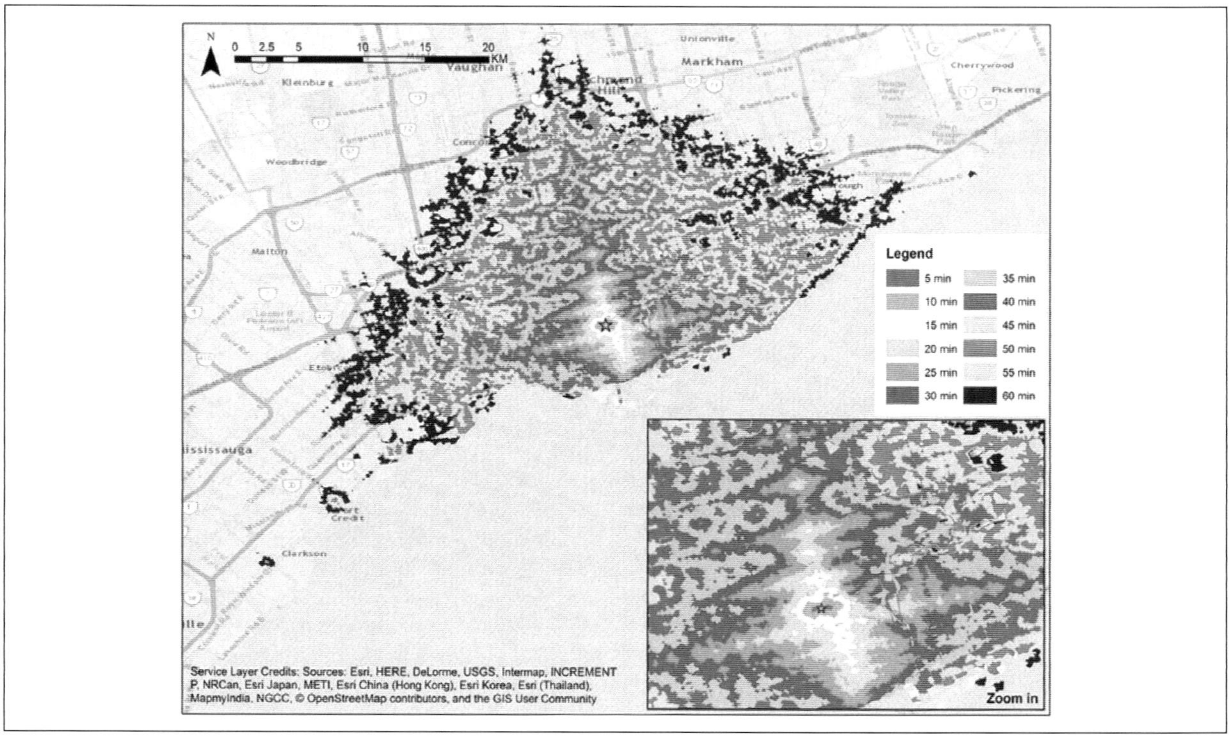

Figure 5. Sample transit isochrones with different cut-off times for a test case.

Toronto than in other regions in the 10~40-minute cut-off range. Also, many inflection points are observed, which indicates the possibility of large differences in job accessibility due to a small change in the cut-off time.

Figure 6 illustrates convergence of the isochrones at overly small and overly large cut-off times. Below 10 minutes there is very little variability between the accessibility calculated for each of the test DAs, after 20–30 minutes for all modes studied there is increasing variability between the calculated accessibility for each mode. The spread of calculated accessibility is useful for differentiating between the accessibility of different places. This continues through 60 minutes. For automobiles, the only mode examined for cut-off times above 60 minutes, the calculated accessibility begins to converge around 70 minutes, indicating diminishing returns of larger isochrones as major population and employment centers are already within the isochrone.

Ranking and IQR Analysis

When used as a predictor of other outcomes (e.g., travel mode choice, land use intensification, equity analysis), the relative accessibility between areas is critical. Similarly, when planning transportation infrastructure, understanding which parts of a study area have more (or less) accessibility can be key to allocating new infrastructure to underserved areas.

An ideal cut-off time would be when there was enough spread between different points to differentiate between the accessibility (i.e., higher variability) and when small changes in cut-off time did not meaningfully change the relative accessibility of studied points (i.e., higher stability).

An Interquartile Range (IQR) analysis was used to examine the variability in test cases by computing the vertical spread of curves at different cut-off times. This "spread" was measured as the difference between the 25 and 75 percentile accessibility values where the DAs are ranked from lowest to highest accessibility for a given cut-off time. Table 1 shows the results of the IQR analysis for Figure 6, where the IQR values represent the vertical spread of the curves. A higher IQR indicates a higher variability in accessibility measures at that cut-off time. For walking access to population and jobs, the IQR increases constantly with the cut-off time indicating a higher variability when using larger cut-off times. In the automobile case, the peak IQR values occur at the 40-minute cut-off for population accessibility and at 35 minutes for job accessibility, after which the IQR values keep decreasing until 65 minutes and then increase again. The IQR for public transit access to population increases with the cut-off time, however, the rate of increase is much larger between 20 and 40 minutes than at other cut-off times. Similarly, the job accessibility also increases with the cut-off time having the highest rate of increase at 30 minutes.

A ranking analysis of the 50 test DAs was carried out to investigate if there is a stable relationship between the accessibility of the studied points at a given cut-off time or range of times. Figure 7 illustrates the ranking of each DA, #1 to

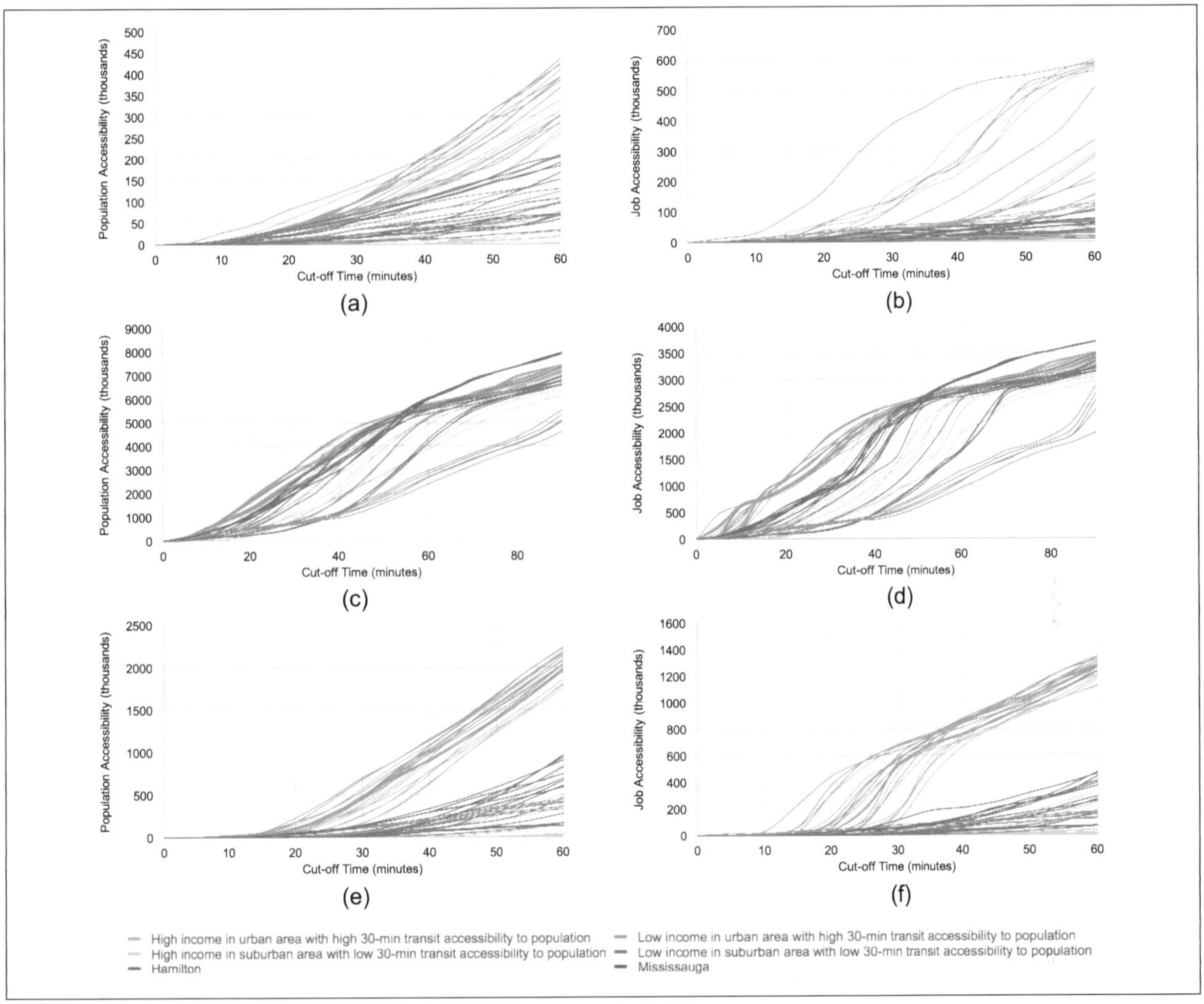

Figure 6. Graphical analysis of population and job accessibility: (*a*) population accessibility of walking; (*b*) job accessibility of walking; (*c*) population accessibility of automobile; (*d*) job accessibility of automobile; (e) population accessibility of public transit; (*f*) job accessibility of public transit.

#50, at each cut-off time, with the most accessible DA for each increment on top. Rankings are graphed for all modes for cut-offs ranging from 0 to 60 minutes. Many changes are observed in the ranking positions of population accessibility (i.e., large increase or decrease of the ranking number for a test case), especially before the 15-minute cut-off time for walking and 25-minute cut-off for public transit. However, for automobile access to population, frequent and significant changes occur at all cut-off times. The job accessibility also shows many changes in the ranking positions, especially before the 20-minute cut-off time for walking and 25 minutes for public transit. Similarly, frequent and significant changes occur for automobile access to jobs.

Accessibility & Mode Share, 50-DA Test Case

Using the equation and steps described in Section 3.3, linear regression models were built to examine the relationship between travel mode shares and accessibility at different cut-off times for the 50 test cases. Table 2 shows the adjusted R-square values and parameters of the regression models.

The adjusted R-square values indicate that public transit mode share is better explained by both the population and job accessibility measures than automobile mode share for cut-off times larger than 20 minutes. For transit mode shares, the model displays a clear optimal predictive power when a 30-minute cut-off time is used, for both the population and job accessibility cases. The automobile mode-share models, on the other hand, display monotonically decreasing adjusted R-square values with increases in cut-off time (i.e., no obvious optimal cut-off time exists). Additionally, the correlations to the transit mode share at different cut-off times are consistently higher with respect to access to population than to jobs, but job accessibility has higher explanatory power for automobile mode shares.

Table 1. IQR Analysis Results for 50 Test Cases

	Population accessibility			Job accessibility		
Cut-off time	Walking	Automobile	Public transit	Walking	Automobile	Public transit
5	958	95,083	1,044	89	38,370	0
10	6,136	325,716	6,242	2,119	232,066	2,050
15	15,455	609,038	23,740	5,955	646,992	9,958
20	28,893	845,524	84,376	12,505	768,225	32,531
25	44,275	1,240,227	185,914	24,572	959,944	119,170
30	58,450	1,477,435	366,436	31,768	1,175,033	495,805
35	81,151	1,727,782	617,638	36,986	1,295,091	634,568
40	105,147	1,793,068	823,918	49,831	1,294,064	728,756
45	125,366	1,416,998	1,003,266	66,747	1,105,904	823,105
50	157,918	1,153,644	1,210,570	71,209	956,264	924,332
55	188,057	826,947	1,415,959	86,829	366,335	996,629
60	230,896	483,326	1,647,500	123,491	200,865	1,098,551
65	NA	330,801	NA	NA	160,858	NA
70	NA	332,187	NA	NA	165,247	NA
75	NA	413,804	NA	NA	186,587	NA
80	NA	483,156	NA	NA	208,124	NA
85	NA	597,421	NA	NA	273,409	NA
90	NA	711,832	NA	NA	312,377	NA

Note: NA = not available.

Extended Examination

To further explore the relationship between accessibility and travel mode shares, the linear regression analysis was extended to include all DAs in the GTHA using the 30- and 45-minute cut-off times. The regression results are shown in Table 3. A large decrease in adjusted R-square value is observed for the public transit models at both the 30- and 45-minute cut-off times for both population and job accessibility relative to the 50-DA test case. This is expected given the much larger dataset. For the automobile models, the full sample adjusted R-square is also somewhat lower for the 30-minute case relative to the 50-DA test case, but is higher for the 45-minute case. The accessibility correlations with automobile mode shares are now higher than the transit mode-share correlations for both population and job accessibility measures for both cut-off times. Presumably, these results reflect the dominance of the automobile mode in the full sample, which adds many suburban and rural DAs to the analysis relative to the 50-DA test case.

Since the adjusted R-square value decreases significantly in the all-DA case, additional regression models were built to investigate potential sources of this poor fit. First, models were run in which DAs with zero automobile and/or transit mode shares were deleted, since the simple linear model being used will not capture zero mode shares well. This results in only a very marginal increase in automobile model fits and decreases in transit model fits. Next, since the dependent variable is mode share and small-sample mode shares can be subject to significant sampling bias, models were estimated excluding DAs with less than 10 and 30 total trips. This improves the model fits marginally, yet overall the conclusions remain the same.

Nevertheless, 30-minute automobile model adjusted R-squares are generally about 2%~5% higher than the 45-minute model values. Yet for transit models, the 30-minute case only has a marginally higher adjusted R-square value than the 45-minute case for population accessibility, while the reverse is true for job accessibility. At both cut-off times, the job accessibility measure has a slightly higher correlation with either automobile or public transit mode shares than the population accessibility measure.

Conclusions

The analysis presented in this paper indicates that different cut-off times do impact the interpretability of accessibility measurements in the isochrone approach, as well as the extent to which such measures can explain travel mode choice.

The effect of picking different cut-off times changes between households because of their distinctive characteristics (e.g., located in urban or suburban area). Different types of accessibility measurements are also differentially affected by the chosen cut-off value. Given the large concentration of jobs in major, discrete employment areas within the urban area (e.g., downtown, near the airport) and the agglomeration of jobs to TAZ the impact of different cut-off times is greater on job-based accessibility measurement than population-based ones; a slightly different cut-off time used for measuring job accessibility may lead to a large difference in the results.

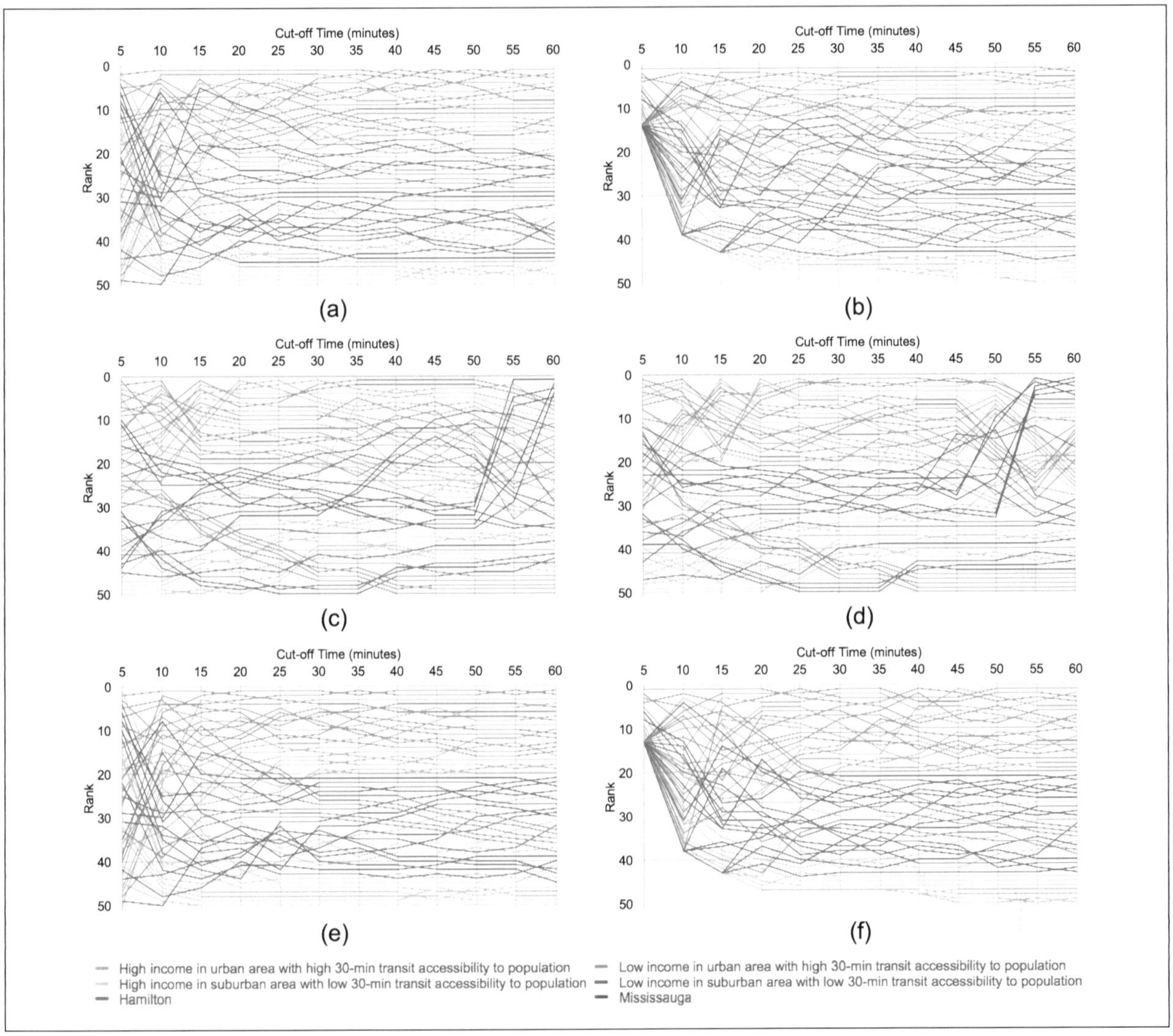

Figure 7. Ranking analysis of population and job accessibility: (*a*) population accessibility of walking; (*b*) job accessibility of walking; (*c*) population accessibility of automobile; (*d*) job accessibility of automobile; (e) population accessibility of public transit; (*f*) job accessibility of public transit.

An ideal cut-off time would permit the analyst to differentiate meaningfully between cases ("variability") while providing stable results for marginal changes in the cut-off value ("stability"). As indicated by the IQR analysis, variability is optimal at a 40-minute cut-off for population accessibility and a 35-minute cut-off for job accessibility in the automobile case. The variability increases with the cut-off time for both walking and public transit accessibility calculations without an obvious optimal value, although the rate of increase is higher between 20 and 40 minutes for transit access to population and 30 minutes for jobs. In the ranking analysis, the relationships between test cases becomes stable when the cut-off time for walking accessibility is larger than 15 or 20 minutes depending on whether a population- or job-based measure is used. In the public transit case, a stable relationship occurs after a 25-minute cut-off time for both types of accessibility measurements. However, there is no stable relationship observed at any cut-off times in the calculation of automobile accessibility.

Based on the linear regression results using the small 50-DA sample, a 30-minute cut-off time is preferred, in the region-wide analysis, the 30-minute cut-off works better than 45 minutes for population accessibility of automobiles, job accessibility of automobiles, and population accessibility of transits. For job accessibility to transit, a 45-minute cut-off time finds a better fit. Although the 50-DA test fails to identify a "best" cut-off time, for automobiles, a 30-minute cut-off works better than 45 minutes

Table 2. Linear Regression Results for 50 Test Cases

		Cut-off time	Adjusted R^2	Intercept	Accessibility parameter
Population accessibility	Automobile	10	0.539	103.015	-8.887×10^{-5}
		15	0.367	99.638	-3.894×10^{-5}
		20	0.254	96.644	-2.086×10^{-5}
		30	0.195	98.408	-1.141×10^{-5}
		40	0.129	98.270	-7.278×10^{-6}
		45	0.099	98.982	-6.262×10^{-6}
		50	0.066	98.971	-5.429×10^{-6}
	Public transit	10	0.380	4.800	2.072×10^{-3}
		15	0.556	3.077	7.073×10^{-4}
		20	0.643	3.561	2.321×10^{-4}
		30	0.672	2.926	6.299×10^{-5}
		40	0.572	2.427	2.954×10^{-5}
		45	0.526	2.355	2.272×10^{-5}
		50	0.497	2.216	1.839×10^{-5}
Job accessibility	Automobile	10	0.576	92.279	-7.608×10^{-5}
		15	0.474	94.784	-4.593×10^{-5}
		20	0.361	94.585	-3.155×10^{-5}
		30	0.301	96.657	-1.986×10^{-5}
		40	0.216	98.388	-1.401×10^{-5}
		45	0.115	95.883	-1.054×10^{-5}
		50	0.069	94.742	-8.835×10^{-6}
	Public transit	10	0.091	12.999	1.129×10^{-3}
		15	0.110	13.596	1.668×10^{-4}
		20	0.300	11.265	9.264×10^{-5}
		30	0.619	5.176	5.136×10^{-5}
		40	0.489	4.758	3.102×10^{-5}
		45	0.470	4.472	2.725×10^{-5}
		50	0.467	4.106	2.462×10^{-5}

in both 50-DA and all-DA cases. In all tested linear regressions, job accessibility is a better predictor of mode choice than population accessibility.

Finally, the impact of different cut-off times on accessibility measurements exists when using the isochrone approach, but the results vary considerably between different transportation modes, population or job accessibility, and household characteristics. Comparing two commonly-used cut-off times, 30 minutes appears to be more suitable than 45 minutes for measuring automobile accessibility, but neither cut-off time demonstrates outstanding advantages in measuring public transit accessibility. Thus, a defined cut-off time for general use cannot be determined based on this analysis.

In general, the primary objective of this study was to find the ideal cut-off time for accessibility measures, but the analysis results indicated that it is hard to identify an optimal value for general use as it changes significantly under different conditions. Accordingly, this paper analyzed how the cut-off time varied and affected the measurement of accessibility to provide possible directions for future studies.

Future Work

Although it is found that different cut-off times do affect accessibility measurements, which further affect the predictions of travel behavior, the best general-purpose cut-off time for isochrone-based accessibility measurements is not yet determined, and may not be identifiable. This paper first explored the impacts of a wide range of cut-off times for a selected sample of 50 DAs, and then focused in detail on an analysis of all DAs in the GTHA using the two commonly used cut-off times (i.e., 30 and 45 minutes). It is suggested to continue the analysis for all DAs using more cut-off times and potentially smaller intervals. Also, the predictive power of these selected cut-off times can be further evaluated by comparing the predicted travel behaviors with real-time travel data once it is available in the future. More elaborated models analyzing the relationship between mode choice and accessibility (non-linear specifications, additional explanatory variables, etc.) can also be explored.

Table 3. Linear Regression Results for Extended Examination

30-minute cut-off time			Adjusted R^2	Intercept	Accessibility parameter
Population accessibility	Automobile	50 test cases	0.195	98.408	-1.141×10^{-5}
		All DAs	0.173	93.728	-7.304×10^{-6}
		All DAs without zero mode share	0.175	93.737	-7.272×10^{-6}
		All DAs with 10 or more trips	0.183	93.683	-7.319×10^{-6}
		All DAs with 30 or more trips	0.189	93.380	-7.228×10^{-6}
	Public transit	50 test cases	0.672	2.926	6.299×10^{-5}
		All DAs	0.134	8.973	5.415×10^{-5}
		All DAs without zero mode share	0.118	11.542	4.678×10^{-5}
		All DAs with 10 or more trips	0.123	11.241	4.642×10^{-5}
		All DAs with 30 or more trips	0.162	9.908	4.897×10^{-5}
Job accessibility	Automobile	50 test cases	0.301	96.657	-1.986×10^{-5}
		All DAs	0.218	91.985	-1.225×10^{-5}
		All DAs without zero mode share	0.221	92.001	-1.220×10^{-5}
		All DAs with 10 or more trips	0.230	91.915	-1.223×10^{-5}
		All DAs with 30 or more trips	0.235	91.590	-1.202×10^{-5}
	Public transit	50 test cases	0.619	5.176	5.136×10^{-5}
		All DAs	0.137	9.801	4.881×10^{-5}
		All DAs without zero mode share	0.130	12.279	4.318×10^{-5}
		All DAs with 10 or more trips	0.135	11.979	4.265×10^{-5}
		All DAs with 30 or more trips	0.173	10.661	4.365×10^{-5}
45-minute cut-off time					
Population accessibility	Automobile	50 test cases	0.099	98.982	-6.262×10^{-6}
		All DAs	0.152	95.851	-4.659×10^{-6}
		All DAs without zero mode share	0.154	95.841	-4.636×10^{-6}
		All DAs with 10 or more trips	0.162	95.860	-4.684×10^{-6}
		All DAs with 30 or more trips	0.166	95.596	-4.647×10^{-6}
	Public transit	50 test cases	0.526	2.355	2.272×10^{-5}
		All DAs	0.128	9.044	1.835×10^{-5}
		All DAs without zero mode share	0.110	11.482	1.589×10^{-5}
		All DAs with 10 or more trips	0.114	11.211	1.571×10^{-5}
		All DAs with 30 or more trips	0.149	9.790	1.664×10^{-5}
Job accessibility	Automobile	50 test cases	0.115	95.883	-1.054×10^{-5}
		All DAs	0.169	93.955	-8.165×10^{-6}
		All DAs without zero mode share	0.172	93.958	-8.126×10^{-6}
		All DAs with 10 or more trips	0.181	93.947	-8.203×10^{-6}
		All DAs with 30 or more trips	0.187	93.703	-8.138×10^{-6}
	Public transit	50 test cases	0.470	4.472	2.725×10^{-5}
		All DAs	0.154	9.405	2.516×10^{-5}
		All DAs without zero mode share	0.138	11.756	2.202×10^{-5}
		All DAs with 10 or more trips	0.144	11.478	2.180×10^{-5}
		All DAs with 30 or more trips	0.178	10.098	2.255×10^{-5}

Author Contributions

The authors confirm contribution to the paper as follows: study conception and design: Prof. Shoshanna Saxe, Prof. Eric J. Miller; data collection: Prof. Saxe, Yang Xi; analysis and interpretation of results: Yang Xi, Prof. Saxe, Prof. Miller; draft manuscript preparation: Yang Xi. All authors reviewed the results and approved the final version of the manuscript.

References

1. Chen, Y., S. Ravulaparthy, K. Deutsch, P. Dalal, S. Yoon, T. Lei, K. Goulias, R. Pendyala, C. Bhat, and H. H. Hu. Development of Indicators of Opportunity-Based Accessibility. *Transportation Research Record: Journal of the Transportation Research Board*, 2011, 2255: 58–68.

2. El-Geneidy, A., A. Cerdá, R. Fischler, and N. Luka. The Use of Accessibility Measures to Evaluate the Impacts of Transportation Plans: An Application in Montréal, Québec. *Canadian Journal of Urban Research: Canadian Planning and Policy (supplement)*, Vol. 20, 2011, pp. 81–104.
3. Owen, A., and D. M. Levinson. Modeling the Commute Mode Share of Transit Using Continuous Accessibility to Jobs. *Transportation Research Part A: Policy and Practice*, Vol. 74, 2015, pp. 110–122.
4. Martens, K. *Transport Justice: Designing Fair Transportation Systems*. Routledge, New York, 2016.
5. Rodrigue, J.-P., C. Comtois, and B. Slack. *The Geography of Transport Systems*. Routledge, New York, 2016.
6. Farber, S. *Connecting People to Places: Spatiotemporal Analysis of Transit Supply Using Travel Time Cubes*. NITC-RR-662. Transportation Research and Education Center (TREC), Portland, OR, 2016. http://dx.doi.org/10.15760/trec.143.
7. Federal Highway Administration and Oregon Department of Transportation. *Newberg-Dundee Transportation Improvement Project: Bypass Element Location (Tier 1) Final Environmental Impact Statement*, 2005. www.oregon.gov/odot/regions/pages/region-2-northwest-oregon.aspx. Accessed July 15, 2017.
8. United States. Federal Transit Administration. *Hillsboro Extension of the Westside Corridor Project, Washington County: Environmental Impact Statement*, 1994. https://catalog.hathitrust.org/Record/100980862. Accessed July 15, 2017.
9. Hansen, W. G. How Accessibility Shapes Land Use. *Journal of the American Institute of Planners*, Vol. 25, No. 2, 1959, pp. 73–76.
10. Ben-Akiva, M. E., X. Dong, J. L. Bowman, and J. L. Walker. Moving from Trip-Based to Activity-Based Measures of Accessibility. *Transportation Research Part A: Policy and Practice*, Vol. 40, No. 2, 2006, pp. 163–180.
11. Geurs, K. T., and B. Van Wee. Accessibility Evaluation of Land-use and Transport Strategies: Review and Research Directions. *Journal of Transport Geography*, Vol. 12, No. 2, 2004, pp. 127–140.
12. Harrison, K. Measuring Access to Employment to Guide and Evaluate Public Transit Service Planning in New Orleans. *University of New Orleans Theses and Dissertations*, Vol. 2256, 2016.
13. Saxe, S., E. Miller, and S. Farber. *Regional Accessibility and Mode Choice in the Greater Toronto Hamilton Area*. Presented at World Society of Transport and Land Use Symposium, Brisbane, Australia, July 2017.
14. Metrolinx. *The Big Move: Transforming Transportation in the Greater Toronto and Hamilton Area*, 2008. www.metrolinx.com/thebigmove/Docs/big_move/TheBigMove_020109.pdf. Accessed July 17, 2017.
15. City of Toronto. *2016 Census: Population and Dwelling Counts*, 2017. www1.toronto.ca/City%20Of%20Toronto/City%20Planning/SIPA/Files/pdf/C/2016%20Census%20Backgrounder%20Population%20Dwellings%202017%2002%2009.docx. Accessed July 17, 2017.
16. Data Management Group. *2011 Transportation Tomorrow Survey Version 1.0 Data Guide*, 2011. http://www.dmg.utoronto.ca/pdf/tts/2011/dataguide2011.pdf. Accessed July 17, 2017.
17. Natural Resources Canada. *GeoBase – National Road Network*, 2012. http://geogratis.gc.ca/api/en/nrcan-rncan/ess-sst/c0db1b2e-bdaf-6998-8b8e-e569e5d39d6b.html. Accessed July 18, 2017.
18. Travel Modelling Group. *GTA Model V4.0*, 2015. University of Toronto Transportation Research Institute, Toronto, ON, Canada.
19. Statistics Canada. *Dissemination Area*, 2015. http://www12.statcan.gc.ca/census-recensement/2011/ref/dict/geo021-eng.cfm. Accessed July 18, 2017.
20. Foth, N., K. Manaugh, and A. El-Geneidy. Toward Equitable Transit: Examining Transit Accessibility and Social Need in Toronto, Canada 1996–2006. *Journal of Transport Geography*, Vol. 29, 2013, pp. 1–10.
21. Hess, D. B. Access to Employment for Adults in Poverty in the Buffalo-Niagara Region. *Urban Studies*, Vol. 42, No. 7, 2005, pp. 1177–1200.
22. Hulchanski, J. D. The Three Cities within Toronto: Income Polarization among Toronto's Neighbourhoods, 1970–2000. *Urban and Community Studies (now Cities Centre) Research Bulletin* 41, University of Toronto, Toronto, 2007.
23. Statistics Canada. *2011 National Household Survey Catalogue no. 99-014-X2011048*, 2011. http://www12.statcan.gc.ca/nhs-enm/2011/dp-pd/dt-td/Rp-eng.cfm?LANG=E&APATH=3&DETAIL=0&DIM=0&FL=A&FREE=0&GC=0&GID=0&GK=0&GRP=0&PID=107759&PRID=0&PTYPE=105277&S=0&SHOWALL=Yes&SUB=0&Temporal=2013&THEME=98&VID=0&VNAMEE=&VNAMEF=. Accessed July 18, 2017.
24. ESRI. *ArcGIS - Mapping and Spatial Analysis for Understanding Our World*. http://www.esri.com/software/arcgis. Accessed July 15, 2017.
25. Statistics Canada. *2011 Census of Population - Dissemination Area*, 2012. http://www12.statcan.gc.ca/datasets/Index-eng.cfm?Temporal=2011&Theme=-1&VNAMEE=&GA=53087&S=0. Accessed July 15, 2017.
26. Data Management Group. *2011 Transportation Tomorrow Survey*, 2011. Digital Data. University of Toronto Transportation Research Institute, Toronto, ON, Canada.
27. City of Toronto. *Population and Employment Projections*, 2015. http://www.toronto.ca/legdocs/mmis/2015/ex/bgrd/backgroundfile-84731.pdf. Accessed July 16, 2017.

The Standing Committee on Transportation Demand Forecasting (ADB40) peer-reviewed this paper (18-04640).

Article

TRR
JOURNAL OF THE TRANSPORTATION RESEARCH BOARD

Transportation Research Record
2018, Vol. 2672(49) 125–134

DOI: 10.1177/0361198118797457
journals.sagepub.com/home/trr
SAGE

What Types of Cars Will We Be Driving? Methods of Forecasting Car Travel Demand by Vehicle Type

Tudor Mocanu[1]

Abstract

New technologies are emerging in the private vehicle market. Conventional propulsion systems are set to be replaced by alternative, more environment-friendly ones (e.g., electric vehicles), and certain new features, like autonomous driving, will possibly change the way private cars are employed. To assess the impact of such technologies, one must estimate how often and for which trips these vehicle types will be used. Another issue is the exact localization of certain vehicle types on the network, to assess environmental effects and identify where specific roadside infrastructure (e.g., charging stations) will be required. This paper presents four approaches to forecasting car usage by vehicle type using a macroscopic travel demand model in combination with a vehicle fleet or technology diffusion model. Integrating the two types of models requires tools ranging from assumptions and extrapolation of empirical data to synthetic or incremental discrete choice models. The approaches are employed in a case study forecasting travel demand using privately owned autonomous vehicles (AVs) in Germany in 2030. Despite identical input data, the estimated proportion of vehicle miles traveled (VMT) using AVs varies between 11% and 23% of overall car VMT, depending on the approach chosen. The reasons for this variation in results are investigated and some recommendations are given. To avoid the difficulties of fitting a synthetic model to observed data and to increase the accuracy of the results, the recommendation is to formulate the vehicle type choice as an incremental model added to the travel demand model.

Motorized individual transport is currently the dominant form of transportation in many Western countries and is forecasted to remain at comparably high levels in the coming decades. The expected continuity in car travel demand seems to indicate a status quo, which is far from being true. Several developments with wide-ranging implications can already be observed. The increasing focus on environmental issues spurs on the development of cleaner vehicles (e.g., electric or hybrid cars). Pollution in urban areas caused by traffic-related emissions is also a growing concern and authorities are in search of policy measures to tackle this issue. Finally, a technology-driven evolution will bring new functionalities to the vehicles, the most discussed of these being automation. The emergence of autonomous vehicles (AVs) has the potential to significantly influence the way private cars are used.

Understanding the relationship between the types of vehicles employed and their usage patterns is necessary to enhance fore

casting in some situations in which the differences between vehicles are relevant. For instance, to accurately model local traffic-related emissions, the traffic flows assigned to the network model should be split up according to vehicle emission profiles. Separate traffic flows for each vehicle propulsion system (e.g., petrol, diesel, electric), size, vintage, and so forth, should be calculated. As will be shown in this paper, this modeling step is not trivial. Conversely, the emergence of new vehicle types and technologies might also have an impact on overall car travel demand. In the case of AVs, a decrease in travelers' valuation of travel time might lead to an overall increase in car travel (*1*). Furthermore, governments and local agencies might devise policy measures to encourage (or discourage) the use of specific vehicle types, a pertinent example being the current debate on the negative environmental effects of diesel cars in German city centers (*2*). Restricting diesel vehicle access in city centers would certainly have an impact on (diesel) car traffic in

[1]Deutsches Zentrum für Luft- und Raumfahrt e.V. (DLR)—German Aerospace Center, Institute of Transport Research, Berlin, Germany

Corresponding Author:
Address correspondence to Tudor Mocanu: tudor.mocanu@dlr.de

these areas, but it might also affect overall mode choice, destination choice, and trip generation. The complex relationship between the composition of the vehicle fleet (shares of different vehicle types) and car travel demand (by vehicle type) should ideally be addressed in travel demand models (TDMs).

This paper investigates the possibility of embedding a differentiation by vehicle type into TDMs. The objective of this paper is not to forecast the adoption of different technologies in the vehicle fleet, but rather to examine the interaction between a forecasted vehicle stock and the travel demand associated with it.

First, an overview of past research related to this subject is presented. Then, the overall model framework is described and the target variables are defined. Four different approaches to integrating vehicle type choice into TDMs are proposed, and their underlying assumptions, advantages, and disadvantages discussed. A case study, forecasting the demand for trips using private AVs in Germany in the year 2030, is used to illustrate the different approaches. Finally, some conclusions and recommendations for further research are offered.

Literature Review

Research on vehicle type differentiation has been conducted since the 1970s, when the first vehicle fleet models (VFMs) were estimated (*3*). Such models predict long-term choices made by individuals or households deciding on the number and type of vehicles they own or purchase. VFMs, also known as vehicle stock, holdings, or car ownership models, can vary significantly in complexity, ranging from simple time series to multivariate discrete choice models (*4*). For an excellent review of state-of-the-art VFMs pertaining to the adoption of electric and hybrid vehicles, see the work of Al-Alawi and Bradley (*5*). Other recent studies by Litman (*6*) and Bansal and Kockelman (*7*) have forecasted the long-term adoption levels of AVs in the United States, and Kröger et al. have estimated AV adoption rates in the German car fleet using an s-shaped market-take-up vehicle technology diffusion model (*8*).

VFMs forecast the number and type of vehicles owned, but only rarely address vehicle usage metrics (e.g., the average number of trips per day and car, average mileage per year and car, etc.). An example of a VFM also considering vehicle use is Bhat and Sen's multiple discrete-continuous extreme value (MCDEV) model (*9*), employed to predict both the vehicle holdings and the yearly vehicle miles of travel (VMT) per vehicle type. Bhat and Sen use the demographic characteristics of the household, vehicle specifications, and land-use attributes as explanatory variables for both vehicle type ownership *and* use.

Although such approaches may deliver aggregate results pertaining to vehicle type specific use, they are limited in several ways. First, as they do not operate with single trips or tours, they cannot link relevant trip or route attributes (e.g., distance, tolls required, available infrastructure, etc.) to the use of vehicles. Second, they are responsive only to a limited range of policy measures, like changes in the vehicle registration fees or fuel taxes. The impact of highly localized policy tools, targeting only specific areas (e.g., city centers) or roads (e.g., special toll roads), cannot be assessed using such models. Finally, the total VMT estimated with such a model does not necessarily correspond to the VMT outputted by a dedicated TDM, making the interpretation of results difficult.

More recent studies have incorporated specific types of privately owned vehicles as an extra mode of transport in TDMs. Auld et al. (*10*) and Childress et al. (*11*) both explored the impacts of AVs on travel demand using activity-based models for the Chicago and Puget Sound regions respectively. Zhao and Kockelman (*12*) employed a macroscopic travel demand model to anticipate the impact of automated vehicles in Austin, Texas. However, the integration of VFM results into the TDM was not discussed in these approaches, with the AV penetration rates being scenario-based assumptions rather than model results.

To the author's knowledge, the integration of vehicle type differentiation into TDMs has not been thoroughly discussed before. Even in studies on the traffic-induced environmental effects, travel demand shares by vehicle type are either assumed (*13*) or not differentiated at all (*14*). The modeling framework proposed in this paper attempts to rectify these shortcomings.

Modeling Framework

Overview

The aim of this paper is to present an approach to estimating car travel demand by vehicle type by integrating results from a VFM into a TDM. This implies splitting up the overall car travel demand for each origin-destination (OD) pair in the study area depending on attributes of the vehicles, the trip-maker, and the trip itself. Using this setup, the model can assess the impact of a wide range of policy measures on both overall car travel demand and vehicle type specific demand. Furthermore, as separate OD matrices for each vehicle type are generated and can be assigned to the network, this approach can also be used to forecast traffic counts by vehicle type on each road section, which is essential, for example, for fine-grained emission modeling.

For the purpose of this study, the structure and setup of the TDM are irrelevant. The model framework shown here is compatible with various macroscopic TDMs, whether set up as traditional four-step, gravity, logit, hierarchical, or simultaneous models, or as some other model. More differentiation in the TDM (e.g., by population groups/demographics, trip purposes, time of day, etc.) will lead to more detailed and possibly more accurate results by vehicle type. It is worth noting that the approach shown here is only required when using a macroscopic TDM dealing with "person groups" or "representative persons" and not individuals or individual households. In microscopic simulations (or multi-agent tools), each individual is assigned a specific vehicle type and thus results by vehicle type can easily be aggregated rather than split up from overall car demand. However, as such models are still difficult to implement in certain large-scale applications (e.g., statewide or national models, long-term forecasting) because of computing limitations and data availability, the approach shown in this paper is (still) considered to be relevant. In the context of this paper, results from the TDM shall be notated as T_{ij}, indicating a matrix with the number of car trips from origin i to destination j.

The methodology and functionality of the VFM will not be discussed, and only its results will be considered here. Let $C_1, C_2, \ldots, C_n$ be vehicle categories (types) in which cars are grouped according to their attributes. Results from the VFM can be notated as the following vector of vehicle fleet shares **vs**

$$\mathbf{vs} = \{\mathrm{vs}_{C_1}, \mathrm{vs}_{C_2}, \ldots, \mathrm{vs}_{C_n}\} \tag{1.1}$$

where the sum of all vehicle fleet shares equals unity.

$$\sum_C \mathrm{vs}_C = 1 \tag{1.2}$$

Depending on the granularity of the VFM, results may be available on an individual/household level or at any superior level of aggregation (e.g., traffic zones, counties, regions, etc.). For the sake of simplicity, the indices denoting the spatial distribution of the vehicle fleet shares will be omitted in this paper.

The target variable is defined as the proportion of car trips using a vehicle from category C, and will be noted as the (matrix) vector **us**

$$\mathbf{us}_{ij} = \{\mathrm{us}_{ij,C_1}, \mathrm{us}_{ij,C_2}, \ldots, \mathrm{us}_{ij,C_n}\} \tag{2.1}$$

where the sum of all usage shares also equals unity.

$$\sum_C \mathrm{us}_{ij,C} = 1 \tag{2.2}$$

Note that unlike the vehicle fleet shares **vs**, which denote attributes of the population and are only influenced by the home location of the travelers, the usage shares might be different for each OD pair ij and should therefore be noted as a vector $\mathbf{us}_{ij}$. In other words, vehicles will not automatically be used in equal proportion for all trips.

Multiplying the usage shares of a vehicle category with the total car trip matrix T_{ij} then gives the trip matrix for that vehicle category

$$T_{ij,C} = \mathrm{us}_{ij,C} T_{ij} \tag{3.1}$$

with

$$\sum_C T_{ij,C} = T_{ij} \tag{3.2}$$

The key in determining car travel demand by vehicle type is forecasting the usage shares vector $\mathbf{us}_{ij}$, as this information is not included in either the TDM or the VFM alone. Using the vehicle fleet shares **vs** from a VFM as input, there are several methods of estimating $\mathbf{us}_{ij}$. Four of these, numbered A1 to A4, will be discussed in the following section. Depending on the model application and data availability, each one of them might seem attractive to use at some point. However, as will be shown in the case study, results might greatly differ, so the adequacy of the method employed must be well weighed.

Approaches to Estimating Vehicle Usage Shares

A1. Trivial Approach. A simple approach to link the results from the TDM and VFM would be to consider that the vehicle usage shares are equal to the vehicle fleet shares (obtained from the VFM) for all OD pairs:

$$\mathbf{us}_{ij} = \mathbf{vs} \tag{4.1}$$

$$T_{ij,C} = \mathrm{us}_{ij,C} T_{ij} = \mathrm{vs}_C T_{ij} \tag{4.2}$$

A1 works under the assumption that all vehicles are used equally often, independent of any attribute of the vehicle itself (e.g., size, comfort, fuel costs, etc.) or of the trip undertaken (e.g., trip purpose, distance, travel time, etc.). The obvious advantage of A1 is its simplicity and ease of use. This approach does not require any additional data and there are no model parameters to be estimated or calibrated.

Results using A1 will be unrealistic whenever there is good reason to assume that certain car attributes are correlated to specific usage statistics. For instance, the average annual mileage of a petrol-fueled car in Germany amounts to circa 10,000 km, whereas diesel vehicles are driven on average approximately 17,000 km (own analyses on national household survey data). In this case, the assumption of equal usage would obviously lead to wrong forecast results. An argument could be made that the independence of the choice outcome with respect to vehicle and trip attributes is indeed given for a large proportion of the population—those individuals or households that only possess one vehicle and thus cannot choose between multiple vehicles for each trip. However, it is not unreasonable to assume that these decision makers will consider their predicted annual mileage when buying a car, thus correlating the available vehicles to their travel patterns. Therefore, although A1 does deliver a very rough estimation of vehicle usage shares, it is not suited for precise forecasting and should rather be used as a benchmarking tool for the more complex approaches shown in the following sections.

A2. Trip Segmentation Approach. The Trip Segmentation Approach (A2) derives the usage shares from observed data in the base year. The concept revolves around the idea that, by segmenting trips into homogenous groups, all relevant differences in the car usage patterns can be identified. Assuming that all relevant explanatory factors have been identified in the base year analysis, the empirically derived usage shares can then also be used for forecasting.

To provide statistically reliable results, comprehensive data on the vehicle usage in the base year is required. Ideally, a trip database should be available containing information on the trip itself, as well as on the characteristics of the vehicle used. Regional or national household surveys should provide this information for a sufficiently large number of trips. For instance, the German national household survey *(15)* reports on approximately 70,000 car trips and the corresponding approximately 35,000 vehicles used on these trips.

Table 1 exemplifies an analysis using the Trip Segmentation Approach on German data. Car trips are grouped according to four distance categories and three trip purposes, and vehicles are differentiated according to their size and age. Each row shows the vehicle usage shares $\mathbf{us}_{ij}$ for that particular trip category. The vehicle fleet shares $\mathbf{vs}$ (proportion of vehicles in a category from the total car fleet) are also given at the bottom of the table for comparison.

The empirical data in Table 1 indicate that trip purpose and distance are not independent of the size and age of the vehicle employed. For short trips, the usage shares closely resemble the vehicle fleet shares, suggesting that for these trips, car attributes have little explanatory power. This is exactly the assumption followed by the Trivial Approach (A1). For longer trips, however, the usage shares of both larger and newer vehicles increase. This is particularly evident for business trips, where the disparity between the usage and vehicle fleet shares is the largest. As long trips have a significantly higher impact on the overall VMT, this also shows the limitations of A1.

Table 1. Observed Vehicle Usage Shares in Germany

		Small			Compact			Mid-Size			Large			
Distance	Purpose	< 4 yrs	4–8 yrs	> 8 yrs	< 4 yrs	4–8 yrs	> 8 yrs	< 4 yrs	4–8 yrs	> 8 yrs	< 4 yrs	4–8 yrs	> 8 yrs	Σ
< 20 km	Commute	0.072	0.078	0.133	0.104	0.096	0.167	0.092	0.089	0.154	0.005	0.002	0.010	1
	Business	0.059	0.051	0.089	0.112	0.116	0.159	0.148	0.117	0.120	0.006	0.009	0.015	1
	Other	0.067	0.069	0.097	0.117	0.104	0.162	0.098	0.107	0.165	0.003	0.003	0.007	1
20–50 km	Commute	0.085	0.081	0.104	0.115	0.118	0.155	0.097	0.099	0.131	0.004	0.004	0.006	1
	Business	0.057	0.075	0.057	0.159	0.062	0.115	0.229	0.097	0.137	0.009	0.000	0.004	1
	Other	0.070	0.069	0.080	0.119	0.102	0.172	0.114	0.114	0.151	0.002	0.002	0.005	1
50–100 km	Commute	0.062	0.057	0.084	0.160	0.135	0.116	0.114	0.113	0.138	0.000	0.003	0.017	1
	Business	0.060	0.060	0.036	0.131	0.071	0.083	0.262	0.119	0.119	0.012	0.036	0.012	1
	Other	0.076	0.069	0.057	0.116	0.079	0.169	0.113	0.143	0.146	0.004	0.008	0.019	1
100–200 km	Commute	0.107	0.019	0.049	0.107	0.097	0.136	0.252	0.087	0.136	0.000	0.010	0.000	1
	Business	0.050	0.033	0.017	0.150	0.083	0.050	0.283	0.083	0.183	0.050	0.017	0.000	1
	Other	0.060	0.060	0.044	0.129	0.096	0.107	0.151	0.159	0.173	0.011	0.005	0.005	1
> 200 km	Commute	0.057	0.068	0.034	0.080	0.148	0.136	0.239	0.068	0.125	0.034	0.000	0.011	1
	Business	0.000	0.000	0.022	0.200	0.000	0.044	0.467	0.089	0.156	0.000	0.022	0.000	1
	Other	0.057	0.049	0.036	0.166	0.117	0.101	0.178	0.117	0.162	0.004	0.000	0.012	1
Vehicle fleet shares		***0.067***	***0.075***	***0.121***	***0.104***	***0.097***	***0.185***	***0.083***	***0.089***	***0.160***	***0.004***	***0.003***	***0.011***	**1**

Note: yrs = years.

Using the Trip Segmentation Approach for forecasting is straightforward. Each OD relation from the forecasted overall car travel matrix is split up according to the corresponding trip category usage shares $\mathbf{us}_{ij}$ from Table 1. If the trip structure in the forecast is different from the base year (e.g., higher proportion of long trips, other trip purposes, etc.), then different overall car usage shares will result.

The Trip Segmentation Approach is more complex than A1 and delivers more accurate results. This improvement comes at the price of increased data requirements. A comprehensive database containing observed trips and vehicles is necessary. In the absence of sufficient trip observations in the database, segmentation by too many criteria can lead to statistically unreliable usage shares.

A difficulty using A2 arises when forecasting the usage of vehicle types that will only become available in the future and for which there is no empirical usage data available. Another drawback of A2 is that the future usage shares are not responsive to changes in the forecasted vehicle fleet shares (distribution of car types by age, size, etc.). For instance, with this approach, an increase in the proportion of "large and new" vehicles would not lead to higher vehicle usage shares for these cars. To make the results responsive to multiple factors, a more complex modeling approach is required.

A3. Choice Model Approach. Mocanu and Winkler (*16*) present an approach to differentiate car travel demand using the discrete choice model framework. They propose splitting results from the TDM into different vehicle categories using a nested logit-structured model extension. This model setup adds car type choice at the lowest level of the TDM choice hierarchy. Thus, a two-level nested logit is created, in which destination and mode choice are calculated in the upper nest and vehicle type choice is calculated in the lower nest.

The choice of vehicle type in the lower nest is formulated as a standard multinomial logit (MNL), with the usage shares vector $\mathbf{us}_{ij}$ resulting from the individual utilities V of each vehicle category C:

$$\mathrm{us}_{ij,C} = \frac{\exp(\mu V_{ij,C})}{\sum_{C} \exp(\mu V_{ij,C})} \qquad (5)$$

where μ defines the logsum scale parameter of the (lower) vehicle type choice nest.

The utilities V are given as a function of the vehicle fleet shares $\mathbf{vs}$ and all other relevant variables, such as travel time (tt), access and egress time, cost, and so forth.

$$V_{ij,C} = f(\mathrm{vs}_C, \mathrm{tt}_{ij,C}, c_{ij,C}, \ldots) \qquad (6)$$

Using the model structure proposed by Mocanu and Winkler, it is possible to fully embed the vehicle type choice in the TDM and also to create a link back from the vehicle type choice to the TDM using the logsum (also called inclusive value or maximum expected utility) over all available alternatives. The authors show that, despite linking up the two models using the nested logit framework, the model can be implemented independently of the original TDM setup and the required modifications to the TDM are minimal.

The Choice Model Approach enables the entire TDM to be responsive to measures that only affect some vehicle types. Imagine a policy measure like a pollution tax for city centers, applicable only to conventionally-fueled vehicles (but not to electric cars). The model should forecast not only a decreasing number of car trips using conventional vehicles in city centers, but also a general reduction of car traffic (in those areas), changes in the destination choice, and possibly also changing trip rates. Conversely, adding a very attractive vehicle type (e.g., AVs) will increase the car logsum and influence mode choice, destination choice, and trip generation, thus creating the expected induced traffic. Using forms of composite costs over all vehicle types other than the logsum (e.g., simple or weighted averages) might lead to undesired model results, as was shown for instance by Williams as early as 1977 (*17*).

One issue with A3 is that this choice model is only a hypothetical construct, for two reasons. On the one hand, vehicle purchase and ownership are long-term decisions, whereas vehicle type usage for each trip is a short-term decision. On the other hand, not all individuals have the freedom to choose between all vehicle types for each of their trips. As each household owns only a limited number of cars, the long-term ownership decision restricts the available choice set in the short-term decision. Consequently, separate models should be set up for each combination of available vehicles. Because of the large number of possible vehicle type ownership combinations and the ensuing computing time and number of parameters required, such a model setup is not feasible. Instead, restrictions in the available choice set have to be applied aggregately on only one model using the vehicle fleet shares $\mathbf{vs}$, as discussed by Mocanu and Winkler (*16*).

This implementation of a discrete choice model has wide-ranging implications for the estimation of preferences and model parameters. Because individual choices, as modeled, cannot be observed in reality, a model estimation using standard techniques is virtually impossible. Stated preference experiments and revealed preference data cannot simulate sufficient variation in the input variable $\mathbf{vs}$, while at the same time keeping all other variables unchanged, in order to estimate the impact of this variable on utility. Therefore, the calibration of the vehicle type choice parameters, and also the logsum scale parameter μ, must occur heuristically.

Because of the difficulties in estimating the vehicle type choice parameters, it becomes evident that the more vehicle categories and differentiation criteria are considered, the harder it is to fit such a model to observed data. Furthermore, it is important not only to replicate observed car usage patterns, but also to ensure that the choice model reacts in a predictable manner to variable input and exhibits plausible elasticities and substitution patterns. This is especially difficult when differentiating vehicles by multiple criteria (e.g., propulsion system and size), as this setup obviously violates the MNL assumption of independently distributed error terms. Under these circumstances, formulating and estimating a vehicle type choice model to fit the data in Table 1 is no trivial task.

A4. Incremental Approach. Both the Trip Segmentation Approach (A2) and the Choice Model Approach (A3) can deliver plausible results for the differentiation of car travel, but both also have significant drawbacks and weaknesses. A2 is based on the assumption that the observed vehicle type usage will remain unchanged for similar trips, whereas A3 has difficulties in reproducing the observed data if the vehicle classification is too complex.

The goal of the Incremental Approach (A4) is to enhance the Trip Segmentation Approach with the advantages offered by the Choice Model Approach. Observed vehicle type usage preferences should be reproduced in the base year, while at the same time making the choice sensitive to variable input. This is possible by setting up the vehicle type choice as an incremental (or pivot-point) choice model, as described for instance in (*18*). The forecasted vehicle type usage shares vector $\mathbf{us}_{ij}^{F}$ will then be

$$\mathrm{us}_{ij,C}^{F} = \frac{\mathrm{us}_{ij,C}^{\mathrm{BY}} \exp(\mu(V_{ij,C}^{F} - V_{ij,C}^{\mathrm{BY}}))}{\sum_{C} \mathrm{us}_{ij,C}^{\mathrm{BY}} \exp(\mu(V_{ij,C}^{F} - V_{ij,C}^{\mathrm{BY}}))} \quad (7)$$

where $\mathrm{us}_{ij,C}^{\mathrm{BY}}$ denotes the observed vehicle type usage shares from the base year, as exemplified in Table 1, and $V_{ij,C}^{F}$, $V_{ij,C}^{\mathrm{BY}}$ the utilities of vehicle category C in the forecast and base year respectively.

This model formulation presents several advantages compared with A3. First, the observed usage shares can be replicated exactly, independent of the formulation of utility $V_{ij,C}$. Running a "forecast" under the assumption that $V_{ij,C}^{F} = V_{ij,C}^{\mathrm{BY}}$ is exactly what is assumed in A2 and leads to $\mathrm{us}_{ij,C}^{F} = \mathrm{us}_{ij,C}^{\mathrm{BY}}$. Therefore, A4 is a more general formulation of A2. Second, if the vehicle fleet share vs_C is incorporated into the utility $V_{ij,C}$, the model results become responsive to changes in the vehicle fleet structure. The lack of such sensitivity was mentioned above as one of the major limitations of A2. Finally, as only differences in utility are relevant, the model can be specified with fewer variables than A3. This should simplify the process of parameter estimation and make the assessment of elasticities and overall model responsiveness easier to handle. Nevertheless, the difficulties in estimating the model parameters mentioned in the previous section also apply to A4.

Case Study

Scope

The following case study will be used to exemplify the modeling framework proposed in this paper. The primary goal of the study is to compare the differences in the four approaches presented above. The object of the study is a forecast of travel demand by AVs in Germany in 2030. The relevant measure of travel demand considered here is VMT per average working day.

The aim of the case study is not to comprehensively analyze all aspects related to the future development of demand for AVs in Germany. There are many factors of a technical, behavioral, and legal nature not considered in this case study. The focus here lies more on the methodology and less on the actual results. To simplify the model implementation and the interpretation of results, the following restrictions apply:

- Only commuting trips are considered.
- Only privately owned vehicles are considered (no car sharing, vehicles on demand, robotaxis, car rentals, etc.).
- AVs are considered to be only those vehicles with full automation (e.g., Highway Pilot technology, Level 4 and above); all other vehicles are defined as non-autonomous (non-AV).
- Empty AV trips are not considered in the VMT calculations.
- Overall mode and destination choice is not affected by the vehicle type choice; therefore the overall number of car trips and car VMT is equal in all approaches.

The last bullet point on the list is necessary to enable a comparison between those approaches that can model the effects on mode and destination choice (A3 and A4) and those that cannot (A1 and A2). Furthermore, to exemplify the impact that user preferences have on the vehicle type usage, it is also assumed that AVs lead to a decrease in the users' valuation of travel time savings (vtts), as the in-vehicle time can also be used for other activities instead of driving. The vtts reduction assumed

Table 2. Forecasted Overall Car Travel Demand in Germany, 2030

Indicator	Forecasted value (commuting trips only)
Total number of car trips per average working day	30.5 million trips
Total car VMT per average working day	568.1 million vehicle km
Average trip length	18.6 km per trip

is 25% starting from the eleventh minute of travel on, following the results of an extensive literature study (*8*). Note that the vtts reduction only affects the choice between AV and non-AV and not the overall mode and destination choice, once again in order to enable the comparison between the four approaches.

TDM and VFM Input

Total car travel demand for Germany was leveraged from the DEMO national travel demand model (*19*) in form of trip matrices T_{ij}.

A VFM was used to forecast the shares of AVs in the German fleet. The model setup is briefly presented in (*8*) and uses an s-shaped market-take-up curve to model the diffusion of automation technologies. Vehicles were grouped into four size categories to enable different technology diffusion rates, and vehicle vintage is a model result. This case study assumed that AV technology will first be available for large vehicles, starting in 2020, and subsequently also for the smaller car segments. Table 3 shows the main results from the VFM.

Table 3. Forecasted Vehicle Fleet Shares in Germany, 2030

Vehicle size category	AV	Non-AV	Σ	AV available starting
Small	0.000	0.227	0.227	na
Compact	0.033	0.310	0.343	2024
Mid-Size	0.075	0.346	0.421	2021
Large	0.004	0.005	0.009	2020
Σ	0.112	0.888	1.000	na

Note: na = not applicable.

Implementation of the Linking Approaches

A1. The implementation of the Trivial Approach is straightforward. AV usage shares for all trips result from the vehicle fleet shares given in Table 3, meaning that for any trip purpose and distance the share of AV trips from all car trips will be 11.2%.

A2. The Trip Segmentation Approach used the analysis shown in Table 1. The German car fleet was segmented into the 12 categories shown (four size × three age categories). The vehicle usage shares given in Table 1 were applied to the forecasted car travel demand matrix, thus obtaining the usage share of each of the 12 vehicle categories for every OD relation.

In order to estimate demand for autonomous vehicles, the assumption was made that for each of the 12 vehicle categories, AV and non-AV usage shares are equal to the vehicle fleet shares in that category. This implies that usage preferences for new and large AV are equal to those for new and large non-AV, and so on. Thus, differences in the utilization of AV and non-AV only result from the car size and age.

A3. To illustrate the Choice Model Approach, a simple model was set up with two vehicle categories: autonomous (AV) and non-autonomous (nAV) vehicles. The utility function for these vehicle categories was adopted from the German National TDM (*9*). This simplified utility function takes the linear form

$$V_{ij} = \beta_{vs}\ln(\mathrm{vs}) + \beta_{tt}\mathrm{tt}_{ij} + \beta_{aet}\mathrm{aet}_{ij} + \beta_{c}(\mathrm{fc}_{ij} + \mathrm{pc}_{ij}) \quad (8)$$

where

tt_{ij} denotes travel time,
aet_{ij} denotes access and egress time,
fc_{ij} denotes fuel costs, and
pc_{ij} denotes parking fees.

Note that car size and age do not enter the utility function. The differences in AV and non-AV utility are given only by the vehicle fleet shares and the reduced valuation of travel time. As only differences in utility are relevant in the logit model setup, with the logsum scale parameter $\mu = 1$ and the vehicle fleet shares parameter $\beta_{\mathrm{vs}} = 1$ the resulting vehicle usage shares can be formulated as

$$\mathrm{us}_{ij,\mathrm{AV}} = \frac{\mathrm{vs}_{\mathrm{AV}}\exp(f(\mathrm{tt}_{ij}))}{\mathrm{vs}_{\mathrm{AV}}\exp(f(\mathrm{tt}_{ij})) + \mathrm{vs}_{\mathrm{nAV}}\exp(\beta_{\mathrm{tt}}\mathrm{tt}_{ij})} \quad (9.1)$$

$$\mathrm{us}_{ij,\mathrm{nAV}} = \frac{\mathrm{vs}_{\mathrm{nAV}}\exp(\beta_{\mathrm{tt}}\mathrm{tt}_{ij})}{\mathrm{vs}_{\mathrm{AV}}\exp(f(\mathrm{tt}_{ij})) + \mathrm{vs}_{\mathrm{nAV}}\exp(\beta_{\mathrm{tt}}\mathrm{tt}_{ij})} \quad (9.2)$$

As the utility function V is linear in travel times and costs, the vtts reduction by 25% from the eleventh minute on can be formulated as

Table 4. Overview of the Case Study Results

	Total number of car trips (millions)		Total car VMT (million veh. km)		Average car trip length (km)		Considers higher AV usage because of	
Model	AV	Non-AV	AV	Non-AV	AV	Non-AV	AV size and age	vtts Reduction
A1	3.4	27.1	63.7	504.4	18.6	18.6	Not	Not
A2	4.8	25.7	102.1	466.0	21.1	18.1	Yes	Not
A3	3.9	26.7	84.5	483.6	21.9	18.1	Not	Yes
A4	5.5	25.1	130.5	437.6	23.9	17.5	Yes	Yes

Note: veh. = vehicle.

$$f(\mathrm{tt}_{ij}) = \max \left\{ \beta_{\mathrm{tt}}\mathrm{tt}_{ij}, \frac{\beta_{\mathrm{tt}}}{1.25}\left(\mathrm{tt}_{ij}[\mathrm{min}] - 10\,\mathrm{min}\right) + \beta_{\mathrm{tt}} 10\,\mathrm{min} \right\} \tag{10}$$

The travel time parameter used for this case study was $\beta_{\mathrm{tt}} = -0.06$.

A4. The Incremental Approach employs the vehicle usage shares from Table 1 as base year values in an incremental (pivot-point) model. Each of the vehicle categories was split into two corresponding vehicle types, with and without autonomous technology respectively. The vehicle fleet shares $\mathbf{vs}^{\mathrm{BY}}$ and usage shares $\mathbf{us}^{\mathrm{BY}}$ in the base year are considered to be 0 for AV and 1 for non-AV for all vehicle size and age categories, as this technology does not yet exist in the base year. For the forecast, the VFM provides the vehicle fleet shares $\mathbf{vs}^{F}$, as given in Table 3. The utility formulation is identical to the implementation of A3, including the parameters and the reduction of travel time valuation shown in Equation 10. Thus, the incremental model for each vehicle category can be formulated as

$$\mathrm{us}^{F}_{ij,AV} = \frac{\mathrm{vs}^{F}_{\mathrm{AV}} \exp\left(f(\mathrm{tt}_{ij}) - \beta_{\mathrm{tt}}\mathrm{tt}_{ij}\right)}{\mathrm{vs}^{F}_{\mathrm{AV}} \exp\left(f(\mathrm{tt}_{ij}) - \beta_{\mathrm{tt}}\mathrm{tt}_{ij}\right) + \mathrm{vs}^{F}_{\mathrm{nAV}}} \tag{11.1}$$

$$\mathrm{us}^{F}_{ij,\mathrm{nAV}} = \frac{\mathrm{vs}^{F}_{\mathrm{nAV}}}{\mathrm{vs}^{F}_{\mathrm{AV}} \exp\left(f(\mathrm{tt}_{ij}) - \beta_{\mathrm{tt}}\mathrm{tt}_{ij}\right) + \mathrm{vs}^{F}_{\mathrm{nAV}}} \tag{11.2}$$

Note the simplifications

$$\lim_{\mathrm{us}^{\mathrm{BY}}_{\mathrm{AV}},\, \mathrm{vs}^{\mathrm{BY}}_{\mathrm{AV}} \to 0} \frac{\mathrm{vs}^{F}_{\mathrm{AV}} \mathrm{us}^{\mathrm{BY}}_{\mathrm{AV}}}{\mathrm{vs}^{\mathrm{BY}}_{\mathrm{AV}}} = \mathrm{vs}^{F}_{\mathrm{AV}} \tag{11.3}$$

and

$$\frac{\mathrm{vs}^{F}_{\mathrm{nAV}} \mathrm{us}^{\mathrm{BY}}_{\mathrm{nAV}}}{\mathrm{vs}^{\mathrm{BY}}_{\mathrm{nAV}}} = \frac{\mathrm{vs}^{F}_{\mathrm{nAV}} 1}{1} = \mathrm{vs}^{F}_{\mathrm{nAV}} \tag{11.4}$$

in Equations 11.1 and 11.2.

Results and Interpretation

Table 4 gives an overview of the aggregated results obtained with the four segmentation approaches. The number of AV and non-AV trips is given, as well as overall VMT and the ensuing average trip distances. The differences in results are significant, particularly for the VMT shares.

A1 forecasts the lowest share of trips and VMT using AV. In contrast to the other approaches, with A1 there is no reason why autonomous cars should be used more often than non-AVs. The vehicle usage shares are equal for all trips, thus leading to an equal average trip distance for AVs and non-AVs. Furthermore, the assumed reduction in the valuation of travel time cannot be considered using this approach.

A2 forecasts a significantly higher share of AV trips and VMT. This result is caused by VFM results showing that autonomous cars are larger and newer than the average car, and observed data shows such vehicles are being used more often and for longer trips. This also explains the longer average AV trip distances. The Trip Segmentation Approach, however, cannot consider the additional assumption of reduced vtts for AVs. If differentiated results for each vehicle category (grouped by size and age as in Table 1) were shown, usage shares and trip distances for autonomous and non-autonomous vehicles would be equal inside each category.

A3 also predicts more AV trips and VMT than A1. This approach considers the relative advantage of AVs caused by the assumed vtts reduction, thus leading to more AV trips and longer average trip distances. However, A3 does not take into consideration that AVs are more often larger and newer vehicles, which have been shown to be used relatively more often, whether or not they have any automation technology onboard.

Finally, A4 attempts to consider both AV comparative advantages and thus leads to the highest number of AV trips and VMT. Compared with A2, average AV trip distances increase by a further 2.8 km because of the vtts assumptions.

The results in Table 4 suggest that A1 is only suited for very rough first estimations. If the vehicle usage differs considerably between the vehicle categories, then A1 cannot deliver accurate results. A comparison between A2 and A3 is difficult, as in their implementation in the case study shown above they focus on different usage aspects. Ideally, a discrete choice modeling approach could and should also incorporate vehicle size and age in the utility function, which would make the results easier to compare with A2. If all relevant input data was captured in such an ideal model, one would expect the results to be similar to those obtained using A4. The problem is that such a model is difficult to formulate and estimate, as discussed in the section pertaining to the Choice Model Approach. Thus, further research is required to enhance the discrete choice modeling approach.

A4 offers an attractive and apparently easy to implement option of incorporating the advantages of the other two approaches, and would seem to deliver the most plausible results in this case study. However, this hybrid empirical and choice model construct should be studied further and implemented in other case studies as well to better understand the overall model responsiveness and the predictability of results.

The case study shown here does not contain any spatially heterogeneous policy measures, which would not affect car travel demand uniformly over the entire OD matrix, but only influence certain OD pairs. Such measures cannot be adequately assessed using A1 or A2. A modeled approach like A3 or A4 would be necessary to assess such scenarios. Furthermore, overall mode and destination choice was kept unchanged in the four approaches even though A3 and A4 offer the possibility of also estimating induced traffic. Enabling this feature will likely further increase the AV VMT shares using A3 and A4.

Conclusions and Outlook

The work presented in this paper is driven by the need to better forecast the impact of new technologies on motorized individual transport demand. As vehicle attributes are correlated to usage patterns, it becomes necessary to differentiate car travel demand according to these attributes. This becomes even more important when assessing the environmental effects of car traffic.

The aim of this paper was to present a modeling framework for splitting up overall car travel demand originating from a macroscopic travel demand model. A necessary prerequisite for this method is the existence of an independent vehicle fleet model forecasting the composition of the car fleet. Four approaches to linking the results from these two models were discussed.

Apart from the trivial solution of splitting up demand according only to the shares of respective vehicles in the fleet, the other three approaches all have their merits, but also their shortcomings. The Trip Segmentation Approach relies on observed usage patterns and assumptions and is therefore not sensitive to modifications in the input parameters. The Choice Model Approach is versatile, but the logit model parameters leading to a good fit to observed data are difficult to estimate. Finally, the Incremental Approach attempts to combine the advantages of the other approaches, but its responsiveness to variable input is somewhat difficult to ascertain.

Results obtained using the four approaches were compared in a case study forecasting demand for travel with private AVs in Germany in the year 2030. The results showed significant differences, particularly in the prediction of AV VMT. As expected, the forecasted demand for AV travel was lowest using the Trivial Approach and highest using the Incremental Approach.

In the case study shown, the Discrete Choice Approach was implemented using a very simple model formulation. Further work is required to enhance the utility function by also incorporating other vehicle attributes. It is expected that an ideally formulated and calibrated Discrete Choice Approach should lead to similar results to the Incremental Approach. The theoretical and practical difficulties related to the model estimation require further analyses.

Until a more complex Discrete Choice Approach can be implemented, the Incremental Approach seems to be a better suited method of linking macroscopic travel demand and VFMs. The approach should be put to use in other case studies as well, to better understand the way it responds to different input data.

References

1. Fagnant, D. J., and K. Kockelman. Preparing a Nation for Autonomous Vehicles: Opportunities, Barriers and Policy Recommendations. *Transportation Research Part A: Policy and Practice*, Vol. 77, 2015, pp. 167–181.
2. Reuters. *German Court Backs Bid to Ban Diesel Cars in Stuttgart*. https://www.reuters.com/article/us-germany-emissions-idUSKBN1AD10YAccessed August 1, 2017.
3. Lave, C. A., and K. Train. A Disaggregate Model of Auto-Type Choice. *Transportation Research Part A: General*, Vol. 13, No. 1, 1979, pp. 1–9.
4. Zhao, Y., and K. M. Kockelman. Household Vehicle Ownership by Vehicle Type: Application of a Multivariate Negative Binomial Model. Presented at 81st Annual Meeting of the Transportation Research Board, Washington, D.C., 2002.
5. Al-Alawi, B. M., and T. H. Bradley. Review of Hybrid, Plug-In Hybrid, and Electric Vehicle Market Modeling

Studies. *Renewable and Sustainable Energy Reviews*, Vol. 21, 2013, pp. 190–203.
6. Litman, T. Autonomous. Vehicle Implementation Predictions: Implications for Transport Planning. Presented at 94th Annual Meeting of the Transportation Research Board, Washington, D.C., 2015.
7. Bansal, P., and K. M. Kockelman. Forecasting Americans' Long-Term Adoption of Connected and Autonomous Vehicle Technologies. *Transportation Research Part A: Policy and Practice*, Vol. 95, 2017, pp. 49–63.
8. Kröger, L., T. Kuhnimhof, and S. Trommer. Modelling the Impact of Automated Driving: Private Autonomous Vehicle Scenarios for Germany and the US. Presented at European Transport Conference, Barcelona, 2016.
9. Bhat, C. R., and S. Sen. Household Vehicle Type Holdings and Usage: An Application of the Multiple Discrete-Continuous Extreme Value (MDCEV) Model. *Transportation Research Part B: Methodological*, Vol. 40, No. 1, 2006, pp. 35–53.
10. Auld, J., V. Sokolov, and T. S. Stephens. Analysis of the Effects of Connected–Automated Vehicle Technologies on Travel Demand. *Transportation Research Record: Journal of the Transportation Research Board*, 2017. 2625: 1–8.
11. Childress, S., B. Nichols, B. Charlton, and S. Coe. Using an Activity-Based Model to Explore the Potential Impacts of Automated Vehicles. *Transportation Research Record: Journal of the Transportation Research Board*, 2015. 2493: 99–106.
12. Zhao, Y., and K. M. Kockelman. Anticipating the Regional Impacts of Connected and Automated Vehicle Travel in Austin, Texas. *Journal of Urban Planning and Development*, Vol. 144 Issue 4, 2018.
13. Stephens, T., J. Sullivan, and G. A. Keoleian. A Microsimulation of Energy Demand and Greenhouse Gas Emissions from Plug-In Hybrid Electric Vehicle Use. Presented at 26th Electric Vehicle Symposium, Los Angeles, California, United States, 2012.
14. Wang, G., S. Bai, and J. M. Ogden. Identifying Contributions of On-Road Motor Vehicles to Urban Air Pollution Using Travel Demand Model Data. *Transportation Research Part D: Transport and Environment*, Vol. 14, No. 3, 2009, pp. 168–179.
15. infas, DLR: Mobilität in Deutschland 2008—Ergebnisbericht; Bonn und Berlin, 2010.
16. Mocanu, T., and C. Winkler. Petrol, Diesel or Electric? An Extension of Passenger Transport Models for Differentiating Car Travel. Presented at 6th Transportation Research Board Innovations in Travel Modeling Conference, Denver, Colorado, 2016.
17. Williams, H. C. W. L. On the Formation of Travel Demand Models and Economic Evaluation Measures of User Benefit. *Environment and Planning A*, Vol. 9, No. 3, 1977, pp. 285–344.
18. Daly, A., J. Fox, and B. Patruni. Pivoting in Travel Demand Models. Presented at European Transport Conference, Glasgow, 2011.
19. Winkler, C., and T. Mocanu. Methodology and Application of a German National Passenger Transport Model for Future Transport Scenarios. Presented at European Transport Conference, Barcelona, Spain, 2017.

The Standing Committee on Transportation Demand Forecasting (ADB40) peer-reviewed this paper (18-05027).

Article

TRR
JOURNAL OF THE TRANSPORTATION RESEARCH BOARD

Transportation Research Record
2018, Vol. 2672(49) 135–145

DOI: 10.1177/0361198118798288
journals.sagepub.com/home/trr
SAGE

Impact of Travel Time Constraints on Taste Heterogeneity and Non-Linearity in Simple Time–Cost Trade-Offs

Jeff Tjiong[1], Stephane Hess[1], Thijs Dekker[1], and Manuel Ojeda Cabral[1]

Abstract

Discrete choice models are a key technique for estimating the value of travel time (VTT). Often, stated choice data are used in which respondents are presented with trade-offs between travel time and travel cost and possibly additional attributes. There is a clear possibility that some respondents experience time constraints, leaving some of the presented options unfeasible. A model not incorporating information on these constraints would explain choices for faster and more expensive options as an indication that those respondents have a higher VTT when in reality they may be forced to select the more expensive option as a result of their personal constraints. This paper puts forward the hypothesis that this can have major impacts on findings in terms of heterogeneity in VTT measures. This paper examines via simulation the bias in VTT estimates and especially preference heterogeneity when such constraints are (not) accounted for. Empirical evidence is provided that preference heterogeneity is confounded with the travel budget impact on the availabilities of alternatives, and it is shown that there is a risk of producing biased estimates for appraisal VTT if studies do not explicitly model choice set formation. The inclusion of an opt-out alternative could be an effective measure to reduce the bias. This paper also explores the potential use of non-linear functional forms to capture the time budget impacts.

Economic theory and empirical findings support the argument that the value of travel time (VTT) is directly related to the stringency of time and money (budget) constraints. Recent empirical advances explicitly model the impact of constraints through the use of choice set formation (*1–3*). However, such constraints are typically not taken into account in some recent national VTT studies (*4*). This paper hypothesizes that not accounting for constraints could create significant risks in producing biased VTT estimates based on stated choice (SC) data. In particular, let us contrast two situations. If a traveler chooses a faster and more expensive option for the reason of wanting to save time for other activities, then this should reasonably be seen as this traveler doing so as his/her VTT is high enough to warrant paying the difference. If, on the other hand, the traveler is faced with two options departing at the same time and one being faster than the other, then he/she might simply be choosing the more expensive and faster option because of a constraint on needing to arrive by a specific time. In the majority of SC studies, the respondent is not given the option of changing his/her departure time and there is thus a substantial risk of constraints on timing influencing findings on the VTT.

This paper studies the confounding impact on VTT estimates and especially preference heterogeneity findings resulting from unaccounted (travel) time constraints. This confounding becomes, in the view of the authors, even more important given the increasing popularity of mixed multinomial logit (MMNL) models explaining unobserved taste heterogeneity among respondents. This paper argues that the estimated variance of the marginal utilities (and therefore VTT) captured by the MMNL models could in part be an artifact of constraints (or thresholds) rather than preference heterogeneity. This is tested through the use of simulated data to simulate fixed or random VTT among the simulated population, who are subject to either fixed or a mix of time budget constraints. Using simulated data, this paper examines the confounding effect that could happen when the choice model is mis-specified by ignoring the impact of travel

[1]Institute for Transport Studies & Choice Modelling Centre, University of Leeds, Leeds, UK

Corresponding Author:
Address correspondence to Jeff Tjiong: ts12lkjt@leeds.ac.uk

time constraints. In addition, given that the use of non-linear functional forms for utility function does not require any changes to the choice model structure, this paper also illustrates the use of non-linear functional forms to catch the tail of the VTT distributions where attribute levels exceed travel budgets.

This paper is organized as follows. The second section provides a review of existing literature in constrained modeling. The simulated dataset and analytical framework are then described. After this a section summarizes model results, then the implications of including an opt-out option are discussed. The use of non-linearities to capture the time budget effects is described, and the final section concludes.

Literature Review

Budget Constraints in SC Experiments

The empirical measurement of VTT is inextricably linked to the theories of time allocation in economics, as they provide justification for the VTT concept. By implementing the time allocation framework developed by DeSerpa (*5*) in the empirical random utility model (RUM) within the discrete choice setting, the VTT can be estimated as the marginal rate of substitution between travel time and cost in the conditional *indirect* utility function that is linear in income. As the utility which appears in the empirical models is the *indirect* utility, which is a result from decisions about consumption that is subject to both money and time budget constraints, the budget constraints are implicitly accounted for within the discrete choice models in principle. In practice, however, such budget constraints are not observed. This implies that researchers might present unfeasible alternatives to respondents in the SC experiment and thus introduce bias when unfeasible alternatives are modeled with non-zero choice probabilities. This problem is particularly apparent within the SC context because of its hypothetical setting, whereas in revealed preference data the chosen alternatives observed should be within budget unless irrational decisions are made. It is hypothesized in this study that money and time budget constraints are latent by nature as suggested by Ahmed and Stopher (*6*), and so it is inevitable that some attribute levels set out by researchers in the SC experiment might exceed some respondents' budget constraints.

Potential Bias Caused by Budget Constraint

The potential bias resulting from model mis-specifications for ignoring the impact of budget constraints on the availabilities of alternatives was identified soon after the development of the discrete choice modeling framework (*7*). Since then, many studies had provided evidence that suggests ignoring the impact of travel (or budget) constraints may lead to biased estimation. Among these studies, Cantillo and Ortúzar (*8*) and Li et al. (*9*) estimated the mis-specified models which also allow for random taste heterogeneity. Cantillo and Ortúzar found seriously biased estimates for VTT valuation in the presence of random attribute thresholds and concluded that the MMNL model is not capable of capturing the non-compensatory behavior (*8*). Li et al. assumed fixed tastes in simulated data but found welfare measures that are biased even when choice set formation is purposefully treated as taste heterogeneity in random parameter logit models (*9*). However, none of the above have pinpointed the direct confounding issue between the taste heterogeneity and the attribute thresholds. As such, this study aims to fill this research gap by allowing for random VTT that vary across a simulated population to test the impacts on mis-specified models. Although the main focus is on the impact of retrieving heterogeneity, it should be clear that bias can also arise in fixed coefficients models.

Implications of Budget Constraints on Choice Set Modeling Tools

It is anticipated that examination of the potential confounding impacts on taste heterogeneity findings caused by unaccounted budget constraint effects could provide valuable insights into the performance of existing choice set formation models. A full two-stage probabilistic choice set model (*1*, *3*) includes modeling a first stage non-compensatory decision-making process in which travelers restrict their decisions to a particular subset of a full choice set in order to conform to their travel budget constraints. This is followed by a compensatory second stage in which utilities are maximized within each subset of choice set. It is hypothesized that if taste heterogeneity is indeed confounded with the budget constraint effect during the choice set generation stage, then it could also lead to bias in the choice evaluation stage. It is also anticipated that such issues also apply to the single-stage constrained choice set models for approximation of the constrained choice sets (*2*, *10*).

Empirical Setup

Data-Generating Process

Monte Carlo Simulations. Simulated datasets are generated through a Monte Carlo simulation to provide empirical evidence of the impacts of the budget constraints on the availabilities of alternatives. Simulated datasets are used for this application as the data-generating process is fully controlled while the true parameters are available for fair comparisons across different model specifications. In this exercise a simple time–cost trade-off exercise is adopted, which has been used mostly in the national value of time studies in Western Europe (*4*). The use of simple trade-offs has received increasing criticism as the valuations from more complex SC designs are deemed more reliable. More complex choices are also thought to be more comprehensible to respondents (*11*). Nevertheless, such simple trade-offs are useful in this study to enable the disentangling of the confounding effects, which is more difficult under the presence of more than two attributes. Also, it is anticipated that the impact of budget constraints on alternative elimination would be the most severe as only one alternative remains in the choice set when the counterpart gets eliminated for exceeding the time budget thresholds. As such, the impact of budget constraint at its most extreme condition could be explored. The findings from this exercise should provide insights to researchers for further test on designs with more complex choices.

Choice Scenarios. Within the simulated population, all pseudo-observed decision-makers are presented with two alternatives, each with varying levels of travel time between 30 min and 75 min and travel cost between £3 and £7.50. A full factorial design is first generated, with dominated choices removed. A choice-rejection mechanism is enforced such that if the travel time attribute presented exceeds the predefined time budget, the respective alternative is then rejected. To retain the same number of observations across different budget threshold bands for fair comparisons, the design only allows trade-offs between travel times where at most one alternative exceeds the time budget of 45 min. For those choice tasks with both alternatives retained, the simulation assumes that individuals make choices according to the standard random utility maximizing rule. A total of 2,700 choice tasks are generated, and these choice tasks are randomized and organized into 10 choice tasks each for 540 pseudo-observed decision-makers, with each of the original choice tasks used twice.

Generation of Choices. Using a RUM, we write the utility $U_{i,n,t}$ that an individual n obtains from choosing alternative i in choice task t as being decomposed into an observed component $V_{i,n,t}$ and a random component $\varepsilon_{i,n,t}$. The observed component of the utility of the time and cost trade-offs can simply be written as

$$V_{i,n,t} = \beta_t T_{i,n,t} + \beta_c C_{i,n,t}$$

where T is the time attribute and C is the cost attribute, while β_t and β_c refer to the marginal utilities of time and cost, respectively.

Four combinations of specifications for time sensitivities and time budget variations across simulated population are set out to generate choices in this study, and cost coefficients are always kept fixed and linear. Mean time and cost coefficients are set as −0.075 and −0.90, respectively, thus resulting in a "true" mean VTT of £5/hr across the simulated population. The four combinations used are as follows:

- Fixed time budget
  - Set A: Fixed and linear time sensitivities of −0.075
  - Set B: Negative lognormally distributed time sensitivities with an arithmetic mean of −0.075 and standard deviation of 0.038
- Mixed time budget
  - Set C: Fixed and linear time sensitivities of −0.075
  - Set D: Negative lognormally distributed time sensitivities with an arithmetic mean of −0.075 and standard deviation of 0.038

Two sets of time budgets are tested in this study. The fixed budget assumption as in sets A and B assumes all individuals share the same time budget threshold, which varies from the unconstrained case of 75 min to 55 min in the most stringent scenario. The mixed budget assumption as in sets C and D assumes that 50% of the simulated population are 10 min more restricted in terms of the time budget when compared with the rest of the simulated population.

Model Estimations

Several different model specifications were tested on the simulated data.

Fixed and Linear Time Sensitivities. The multinomial logit (MNL) model is used for estimations of the fixed and linear time sensitivities. Let $P_{n,t}(i|\beta)$ give the probability of respondent n (with $n = 1, \ldots, N$) for alternative i (with $i = 1, \ldots, I$) in choice situation t (with $t = 1, \ldots, T_n$), conditional on a vector of taste coefficients β, with $\varepsilon_{i,n,t}$ following a Type I extreme value distribution, distributed identically and independently across alternatives and choice situations. The choice probability given by the MNL model then becomes $P_{n,t}(i|\beta) = e^{V_{i,n,t}} / \sum_{j=1}^{J} e^{V_{j,n,t}}$.

The log-likelihood (LL) function, conditional on β, is given by

$$LL(\beta) = \sum_{n=1}^{N} \sum_{t=1}^{T_n} ln(P_{n,t}(j_{n,t}|\beta))$$

where $j_{n,t}$ is the alternative chosen by respondent n in choice situation t. As time sensitivities are specified as fixed and linear in this set of scenarios, VTT can be computed by taking the ratio of the partial derivatives of the utility against time and cost, which is the marginal rate of substitution between time and cost, expressed as β_t/β_c.

Negative Lognormally Distributed Time Sensitivities. The MMNL model is used for estimations of the random VTT. In the MMNL model, the vector of the taste coefficients β follows a random distribution across respondents, such that we have $\beta \sim g(\beta|\Omega)$, with Ω representing a vector of parameters of the distribution of β. This study allows tastes to vary across respondents only but stay constant across choice situations (cf. *12*). The choice probability of the chosen alternative given by the MMNL model for respondent n over a sequence of choices he/she faced becomes

$$P_n(\Omega) = \int_{\beta} \prod_{t=1}^{T_n} P_{n,t}(j_{n,t}|\beta) g(\beta|\Omega) d\beta$$

The LL function is given by

$$LL(\Omega) = \sum_{n=1}^{N} ln\left(\int_{\beta}\left[\prod_{t=1}^{T_n}(P_{n,t}(j_{n,t}|\beta))\right] g(\beta|\Omega) d\beta\right)$$

It is assumed that the time sensitivities are negative lognormally distributed in the model estimations in which random VTT are estimated. 200 Halton draws are used to approximate the integral through Monte Carlo simulation for all the MMNL models.

Non-Linear Time Sensitivities. Non-linear functional forms were finally tested to catch the tail of the VTT distributions where attribute levels exceed travel budgets. As such, the 3rd-degree polynomials with the form $\beta_{t1}T + \beta_{t2}T^2 + \beta_{t3}T^3$ specified for time sensitivities are estimated using the MNL models. In terms of the VTT calculations, the partial derivative of the utility also depends on the time attribute owing to the non-linearities. For the time sensitivities formulated in 3rd-degree polynomial form, the VTT becomes $(\beta_{t1} + 2\beta_{t2}T + 3\beta_{t3}T^2)/\beta_c$.

Incorporation of Constraints. All scenarios are tested with and without the knowledge of the availabilities of alternatives (owing to constraints) for each choice task. The model runs with known availabilities of alternatives are used for replicating the true parameters in the unbiased models, and another set of model runs are undertaken for testing the budget constraint impacts in the biased models.

Empirical Results

The results of the various models are now presented, and each simulated data setting is examined in turn. For each model specified for estimation, 100 simulated data sets are drawn randomly. All the model results reported are averages across all 100 simulated data sets.

Linear Time Sensitivities under Fixed Time Budget

Replication of Time and Cost Sensitivities. When the availabilities of alternatives for all the choice tasks in the SC experiment are known to the analyst, the MNL models can replicate the true time sensitivity of −0.075 consistently across different levels of thresholds set out in the unbiased models (A1) as shown in Table 1. It is also shown that MMNL models (B1) are able to retrieve the true arithmetic mean of −0.075 for time sensitivity and standard deviation of 0.038 from the simulated population with negative lognormally distributed time sensitivities. The true cost sensitivity of −0.9 is also consistently retrieved from the models.

Model Fit (ρ^2 and LL). The unbiased models become more deterministic when alternatives are eliminated because of the stringency of the time budgets, as the probability of observing the chosen alternatives becomes one for these choice tasks. It is shown that there is a significant improvement in LL from −2,967 in the unconstrained scenario to −1,400 when time budget is set at 55 min (A1), when fixed tastes are assumed in the simulated data. Similarly, when random time sensitivities are included in the data-generating process, LL is improved from −3,036 to −1,416 (B1) in the unbiased models. As individuals are assumed to make their choices based on RUM-consistent behavior for the remaining choice tasks that are not eliminated, the true VTT of £5/hr can be retrieved from these choice tasks. These results show that the true values can be replicated when the model structure is properly specified and the availabilities of alternatives are known.

Biased Estimates in MNL Models When the Availabilities of Alternatives Are Unknown. The fact that the availabilities of alternatives are unknown to the analyst has several implications for the model estimation. First, respondents whose time constraints leave them with only one viable option are then forced to choose the faster but more expensive alternatives. As the time constraints are unobserved by the analyst, the choice models consequently over-estimate the time sensitivities, given that the

Table 1. Estimation Results for Linear Time and Fixed Time Budget in Data-Generating Process (DGP)

Budget threshold (min)	ρ^2	LL	β_t Mean	β_t *t*-stat	β_t SD	β_t *t*-stat	β_c Est	β_c *t*-stat	VTT (£/hr) Mean	VTT (£/hr) SD
		DGP:	Fixed time budget—Linear time—Fixed β_t							
A1—UNBIASED		EST:	Fixed time budget—Linear time—Fixed β_t (MNL)							
75	0.21	−2,966.57	−0.075	31.1	–	–	−0.905	31.6	4.99	–
70	0.30	−2,619.77	−0.075	26.7	–	–	−0.896	29.4	5.00	–
65	0.40	−2,230.36	−0.075	22.5	–	–	−0.900	26.7	4.99	–
60	0.51	−1,822.36	−0.075	18.0	–	–	−0.899	23.9	5.00	–
55	0.63	−1,399.64	−0.076	13.7	–	–	−0.905	20.7	5.03	–
		DGP:	Fixed time budget—Linear time—Random β_t							
B1—UNBIASED		EST:	Fixed time budget—Linear time—Random β_t (MMNL)							
75	0.19	−3,035.94	−0.075	61.4	0.037	14.3	−0.899	29.0	5.02	2.50
70	0.29	−2,657.06	−0.075	54.0	0.038	12.5	−0.903	27.2	5.02	2.52
65	0.40	−2,254.72	−0.076	46.3	0.037	10.0	−0.908	24.9	5.04	2.46
60	0.51	−1,843.73	−0.074	37.8	0.037	7.6	−0.897	22.3	4.98	2.47
55	0.62	−1,415.61	−0.075	28.9	0.036	5.2	−0.900	19.1	5.03	2.47
		DGP:	Fixed time budget—Linear time—Fixed β_t							
A2—BIASED		EST:	Fixed time budget—Linear time—Fixed β_t (MNL)							
75	0.21	−2970.04	−0.075	31.1	–	–	−0.901	31.4	5.01	–
70	0.24	−2843.25	−0.090	34.2	–	–	−0.917	31.8	5.88	–
65	0.29	−2650.01	−0.108	37.2	–	–	−0.932	32.0	6.94	–
60	0.36	−2382.77	−0.130	38.9	–	–	−0.941	31.1	8.31	–
55	0.45	−2057.62	−0.157	38.9	–	–	−0.914	28.5	10.31	–
		DGP:	Fixed time budget—Linear time—Random β_t							
B2—BIASED		EST:	Fixed time budget—Linear time—Random β_t (MMNL)							
75	0.19	−3035.08	−0.075	61.5	0.037	14.2	−0.900	29.2	5.01	2.5
70	0.22	−2924.34	−0.089	70.8	0.028	10.5	−0.906	29.6	5.90	1.9
65	0.27	−2722.02	−0.106	75.6	0.018	5.7	−0.904	29.7	7.01	1.2
60	0.35	−2431.31	−0.126	76.5	0.004	0.8	−0.899	29.7	8.40	0.3
55	0.44	−2078.23	−0.154	71.5	0.000	0.5	−0.890	28.2	10.40	0.0

observed choice probabilities of the faster alternatives are higher compared with the estimates in the unbiased scenarios when the availability of alternatives are known to analysts. As shown in Table 1, time sensitivities are overstated significantly by 109% from −0.075 to −0.157 when the time budget is restricted to 55 min (A2). VTT is also overestimated to a similar level, from £5/hr in the unconstrained scenario to £10.3/hr when a time budget of 55 min is assumed. These findings of biased estimates for the VTT are in line with the past empirical evidence discussed earlier.

Second, in the unconstrained scenarios when travel times presented are not restricted by time budgets, some respondents would still choose the slower but cheaper alternatives because of any unobserved factors, even when their VTT are higher than the boundary values of time. These unobserved factors, which are represented as random errors in choice models, do not contribute to the randomness of choices anymore once the time budget constraints are applied when faster alternatives become the only viable choices. As such, the choice processes are estimated to be more deterministic when the budget constraint impacts enter the model estimations. This explains the increase of ρ^2 and LL from 0.21 and −2,970 in the unconstrained scenario to 0.45 and −2,058, respectively, when the budget is set at 55 min, even when no choice tasks are omitted from the LL calculations.

Biased Estimates in MMNL Models When the Availabilities of Alternatives Are Unknown. Given the popularity of using MMNL models to capture preference heterogeneity, it is of particular interest to understand whether the MMNL models can fully capture the preference heterogeneity even when some attribute levels exceed the time budget thresholds of respondents. As shown in Table 1, the MMNL models increasingly fail to capture the preference heterogeneity inherent to the true data set when the time budgets become more stringent (B2). The standard deviation of the negative lognormally distributed time sensitivities is reduced from 0.037 in the unconstrained scenario to 0.028 and 0.018 when the time budget thresholds are set at 70 min and 60 min, respectively. At the time budget threshold of 55 min, the MMNL model fails to capture any preference heterogeneity with the arithmetic mean estimated at −0.154. This arithmetic mean estimate is similar to the biased marginal time utility estimated at −0.157 by the MNL model (A2) when the time budget threshold is 55 min. This implies that the MMNL model effectively treats all respondents as having high time sensitivities.

To explain this further, consider the situation in which people have heterogeneous time sensitivities across the sample. When the time budget constraints are stringent, individuals who have low VTT are forced to choose the fast but expensive alternatives as opposed to the slow and cheap alternatives which they prefer. On the other hand, individuals who have high VTT would also choose the fast but expensive alternatives, either because of their high willingness to pay in the unconstrained choice situations, or because of the budget constraints in the constrained situations. If, as a result, the choice outcomes are the same between these two groups of individuals who share distinctly different VTT, the MMNL model cannot detect any differences in tastes between them when the time budget constraints are not accounted for. It demonstrates that the use of MMNL model could potentially produce misleading findings of a lack of preference heterogeneity, when in fact the preference heterogeneity is simply suppressed by the severe time budget constraints in the model estimations which dominate completely. Similarly, it is also shown that the VTT estimates produced by the MMNL models (B2) align closely with the estimates generated by the biased MNL models (A2) at all levels of the time budget constraints. Both the MNL and MMNL models over-estimate the VTT by twofold in the most extreme case, at around £10.3/hr approximately because of the inflated time sensitivities. Cost sensitivities, on the other hand, are not affected by the time budget constraints and the MMNL models are able to retrieve the true value of −0.90.

Linear Time Sensitivities under Mixed Time Budgets

Replication of Parameters When the Availabilities of Alternatives Are Known. It has been shown above that the MMNL models could produce misleading findings with respect to the presence of preference heterogeneity when all respondents share the same time budget thresholds. This section further introduces mixed time budget thresholds to the model estimations, which assumes that two randomly selected groups within the simulated population share distinctly different perceptions of the time budget constraints. Within this setting, half of the respondents perceive their time budget constraints to be 10 min more restrictive in comparison with the rest of the population (e.g., time budget constraints of 60 min and 70 min perceived by half of the respondents, respectively). The objective of this exercise is to examine whether further confounding of preference heterogeneity would occur when budget thresholds are not fixed among individuals. In the unbiased scenarios in which the availabilities of alternatives subject to the budget constraints are known to analyst, all true values assumed in the data-generating process (VTT of £5/hr, mean time sensitivity of −0.075 and cost sensitivity of −0.90) are retrieved (C1 and D1 in Table 2).

Biased Estimates When the Availabilities of Alternatives Are Unknown. Similar to the findings from scenarios where fixed time budgets are assumed among individuals, the biased models over-estimate time sensitivities when the mixed time budget constraints are stringent. In the most restrictive scenario. in which half of the respondents perceive the time constraints to be either 55 min or 45 min, time sensitivity is overestimated by 123%, from −0.075 to −0.165 (C2 vs. C1 in Table 2). Apart from the inflated time sensitivities caused by unaccounted time budget constraints, we again test whether the MMNL specification for the model estimations would lead to biased results. In general, model results show that the mis-specified MMNL models (C3) pick up preference heterogeneity that does not exist in the data-generating process. In the scenario where the mixed time budget constraint is the most stringent, the mis-specified MMNL model estimates the standard deviation of the time sensitivity at 0.037, with a t-statistic that is significant at 6.5. This provides evidence that the MMNL model could potentially misinterpret the effects of mixed budget thresholds as preference heterogeneity. In other words, despite the fact that all individuals share the same VTT, choice probabilities for the chosen alternatives could still vary significantly across the population according to the mixed budget threshold setting.

To put this issue into context, assume a case where all individuals are willing to pay £1.25 to save 15 min of travel time (i.e., a VTT of £5/hr). They are then asked to choose between the free alternative, which requires 60 min of travel time, and the tolled alternative, which costs £2 for a 45-min journey. As the toll charge is higher than the willingness to pay to save 15 min of travel time for all individuals, they are likely to choose the free alternative over the tolled alternative. Now assume that some but not all of these respondents are also subject to a time budget threshold of 55 min; the tolled alternative then becomes the only available option because of the budget constraints, rather than the free alternative that they prefer. As the time budget variations among individuals are unobserved, the choice models thus wrongly attribute such effects to the differences in taste heterogeneity among population instead.

Table 2. Estimation Results for Linear Time and Mixed Time Budget in DGP

Budget threshold (min)	ρ^2	LL	β_t Mean	β_t *t*-stat	β_t SD	β_t *t*-stat	β_c Est	β_c *t*-stat	VTT (£/hr) Mean	VTT (£/hr) SD
		DGP:	Mixed time budget (50–50%)—Linear time—Fixed β_t							
C1—UNBIASED		EST:	Mixed time budget (50–50%)—Linear time—Fixed β_t (MNL)							
75 & 65	0.30	−2,603.35	−0.075	27.2	–	–	−0.900	29.3	4.99	–
70 & 60	0.41	−2,220.83	−0.075	22.9	–	–	−0.894	26.7	5.01	–
65 & 55	0.51	−1,816.17	−0.074	18.7	–	–	−0.898	24.1	4.97	–
60 & 50	0.62	−1,404.34	−0.075	14.7	–	–	−0.901	21.1	4.98	–
55 & 45	0.74	−989.32	−0.075	10.6	–	–	−0.898	17.3	4.98	–
		DGP:	Mixed time budget (50–50%)—Linear time—Random β_t							
D1—UNBIASED		EST:	Mixed time budget (50–50%)—Linear time—Random β_t (MMNL)							
75 & 65	0.29	−2,641.81	−0.076	54.7	0.037	12.4	−0.904	27.2	5.02	2.5
70 & 60	0.40	−2,244.51	−0.075	47.1	0.037	10.2	−0.905	25.0	4.98	2.5
65 & 55	0.51	−1,840.67	−0.074	38.5	0.038	8.3	−0.898	22.4	4.99	2.5
60 & 50	0.62	−1,416.85	−0.075	31.1	0.036	5.7	−0.896	19.6	5.03	2.4
55 & 45	0.73	−998.14	−0.069	23.5	0.034	5.2	−0.905	16.3	4.97	2.5
		DGP:	Mixed time budget (50–50%)—Linear time—Fixed β_t							
C2—BIASED		EST:	Mixed time budget (50–50%)—Linear time—Fixed β_t (MNL)							
75 & 65	0.24	−2,860.52	−0.089	33.2	–	–	−0.895	31.6	5.95	–
70 & 60	0.28	−2,689.35	−0.104	35.1	–	–	−0.892	31.2	7.03	–
65 & 55	0.35	−2,441.55	−0.124	36.7	–	–	−0.877	29.7	8.49	–
60 & 50	0.42	−2,160.86	−0.144	37.0	–	–	−0.819	26.9	10.58	–
55 & 45	0.51	−1,833.58	−0.165	36.1	–	–	−0.684	21.0	14.49	–
C3—BIASED		DGP:	Mixed time budget (50–50%)—Linear time—Fixed β_t							
(MIS-SPECIFIED)		EST:	Mixed time budget (50–50%)—Linear time—Random β_t (MMNL)							
75 & 65	0.24	−2,858.18	−0.091	77.3	0.014	5.0	−0.920	30.4	5.95	0.9
70 & 60	0.28	−2,675.94	−0.109	75.9	0.016	5.0	−0.922	29.8	7.07	1.0
65 & 55	0.35	−2,443.43	−0.129	71.4	0.019	5.4	−0.906	28.3	8.53	1.3
60 & 50	0.42	−2,152.58	−0.153	63.7	0.025	5.9	−0.858	25.7	10.69	1.8
55 & 45	0.51	−1,832.06	−0.180	51.8	0.037	6.5	−0.736	20.2	14.70	3.0
		DGP:	Mixed time budget (50–50%)—Linear time—Random β_t							
D2—BIASED		EST:	Mixed time budget (50–50%)—Linear time—Random β_t (MMNL)							
75 & 65	0.22	−2,929.18	−0.091	66.2	0.036	13.6	−0.909	29.5	6.04	2.4
70 & 60	0.27	−2,739.95	−0.109	70.0	0.031	11.3	−0.910	29.5	7.17	2.1
65 & 55	0.34	−2,481.53	−0.130	67.8	0.030	9.7	−0.902	28.0	8.66	2.0
60 & 50	0.42	−2,169.94	−0.155	61.6	0.034	8.7	−0.859	25.7	10.86	2.4
55 & 45	0.51	−1,829.64	−0.183	51.2	0.042	7.5	−0.737	20.2	14.90	3.4

The difficulties of distinguishing whether the variations in choice probabilities are caused by preference heterogeneity or differences in budget thresholds are further complicated when both the budget thresholds and tastes vary among individuals. On one hand, it would be anticipated that the MMNL model could not fully capture the preference heterogeneity assumed in the simulated data set when the travel budget constraints are applied, as described earlier. On the other hand, it is also expected that the MMNL model would wrongly attribute the mixed budget effects as taste heterogeneity when time budgets are very stringent. As the variations of time budgets among individuals are unknown to the analyst, there is substantial risk that misleading findings of taste heterogeneity can also be attributed to a mix of these two opposite effects. The model results across different levels of stringency of time budgets for simulated data in which random time sensitivities and mixed time budgets are assumed are shown in D2 in Table 2.

Inclusion of an Opt-Out Alternative

The inclusion of an opt-out alternative, or sometimes referred to as the "no choice,""neither," "none of these," or "status quo" alternative in SC scenarios has been widely discussed in the past. It has been argued that the inclusion of an opt-out alternative increases both the realism of the SC choice tasks and the statistical efficiency of model estimations (cf. *13*). Given the aforementioned risk of confounding impacts on the taste heterogeneity findings owing to unaccounted budget constraint effects, it of interest to explore the effectiveness of the opt-out alternative to reduce the bias associated with budget constraints in the valuation of VTT.

Table 3. Estimation Results for Inclusion of the Opt-Out Alternative

Budget threshold (min)	ρ^2	LL	β_t				β_c		VTT (£/hr)		ASC for Opt-out option	
			Mean	*t*-stat	SD	*t*-stat	Est	*t*-stat	Mean	SD	Est	*t*-stat
B1—UNBIASED		DGP:	Fixed time budget—Linear time—Random β_t									
(W/ OPT-OUT)		EST:	Fixed time budget—Linear time—Random β_t (MMNL)									
75	0.23	−4567.86	−0.075	65.1	0.037	20.2	−0.902	34.5	5.00	2.50	−9.027	−36.7
65	0.31	−4108.58	−0.075	52.5	0.038	18.1	−0.903	32.2	5.01	2.52	−9.028	−32.5
60	0.35	−3875.24	−0.075	46.0	0.038	16.8	−0.899	31.0	5.03	2.52	−9.002	−30.2
55	0.39	−3635.90	−0.075	39.5	0.038	15.3	−0.899	30.0	5.00	2.52	−8.988	−28.2
B2—BIASED		DGP:	Fixed time budget—Linear time—Random β_t									
(W/ OPT-OUT)		EST:	Fixed time budget—Linear time—Random β_t (MMNL)									
75	0.23	−4574.10	−0.075	65.1	0.038	20.4	−0.901	34.7	5.03	2.5	−9.01	−36.9
65	0.24	−4495.65	−0.105	80.9	0.033	20.3	−0.929	35.5	6.82	2.1	−10.36	−40.3
60	0.26	−4379.86	−0.124	84.5	0.032	20.2	−0.927	35.2	8.05	2.1	−11.05	−41.6
55	0.29	−4218.61	−0.145	83.1	0.032	19.3	−0.902	33.2	9.68	2.1	−11.67	−41.1

Model specifications including fixed time budget thresholds and negative lognormally distributed time sensitivities (B2 in Table 1) are retained as the basis for the new data-generating process to generate choices for the scenarios that include opt-out alternatives. The utility of the new opt-out alternative is represented by an alternative-specific constant (ASC), in which a value of −9.0 is assigned to the opt-out alternative to represent the dis-benefits from not being able to travel. This results in approximately 25% of individuals choosing the opt-out alternative in the unconstrained scenario, with a choice probability of 37.5% approximately for any of the two travel alternatives. This setting implies that the dis-utilities of not traveling are slightly larger than the dis-utilities of the travel alternatives in the unconstrained situation, ensuring that the opt-out alternative is not overly attractive relative to the two travel alternatives.

Similar to the unbiased model results presented earlier, all the true values including the arithmetic mean and standard deviation parameters of the time coefficients, cost coefficients, and the ASC values of −9.0 for the opt-out alternatives are retrieved when the availabilities of alternatives are known to the analyst (B1 in Table 3). When the availabilities of alternatives are unknown, taste heterogeneity assumed in the data-generating process cannot be retrieved fully (B2 in Table 3), but the level of bias is not as strong as that in the binary choices examined earlier. When the time budget threshold is set to 55 min, the arithmetic mean and standard deviation of the time coefficient change from −0.075 and 0.038 in the unconstrained case to −0.145 and 0.032, respectively, in the model that includes an opt-out alternative. This is compared with the arithmetic mean of −0.154 and a complete loss of taste heterogeneity in binary choices without an opt-out alternative (B2 in Table 1). It is noted that the capability of recovering taste heterogeneity under the presence of the opt-out alternative would depend on both the SC design and the value of the ASC assigned. The SC design implemented in this study only allows one out of two travel alternatives to exceed the budget thresholds. This setting always allows respondents to choose between the opt-out alternative and at least one other travel alternative, which facilitates the retrieval of the true preference from these trade-offs. In practice, the recovery of some taste heterogeneity might be somewhat less effective, as the respondents could be forced to choose the opt-out alternative only when both the travel alternatives presented exceed their budget thresholds. In summary, the inclusion of the opt-out alternative would provide more information to the choice model to explain taste heterogeneity but cannot fully eliminate the confounding issue when the budget constraints are not accounted for in the choice model.

Non-Linearities

Replication of Parameters When the Availabilities of Alternatives Are Known

This section switches focus to the incorporation of non-linearities in the model specifications to capture potential budget constraint effects. It was demonstrated in earlier sections that the confounding of taste heterogeneity findings caused by unaccounted budget constraint effects could potentially lead to significant bias in the VTT estimation. It was also hypothesized that travel budget constraints are latent in nature, and are difficult to measure without the use of more complicated probabilistic choice set formation models. It is thus useful to examine whether non-linear functional forms could capture the kink of travel dis-utilities, which could occur when stringent budget constraints are applied. This could

Table 4. Estimation Results for Linear Time and Fixed Time Budget in DGP but Estimated by Non-Linear Functional Form

Budget threshold (min)	ρ^2	LL	β_t: β_{t1} Est	β_{t1} *t*-stat	β_{t2} Est	β_{t2} *t*-stat	β_{t3} Est	β_{t3} *t*-stat	β_c Est	β_c *t*-stat	VTT (£/hr)
		DGP:	Fixed time budget—Linear time—Fixed β_t								
E1—UNBIASED		EST:	Fixed time budget—Non-linear time (3rd-degree polynomials)—Fixed β_t (MNL)								
75	0.21	−2,968.33	−0.162	8.1	0.0017	8.8	−1.1E-05	45.98	−0.901	29.9	4.79
70	0.30	−2,616.84	−0.129	4.8	0.0012	6.4	−8.1E-06	39.07	−0.896	28.3	4.86
65	0.40	−2,225.71	−0.087	2.8	0.0002	4.4	−1.7E-06	27.19	−0.905	26.3	4.96
60	0.51	−1,819.41	−0.116	1.6	0.0009	1.8	−6.1E-06	8.92	−0.902	23.7	4.92
55	0.62	−1,401.19	−0.021	2.4	−0.0013	5.5	1.1E-05	5.95	−0.898	20.7	4.95
		DGP:	Fixed time budget—Linear time—Fixed β_t								
E2—BIASED		EST:	Fixed time budget—Non-linear time (3rd-degree polynomials)—Fixed β_t (MNL)								
75	0.21	−2,965.66	−0.164	7.6	0.0018	8.2	−1.2E-05	44.5	−0.901	30.1	4.78
70	0.27	−2,728.27	−0.914	112	0.0187	606.5	−1.3E-04	174.9	−0.844	28.0	2.65
65	0.37	−2,347.61	−1.486	103	0.0325	256.4	−2.4E-04	343.6	−0.840	26.0	1.80
60	0.48	−1,943.58	−2.330	222	0.0538	984.8	−4.2E-04	187.0	−0.834	23.5	1.89
55	0.60	−1,507.56	−3.933	349	0.0957	273.7	−7.7E-04	490.4	−0.852	21.2	4.78

potentially provide useful insights to researcher on the possibility that particular attribute levels set out in SC designs are beyond budget thresholds for some decision-makers.

A 3rd-degree polynomial functional form for time sensitivities is adopted for testing the use of non-linear functional forms in this study. Model results show that the 3rd-degree polynomial functional forms produce very similar cost sensitivities, LL and ρ^2 (E1 in Table 4) as in the MNL models (A1 in Table 1) when the availabilities of alternatives are known to the analyst. Overall, it appears that the 3rd-degree polynomial form specified for time sensitivities collapses to a linear form in the unbiased models, as the estimated time coefficients for the second and third polynomial terms are very small. The true VTT of £5/hr, estimated in quadratic forms as described earlier, is retrieved across all levels of the budget thresholds and attribute levels.

Biased Parameters When the Availabilities of Alternatives Are Unknown

Model results estimated when the availabilities of alternatives are unknown to the analyst are summarized in set E2 in Table 4. The VTT estimates produced by the polynomial utility functional forms are shown to be highly sensitive to the attribute levels of the travel time, as opposed to the VTT estimates in unbiased models that are stable across attribute levels. For instance, when the time budget threshold is set at 55 min, the VTT escalates from £91/hr to £184/hr when the journey times increase from 65 min to 75 min, as shown in Figure 1. These exceptionally high VTT estimates show that respondents are highly unlikely to choose the alternatives where travel times presented are beyond the time budgets, and could become useful indicators to highlight the significant impacts of the budget constraints on the VTT valuation.

Now we examine whether the flexible utility functional forms could capture the tail of the VTT distributions where the attribute levels exceed the travel budgets of the respondents. Figure 1 also shows the utility levels that are related to the travel time components only. It can be seen that the time dis-utilities increase significantly only when time attributes presented are beyond the budget thresholds. For instance, when time budget is set at 55 min, the polynomial utility function produces a stable utility level for journeys that last between 30 min and 55 min. Beyond that, the travel time dis-utilities increase significantly as the time attribute values exceed the designated time budget of 55 min. This indicates that the use of the 3rd-degree polynomial utility function could become a convenient and effective approach to detect the potential budget constraint effects in SC data.

Conclusion

This paper has sought to provide a detailed assessment of the impact of time budget constraints on the VTT estimates and the identification of preference heterogeneity, when explicit modeling of choice set formation is not involved. It is first shown that if time budgets are stringent but not accounted for, VTT can be significantly overestimated. Second, this paper has provided a comprehensive set of empirical evidence to understand the confounding impact on preference heterogeneity findings caused by the unaccounted budget constraint effects

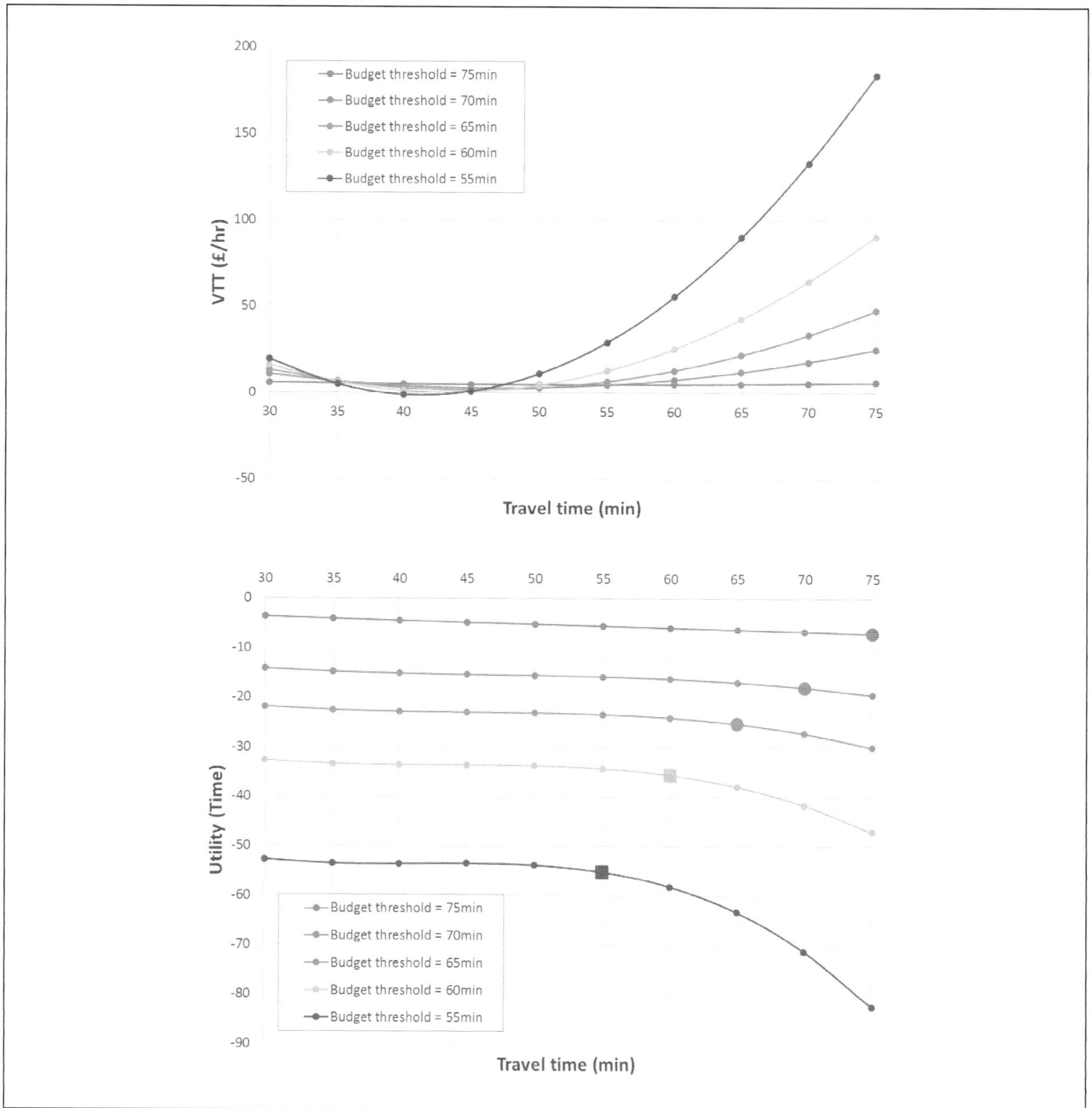

Figure 1. VTT estimates and travel dis-utilities estimated by non-linear functional form.

across a range of time budget stringency. It was found that the MMNL model fails to capture any preference heterogeneity and collapses to a MNL model when the travel budget is very binding within a binary choice and deterministic alternative elimination setting. It was also found that the MMNL model could also wrongly attribute the impacts of the mixed time budget constraints to the findings of preference heterogeneity.

It was found that including an opt-out alternative could potentially help retrieve some but not all preference heterogeneity under the presence of budget constraints. The findings from this study also raise some further questions. First, the question arises whether the confounding issue also occurs at the non-compensatory stage of the choice set formation models (e.g., the Manski-type models). Second, how can the confounding effects be disentangled using real-life SC data? Third, there is a need for a comparative analysis to assess the differences between the single-stage semi-compensatory model for approximation of constrained choice sets (e.g., the constrained multinomial logit model) and the simple non-linear functional forms, given that simple non-linear functions could potentially capture the kink of the time sensitivities when subject to binding budget constraints.

It should be noted that if the SC surveys adequately capture rescheduling by allowing respondents to trade travel time and cost differences against re-timing of their departure and/or arrival times, then many of these aforementioned issues could be avoided or at least reduced (cf. *14*). While there are many VTT studies that have analyzed the impact of trip rescheduling, most appraisal VTT measures for national or regional infrastructure projects are estimated without taking into consideration the possibility of trip rescheduling (*15*). In this context, it can be questioned whether such approaches, especially for the studies which rely on simple time-money trade-offs, could avoid or reduce any potential bias on the VTT estimates that might result from unaccounted for travel budget impacts.

This study represents a key step for extending knowledge of the impact of budget constraints. Future extensions to the simulation work would include varying number of attribute and alternatives, enabling multiple budget constraints and different decision strategies dealing with budget constraints, and improving realism in the assumption of budget constraints.

Acknowledgments

The first author acknowledges financial support by the Economic and Social Research Council [Grant ES/J500215/1]. The second author acknowledges financial support by the European Research Council through the consolidator grant 615596-DECISIONS.

Author Contributions

The authors confirm contribution to the paper as follows: study conception and design: Tjiong, Dekker, Hess; generation of simulated data: Tjiong; analysis and interpretation of results: all authors; draft manuscript preparation: all authors. All authors reviewed the results and approved the final version of the manuscript.

References

1. Manski, C. F. The Structure of Random Utility Models. *Theory and Decision*, Vol. 8, No. 3, 1977, pp. 229–254.
2. Martínez, F., F. Aguila, and R. Hurtubia. The Constrained Multinomial Logit: A Semi-Compensatory Choice Model. *Transportation Research Part B: Methodological*, Vol. 43, No. 3, 2009, pp. 365–377.
3. Swait, J., and M. Ben-Akiva. Incorporating Random Constraints in Discrete Models of Choice Set Generation. *Transportation Research Part B: Methodological*, Vol. 21, No. 2, 1987, pp. 91–102.
4. Hess, S., A. Daly, T. Dekker, M. O. Cabral, and R. Batley. A Framework for Capturing Heterogeneity, Heteroskedasticity, Non-Linearity, Reference Dependence and Design Artefacts in Value of Time Research. *Transportation Research Part B: Methodological*, Vol. 96, 2017, pp. 126–149.
5. DeSerpa, A. A Theory of the Economics of Time. *The Economic Journal*, Vol. 81, No. 324, 1971, pp. 828–846.
6. Ahmed, A., and P. Stopher. Seventy Minutes Plus or Minus 10—A Review of Travel Time Budget Studies. *Transport Reviews*, Vol. 34, No. 5, 2014, pp. 607–625.
7. Williams, H., and J. de. D. Ortúzar. Behavioural Travel Theories, Model Specification and the Response Error Problem. PTRC Summer Annual Meeting, University of Warwick, Coventry, England, United Kingdom, 1979.
8. Cantillo, V., and J. de. D. Ortúzar. A Semi-Compensatory Discrete Choice Model with Explicit Attribute Thresholds of Perception. *Transportation Research Part B: Methodological*, Vol. 39, No. 7, 2005, pp. 641–657.
9. Li, L., W. Adamowicz, and J. Swait. The Effect of Choice Set Misspecification on Welfare Measures in Random Utility Models. *Resource and Energy Economics*, Vol. 42, 2015, pp. 71–92.
10. Swait, J. A Non-Compensatory Choice Model Incorporating Attribute Cutoffs. *Transportation Research Part B: Methodological*, Vol. 35, No. 10, 2001, pp. 903–928.
11. Hess, S., A. Daly, and M. Börjesson. A Critical Appraisal of the Use of Simple Time-Money Trade-Offs for Appraisal Value of Time Measures. Presented at 96th Annual Meeting of the Transportation Research Board, Washington, D.C., 2016.
12. Hess, S., and K. E. Train. Recovery of Inter-and Intra-Personal Heterogeneity Using Mixed Logit Models. *Transportation Research Part B: Methodological*, Vol. 45, No. 7, 2011, pp. 973–990.
13. Kontoleon, A., and M. Yabe. Assessing the Impacts of Alternative 'Opt-Out' Formats in Choice Experiment Studies: Consumer Preferences for Genetically Modified Content and Production Information in Food. *Journal of Agricultural Policy and Resources*, Vol. 5, No. 1, 2003, pp. 1–43.
14. Hess, S., J. W. Polak, A. Daly, and G. Hyman. Flexible Substitution Patterns in Models of Mode and Time of Day Choice: New Evidence from the UK and the Netherlands. *Transportation*, Vol. 34, No. 2, 2007, pp. 213–238.
15. de Jong, G. C., and M. C. Bliemer. On Including Travel Time Reliability of Road Traffic in Appraisal. *Transportation Research Part A: Policy and Practice*, Vol. 73, 2015, pp. 80–95.

The Standing Committee on Transportation Demand Forecasting (ADB40) peer-reviewed this paper (18-05701).

Article

TRR
JOURNAL OF THE TRANSPORTATION RESEARCH BOARD

Transportation Research Record
2018, Vol. 2672(49) 146–157

DOI: 10.1177/0361198118798970
journals.sagepub.com/home/trr
SAGE

Modeling Framework and Implementation of Activity- and Agent-Based Simulation: An Application to the Greater Boston Area

Isabel Viegas de Lima[1], Mazen Danaf[1], Arun Akkinepally[1], Carlos Lima De Azevedo[1], and Moshe Ben-Akiva[1]

Abstract
This paper presents a utility-maximizing approach to agent-based modeling with an application to the Greater Boston Area (GBA). It leverages day activity schedules (DAS) to create a framework for representing travel demand in an individual's day. DAS are composed of a sequence of stops that make up home-based tours with activity purposes, intermediate stops, and subtours. The framework introduced in this paper includes three levels: (1) the Day Pattern Level, which determines if an individual will travel and, if so, what types of primary activities and intermediate stops they will do; (2) the Tour Level, which models the mode, destination, and time-of-day of the different primary activities; and (3) the Intermediate Stop Level, which generates intermediate stops. The models are estimated for the GBA using the 2010 Massachusetts Travel Survey (MTS). They are then implemented in SimMobility, the agent-based, activity-based, multimodal simulator. It run in a microsimulation using a Synthetic Population. Produced results are consistent with the MTS. Compared with similar activity-based approaches, the proposed framework allows for more flexibility in modeling a wide range of activity and travel patterns.

Transportation modeling has played a key role in how planners and politicians make decisions for transport networks and cities. With the increase in available data, the complexity of the designed interventions is growing. To complement and inform these decisions, transportation models must develop to reflect the multidimensional scope of the impact of such interventions and the individualized responses to them. Activity-based models, such as the one presented in this paper, model travelers' behaviors as a sequence of trips originating and ending at home, known as a tour. In addition, many agent-based modeling (ABM) approaches leverage personalized daily schedules for each individual to represent their behavior throughout a day. The generated day activity schedules (DAS) are composed of a sequence of activities, classified as primary activities, intermediate stops, and subtours. In turn, they are used to extract aggregate demand metrics and to simulate travel behavior. By using a variation of the DAS (*1*) approach—a sequence of discrete choice models following utility maximization—we implement a framework that models different travel patterns and can model impacts of behavioral interventions.

In the applications of activity-based travel demand models, Rasouli and Timmermans (*2*) identified three different approaches: (1) constraints-based models; (2) utility-maximizing models and more recently, (3) computational process models. These three approaches differ mainly in the way individual and household activity travel patterns are modeled.

Constraints-based models are the earliest type of activity-based models. They are only designed to check whether any given activity program is feasible given spatial and temporal constraints rather than to predict travel patterns. The constraints in these models originate from Hägerstrand's (*3*) formulation of "time geography," which classifies these constraints into three categories: capability constraints; coupling constraints; and authority constraints. Besides not being able to predict activity travel patterns, these models suffer from major limitations, including the unrealistic assumption of isotropic conditions (*4*), and more importantly, their inability to deal with uncertainty as well as spatial and temporal variability.

[1]Massachusetts Institute of Technology, Cambridge, MA

Corresponding Author:
Address correspondence to Isabel Viegas de Lima: iviegas@mit.edu

On the other hand, utility-maximizing models utilize econometric models—mainly discrete choice models—to model household and individuals' travel. In the earlier applications of these models, Adler and Ben-Akiva (*5*) and Recker and McNally (*6, 7*) proposed a multinomial logit (MNL) model to represent an individual's choice of the optimal travel pattern, defined in terms of the tours traveled on a given day and the number of stops made within each tour.

This approach was further developed by Ben-Akiva and Bowman (*1, 8–12*) who proposed the daily activity schedule model. It utilizes nested logit (NL) models to represent different travel choices as a multidimensional choice with shared unobserved elements. The topmost choice within this approach is an activity pattern choice, in which the primary activity, the primary tour type, and the number and purpose of secondary tours are modeled. The number of secondary tours is restricted between zero, one, or two or more tours. Lower levels—which are conditional on the activity pattern choice—include time-of-day choice for the primary tour, mode and destination choice for the primary tour, time-of-day choice for the secondary tours, and mode and destination choice for the secondary tours. This approach also has a coarse representation of tour time-of-day decisions, which is limited to sixteen alternatives representing different combinations of four time periods: a.m. peak hour, midday, p.m. peak hour, and other. DaySim, the extension to this model implemented in Sacramento, allows for greater variability by estimating tour types and numbers separately (*13*).

The Prism-Constrained Activity-Travel Simulator (PCATS) (*14*) is another utility-based model which utilizes Hägerstrand's time-space prisms. It divides the day into blocked periods in which individuals are engage in fixed activities at specific times and locations. A two-tier nested logit model is used for open periods, whereby individuals first choose between an "in-home activity," an "activity at or near the location of the next fixed activity," and a "general out-of-home activity." Similarly, mode and destination choice models and activity duration models are used to develop the full patterns.

A similar approach is adopted by Bhat et al. (*15*) in the Comprehensive Econometric Micro-simulator for Daily Activity-travel Patterns (CEMDAP), which divides the population into workers and non-workers. The activity patterns of workers are divided into five periods with regards to work while the patterns for non-workers consist of a sequence of home-based tours. The simulator uses a generation-allocation model to determine activity participation which includes: (1) work and school activities; (2) children's travel needs; and (3) independent activities for personal and household needs. Afterwards, a scheduling model is applied to determine the sequencing of the activities produced by the generation-allocation model.

The Comprehensive Utility-based System of Travel Options Modelling (CUSTOM) (*16*) has recently been developed as an approach that considers time budgets for activity scheduling. In the prototype application, trips purpose, mode, location, and time allocation decisions are made sequentially for workers. After each activity, an individual decides to either go home permanently, go home temporarily, or perform another activity.

Several activity-based approaches have also been implemented in different regions of North America. Coordinated Travel-Regional Activity Modeling Platform (CT-RAMP) (*17*), differentiates between non-mandatory and mandatory activities, such as work and education. It determines the frequency of mandatory tours and schedules them, and then uses the remaining time to schedule additional tours. TASHA (*18*), active in the Greater Toronto-Hamilton Area, generates individual projects for each individual including a start time and duration. It schedules household members' days by adjusting the predetermined start times.

More recently, computational process models have been utilized in developing heuristics that represent activity and travel patterns. Some examples of these applications are SCHEDULER (*19*), AMOS (*20, 21*), and ALBATROSS (*22–24*). On the other hand, MATSIM (*25*) draws agents' activity chains conditional on their socio-demographics from the observed distribution in the population. ActivitySim, the open-source agent-based simulator developed by Los Alamos National Laboratory, uses optimization to choose the best schedule at any point based on utilities, priority functions, and location functions (*26*). Other models assume a planning horizon and scheduled and unscheduled trips to be generated, such as ADAPTS (*27*). However, most of these models lack behavioral interpretability compared with utility-maximizing models. For a detailed description of these three approaches, we refer the reader to Rasouli and Timmermans (*2*).

Even though utility-based approaches—particularly the DAS approach—allow for high flexibility in modeling travel patterns, their applications were sometimes restrictive. For example, the approach in Ben-Akiva and Bowman (*1*) did not account for all possible activity patterns; only a finite set of predefined patterns was considered, such as home-work-home (hwh), home-other-home (hoh), etc. Later applications of the model in DaySim utilize a similar approach to this paper. CEMDAP (*15*), on the other hand, divides workers' time periods into: (1) before work; (2) the to-work commute; (3) during work; (4) from work; and the (5) after work, imposing scheduling behavioral assumptions in how an individual's pattern is built. In addition, these approaches suffered from a tradeoff between the computational complexity of the model and its ability to model a wider range of activity patterns.

The remainder of this paper is structured as follows: the next section presents the modeling framework used and explains its different components. An application of this framework to the Greater Boston Area (GBA) and the illustration of how this framework was implemented in the activity-based simulator, SimMobility (*28*) is presented in the next section, followed by a presentation of the estimated significant variables for key models and model groups and simulation results. The final section presents the limitations and future extensions and concludes the paper.

Model Framework

Model Overview

The activity-based modeling framework presented was proposed by Siyu (*29*) for Singapore. The framework leverages utility-maximization to create a complex variety of DAS. It is built with a sequence of discrete choice models, which are functions of different personal characteristics and network parameters.

The framework for each individual's day schedule is divided into three sections: (1) the Day Pattern Level; (2) the Tour Level; and (3) the Intermediate Stop Level. Each of these levels is conditional on the ones above and they are modeled as a sequence of discrete choice models, linked via inclusive values.

The *Day Pattern Level* determines if the individual will make tours, what activity purposes those tours will have, and what types of intermediate stops those tours may contain. Furthermore, it determines how many tours of each chosen tour purpose an individual will execute.

The *Tour Level* is then performed for each chosen tour. It determines the mode, the destination, and the start and end time of each activity. For work tours, a work-based subtour—originating and ending at work during the allocated duration of the work activity—may be modeled. The mode, destination, and time-of-day of those subtours are modeled as well.

Finally, the *Intermediate Stop Level* generates stops for trips to and from the primary activity. Each intermediate stop has a purpose, a mode, a destination, and a time-of-day.

Inclusive values from lower models are included in higher models to represent the nested nature of the decision-making process, in which higher level decisions are affected by the expected utilities of lower level ones. The general framework presented above is summarized in Figure 1. All models are further discussed and can be estimated using travel diaries with individual and household characteristics.

The output of the sequence of models is a DAS, which includes a sequence of stops that make up tours. Each tour has a primary activity and every stop has an arrival time, departure time, purpose, and mode taken to it.

Model Components

This section explains each of the models presented in Figure 1 in detail: their structure, inputs, and outputs.

Day Pattern Level

Day Pattern Travel. The Day Pattern Travel model determines if the individual makes any tours in a given day. The model takes the form of a binary logit (BL), where the choices are either to travel or to stay at home. It takes personal characteristics of the individual and the inclusive value of the Day Pattern Tours model, explained next, as inputs. A decision to travel will lead to Day Pattern Tours model.

Day Pattern Tours. The Day Pattern Tours model determines which activity purpose(s) the individual will pursue as their primary purpose(s) of the tour(s) they make. The model is a MNL where the choice set includes a combination of each available activity purpose. The limit on the number of activity purposes in the combinations is empirically determined. Each choice's utility includes pattern-specific parameters, such as purpose-specific constants; individual-specific parameters such as sociodemographics; and inclusive values from the tour-level models of each activity type. Furthermore, the formulation includes purpose-interaction parameters, which capture the added utility of performing two or more purposes in combination. Work tours are only available for individuals who are employed and education tours are only available for individuals who are enrolled in an educational institution.

Day Pattern Stops. Similar to the Day Pattern Tours model, the Day Patten Stops models the choice between the allowed combination of purposes, yet for for stops performed before or after the primary activity of the tour. It includes the same variables as the Day Pattern Tours model, excluding the inclusive values. The availabilities are also determined in the same manner. The output of the Day Pattern Stops model is used to determine the availability of stops in the Intermediate Stop Generation Model, explained below.

Number of Tours. Using the output of the Day Pattern Tours model, the Number of Tours is then determined for each of the selected tour purposes. The models are

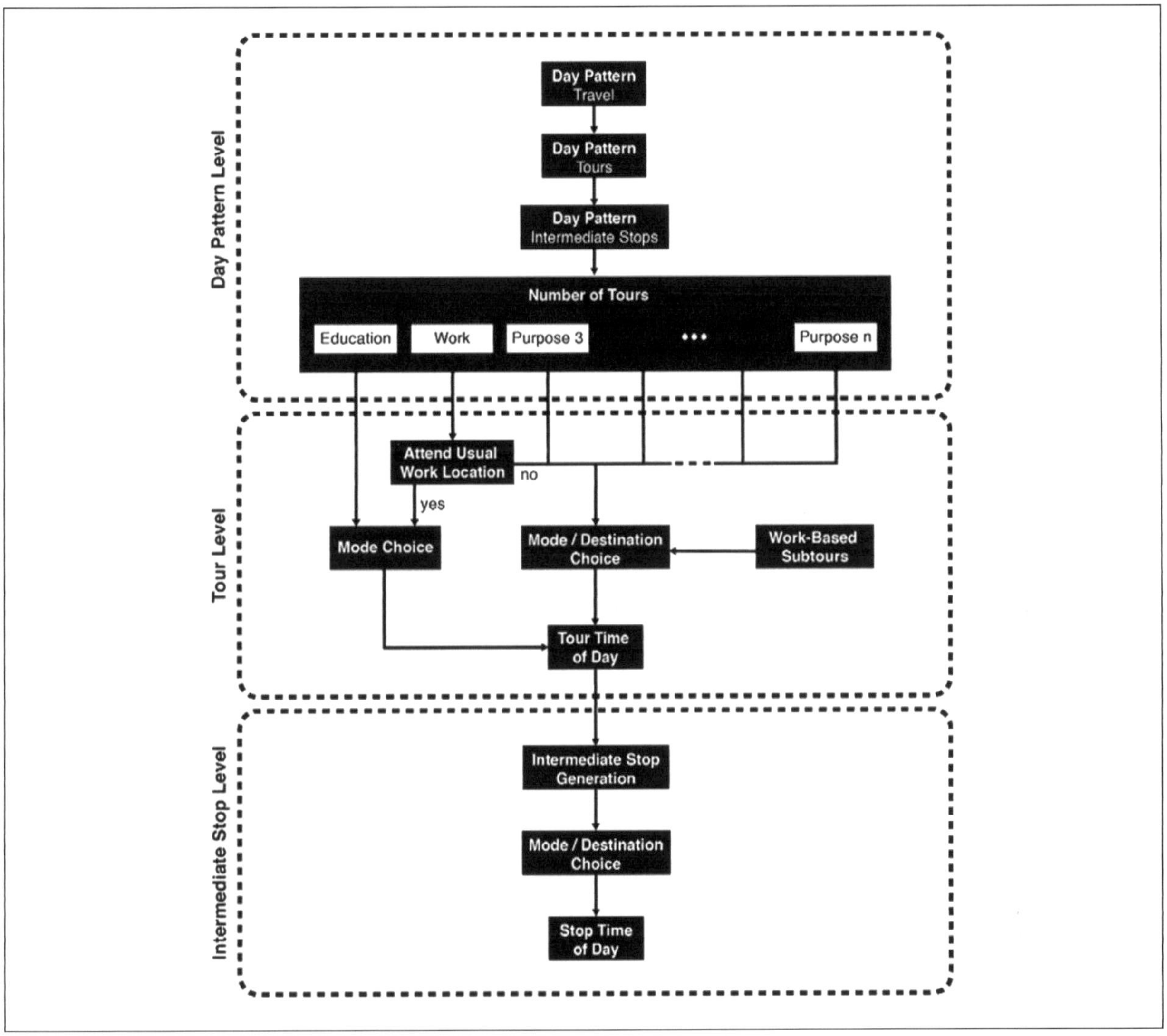

Figure 1. Model framework schematic.

either BL—if the choice is in between one or two tours—or MNL if the choice includes more tours. The number of tours in the choice set for each purpose is determined from the data. Variables in the model include individual characteristics and inclusive values from tour-level models.

Tour Level. Each of the tours, determined after the Number of Tours model, is processed through the Tour Level and then the Intermediate Stop Level.

Usual Work. The Usual Work model is performed for work tours only. It determines whether or not the individual is going to their usual work location. The decision is thus modeled as a BL. The model includes individual characteristics as well as the inclusive value of either the Mode model or the Mode and Destination model for work. If the usual location is chosen, then the Mode is modeled; otherwise, the Mode and Destination model is used.

Tour Mode. The Tour Mode models determine the mode for work tours with a usual location and any other tours which have fixed locations, such as education. The mode choice is modeled as an MNL (or NL) with the choice between the available mode alternatives. This structure allows for the inclusion of new modes, such as on-demand mobility. The models include individual parameters as well as travel time and cost. Alternative availabilities are determined based on distance thresholds for walking and biking, transit availability between different origin-destination pairs, and household characteristics, such as owning a car or bicycle. The usual location of the tour is determined in the synthetic population, such as a worker's work location or a student's school location.

Tour Mode and Destination. Mode and destination are modeled jointly for any tour that has a non-fixed destination. The final choice set includes all combinations of modes and destinations. Destinations are usually the traffic analysis zones (TAZ) in the modeled region. The model includes travel times and costs for each mode when relevant. For each possible destination, size variables, such as area, employment, population, and central business district (CBD) dummy, are included. They are incorporated using the aggregate spatial method outlined by Ben-Akiva and Watanatada (*30*). The model is estimated as a MNL or NL, where the order of nesting is determined empirically.

Tour Time-of-Day. Once the tour destination and mode are chosen, the Tour Time-of-Day models determine the start time and end time of the primary activity. The continuous 24-hour day is discretized into 48 half-hour segments. The full choice set includes all of the possible combinations of start and end times, leading to 1,176 alternatives. The alternative specific constants follow a continuous and cyclical form outlined by Ben-Akiva and Abou-Zeid (*31*) using a trigonometric utility functional form. Activity duration, travel times, and travel costs are also included. If there is no available time, the tour is removed. The number and duration of the time of day segments can be varied both in the simulation and in the model estimation to achieve accurate results. Ben-Akiva and Abou-Zeid, for example, use 35 time intervals consisting of half-hours in peak hours and longer intervals in off-peak hours. For simulation, departure times are randomly generated within the 30-minute interval selected. Logsums from the Tour Time-of-Day model can be included in the Tour Mode and Destination Model to capture the nested structure of the choice.

Work-Based Subtour Generation. Subtours may be generated for work tours. The Work-based Subtour Generation is modeled as a quit/no-quit BL. While the individual chooses to not quit, they continue to schedule subtours. Individual characteristics and the tour mode are parameters in the model.

Work-based Subtour Mode and Destination. The Work-based Subtour Mode and Destination is modeled in the same manner as the primary activity Mode and Destination model. The modes, however, are limited based on the hierarchy described by Siyu (*29*), dictated by the work tour mode. In this hierarchy, an individual can only take the mode of their tour or others more readily available. Therefore, a person may not, for example, take a personal vehicle if they did not drive to work.

Work-Based Subtour Time-of-Day. The Work-based Subtour Time-of-Day is modeled in the same manner as the Tour Time-of-Day. However, the availabilities of alternatives are determined by the duration of the primary work activity.

Intermediate Stop Level. In the Intermediate Stop Level, tours are divided into the inbound and outbound parts—before and after arriving at the primary activity respectively—and each of these parts is processed separately.

Intermediate Stop Generation. The Intermediate Stop Generation model is an NL quit/no-quit model, whereby a no-quit choice results in a new intermediate stop. While one of the nests includes only the quit option, the other includes the other activity purposes. Individual characteristics, tour purpose, and remaining time window—determined based on the start or end time of the primary activity, and preceding or successive stops—are included as variables in the model. Availability of stop purposes is determined based on the outputs of the Day Pattern Stops model. Stops are scheduled sequentially.

Intermediate Stop Mode and Destination. The Intermediate Stop Mode and Destination is modeled like the Tour Mode and Destination. For inbound and outbound stops, the travel time and travel cost being considered are those between the planned stop and the one succeeding or preceding it respectively. The mode availability follows the same hierarchy as that described in the Work-based Subtour Mode and Destination model.

Intermediate Stop Time-of-Day. Unlike the tour and subtour activities, intermediate stops are bound by adjacent activity start and end times. Therefore, their choice set is constructed out of the 48 half-hour time periods. The Intermediate Stop Time-of-Day still has the trigonometric functional forms, but are interacted for the different stop purposes. Travel time and travel cost are also included. The model is MNL and alternatives are defined similar to the window for generation.

Estimation and Simulation in the Greater Boston Area

The modeling framework defined above is applied to the GBA. The region has a population of 4.5 million individuals within 1.6 million households. Data from the 2010 Massachusetts Travel Survey (MTS) (*32*) and skim matrices for car and transit travel times and costs, provided by Boston's Central Transportation Planning Staff (CTPS), are used to estimate the models. The MTS includes activity diaries for 33,000 individuals belonging to 15,000 households. The data were collected between June 2010 and November 2011. Individuals were asked to fill out all activities they performed in a designated weekday (24 hours, Monday to Friday) in chronological order, and to provide the activity location, the transport mode used to arrive at this location, the arrival and departure times, and accompanying household or non-household members. The survey also collected individual and household characteristics for participants.

Estimation for the models was done using PythonBiogeme (*33*). It was performed in two phases. First, all models were estimated using only the processed MTS and CTPS data. Before the second round of estimation, inclusive values from the lower level models were calculated. Models that included these were re-estimated in the second round. This estimation procedure allowed for partial simulation of the topmost levels to determine performance before continuing to lower levels.

Data Preparation

Tours, Subtours, and Intermediate Stops. MTS trip diaries were processed to identify tours, subtours, and intermediate stops. Activities in the MTS are classified into work, education, shopping, personal, recreation, or escort activities. The tour primary activity is determined based on the following hierarchy: (1) work; (2) education; (3) personal activities; (4) shopping; (5) recreation; and (6) escort. For full-time students, education is prioritized over work. If a tour includes two or more highest-priority activities of the same type, the primary activity is the one with the longest duration. All other activities within the tour are classified as intermediate stops. Work-based subtours are defined as trip chains starting and ending at the work location. Work activities before and after subtours are merged and considered as a single activity.

Day Patterns Models. Sixty-four activity patterns can be defined with the possible combinations of the six activity purposes. For example, a possible activity pattern is "work, recreation, and shopping." After examining the MTS data, it was concluded that most individuals participated in at most four different primary activity purposes in a day, reducing the number of possible activity patterns from 64 to 56. Similarly, 56 activity patterns are identified from the data for intermediate stops, representing the different combinations of activities an individual engages in during the stops. For each individual, the number of tours for each activity type is determined. It was observed from the MTS data that individuals perform at most two tours per activity type per day—except for recreational tours, for which the maximum was three.

Tour Mode and Mode Destination Models. The tour mode is determined based on a hierarchy similar to that defined by Siyu (*29*): public transit, carpooling, driving alone, on-demand modes, motorcycle, biking and bike-sharing, and walking. Trips and tours with rare modes (e.g., ambulance, boat, airplane, etc.) are excluded. Walking, biking, driving alone, carpooling with two people, carpooling with three or more people, and public transit were empirically chosen as the available travel modes.

In the Mode and Mode and Destination choice models, walk and bike availability are determined based on the maximum walking and biking distances observed in the MTS: 5 and 15 miles respectively. Biking is available only for individuals whose households own at least one bike. Driving alone is available only for individuals who have a driver's license and at least one car in their household. Public transit availability is obtained from the CTPS skim matrices. Carpooling is assumed to be available to all individuals.

Time-of-Day Models. In Time-of-Day models, time availabilities are assigned in the order of activity priority. For example, a primary work activity is scheduled first, and all possible arrival and departure times are available for that activity. The second most important primary activity has all the times available, except those that are occupied by the primary work activity, and so on. Time availability for intermediate stops and work-based subtours is defined as specified in the framework.

Simulation Implementation

The models estimated in this project follow the structure designated by Siyu (*29*) for integration into SimMobility. The software includes a Long-term Level, which deals with vehicle ownership and land-use; a Mid-term Level, which deals with daily travel patterns; and a Short-term Level, which is a microscopic traffic simulator (*28*).

The Mid-term Level of SimMobility is composed of three components: Pre-day, Within-day, and Supply (*34*). The Pre-day, which is presented in this paper, provides the initial demand for the simulator. The Supply Simulator contains the road and transit network and the capacities for each mode. It is interacted with the Within-day, which contains models that allow users to change their defined route and change their mode according to real-time network conditions, accounting for mode choice as a function of a set departure time.

Developed in C++, the simulator leverages performance improvements by relying on parallel processing and distributed computing. The lightweight, embeddable scripting language Lua is used to implement the above specifications (*35*). Implementation in Lua, combined with the model framework's treatment of activity purposes, allows for the easy addition of new modes and purposes. Furthermore, it makes implementing estimated models for different cities straightforward.

Synthetic Population. As an agent-based simulator, SimMobility requires detailed individual information as an input. Fournier and Christofa (*36*) describe the process through which the synthetic population is generated. Several sources are used to produce and validate the synthesis:

1. Decennial data from the U.S. Census Bureau (2010)
2. 5-year American Community Survey (2010)
3. MTS (2010)
4. Public Use Microdata Samples (PUMS, 2011)

While the census data (1 and 2) are aggregated totals, the PUMS and MTS data are disaggregated samples. A variety of methods were also used to generate the Synthetic Population, including different seeding approaches, iterative proportional fitting, integerization, iterative proportional updating, and Monte Carlo sampling. The developed synthetic population includes 1.7 million households with over 4.5 million individuals, which are inputs to SimMobility.

Estimation and Simulation Results

Estimation Results

Twenty-nine models are estimated for the GBA implementation. The significant values for the key models and model groups are discussed in this sub-section. For the Day Pattern Travel model, age, gender, education and employment, and inclusive values were significant. For the Day Pattern Tours and Day Pattern Stops models, in addition to the aforementioned variables, owning a transit pass, having a driver's license, income, and purpose-interaction terms were also significant. In the majority of the Number of Tours models, the purpose-specific inclusive value of was significant. For the Mode and Mode-Destination models, travel time and costs, as well as zone-specific population, area, and employment tended to be significant. While each Time-of-Day model had a different number of trigonometric components, travel time and cost were significant. Finally, for Intermediate Stop Generation, the remaining time window in the day, combined with how many stops were already performed for that half-tour, was significant. The tour purpose was also significant. For full estimation results, model specifications, and Lua implementation files, please refer to the GitHub repository (*37*).

Simulation Results

The value of the ABM approach is twofold: it can be used to get aggregate metrics of the demand and to simulate demand on a network. While the latter is outside the scope of this paper, the former is presented below. For all figures, simulated results to the MTS are compared.

The simulation was done in two phases. The calculation of the inclusive values for all individuals was done on 20 cores within 21 hours and the simulation of the DAS itself was done on 20 cores in 23 hours. In all, 5,096,606 tours and 10,548,458 stops—including primary activities, subtours, and intermediate stops—were simulated for 4,142,068 people, of which 720,924 did not travel. Note that, since SimMobility allows for parallelization and distributed processing, the run-time can be improved significantly by allocating more cores, or by using High Performance Computing (HPC).

Figure 2*a* shows the split of the tour purposes. The models are overall consistent with the MTS. While work is slightly overestimated, personal, recreation, and escort are slightly underestimated. Education, however, is overestimated by 5%. This can be explained by the underrepresentation of students in the MTS. Of the existing students in the MTS, 90.8% do at least one education tour in a day. Therefore, the Synthetic Population leads to more education tours.

Similarly, Figure 2*b* shows the comparison between the modal split of the MTS and the simulated DAS. Overall, the framework properly models the tour modes. Driving alone is underestimated by 5% while two-person carpool and three or more-person carpool are both overestimated by only 1%. Public transit was as prevalent in the DAS as in the MTS. 10% of the simulated tours have walking as the primary mode, while only 7% of the tours in the MTS are walking tours. Biking, although selected for 1,449 tours, is underestimated. The differing mode share is a result of the differing purpose share: since each tour purpose has its own mode (or mode and destination) choice model, the produced modal splits are different.

Figure 2*c* displays how many trips of a given purpose each individual performs. It shows that the Synthetic Population and models, despite reproducing modal trends well, are slightly underrepresenting work, shop, and recreational tours. More importantly, it shows that traveling is underrepresented. This corresponds to the difference of 1.12 tours per capita in the simulation, versus 1.21 tours per capita observed in the MTS.

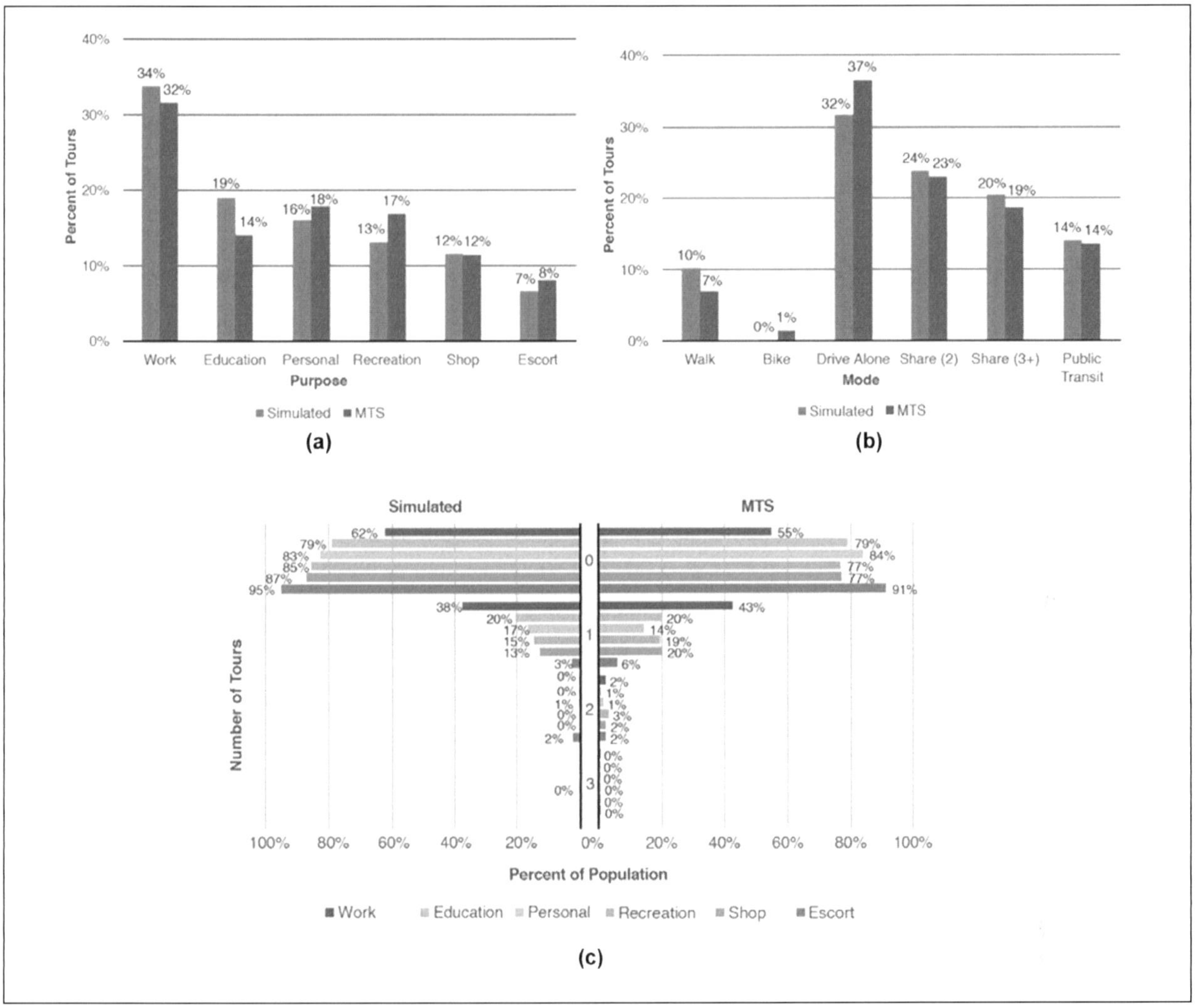

Figure 2. Tours by (*a*) purpose; (*b*) mode; and (*c*) number of tours per person per mode.

Figure 3 shows the times at which people travel to and from their primary activity. The temporal peaks for work and education are clearly visible in both the simulation and the MTS. It has been broken down by activity purpose. As mentioned, education is higher in the simulated results because of the underrepresentation of students in the MTS. It is also noticeable that, as a consequence of the hierarchy of tour purposes for simulation, lower tours are being pushed to later in the day, such as recreation and escort.

Finally, Figure 4 shows the special distribution of the tour destinations. As expected, the CBD—located at the center of the figure—has the highest tour density.

Overall, the framework and the Synthetic Population successfully replicate the major trends in travel demand in the GBA.

Conclusion and Discussion

This paper presents an application of an updated approach for modeling activity patterns in an activity-based, agent-based simulator. Compared with the existing approaches, this framework allows for more flexibility in representing different day patterns. In line with the DaySim modifications to the Bowman structure, the proposed structure allows for a larger set of combinations of different tour purposes and numbers of tours. By including the interaction terms at the Day Pattern Level in the Day Pattern Tours and Day Pattern Stops models, the model captures the added utility of performing different activities in combination—a behavioral trait often ignored. This is further enforced in the Intermediate Stop Generation model, which takes in the primary purpose as one of the inputs.

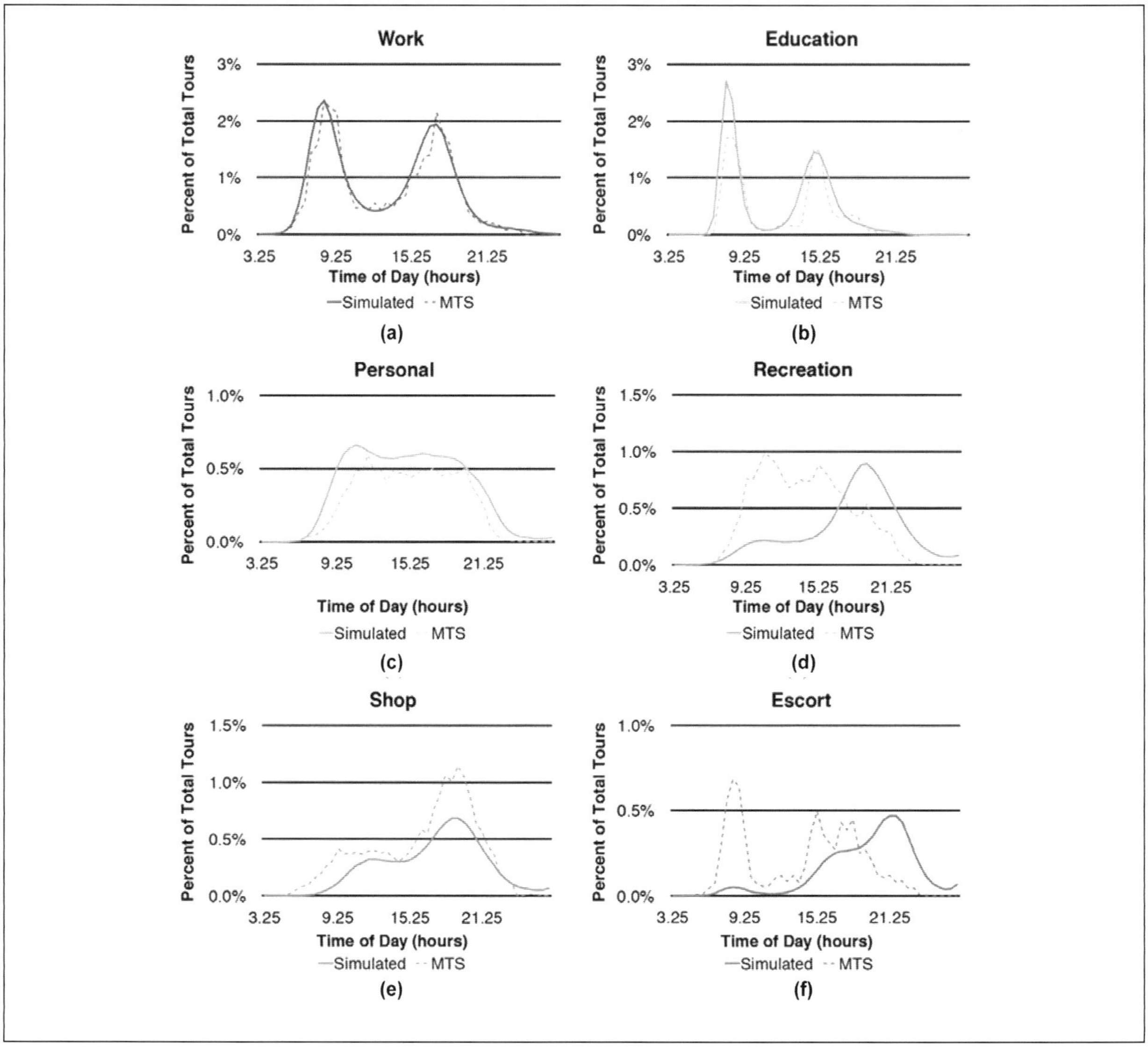

Figure 3. Tours by time-of-day for (*a*) work; (*b*) education; (*c*) personal; (*d*) recreation; (*e*) shop; and (*f*) escort.

Furthermore, the inclusion of the trigonometric functional forms in the Time-of-Day models allows for smoother scheduling of activities. This method was chosen over the use of multiple discrete-continuous models, which would have been combined with the Day Pattern Level as generation models. This Time-of-Day formulation can account for time periods where travel data are not available, and can be modified to model time periods of unequal length. More importantly, the structure captures the cyclical properties of time preferences (*31*). The current framework allows for the effective implementation of interaction terms between tour purposes to create diversified activity patterns and for the consideration of the time-of-day dependence of the value of an activity duration.

However, there are several improvements that can be made to this application. First of all, the data used in this paper were collected in 2010–2011, and thus are relatively old. Vehicle ownership, as well as modal shares in MTS are different from those observed today due to the pervasiveness of on-demand services, such as Uber, Lyft, ZipCar, and so forth. Additionally, the MTS lacks information on weekend travel and undersamples university students, even though they represent a significant fraction of the GBA population.

To correct for all of these data limitations, a data collection effort is planned in the GBA. During the data collection, 500 subjects are expected to validate their trips for multiple weeks and fill out the stated preference (SP) surveys using the Future Mobility Sensing (FMS)

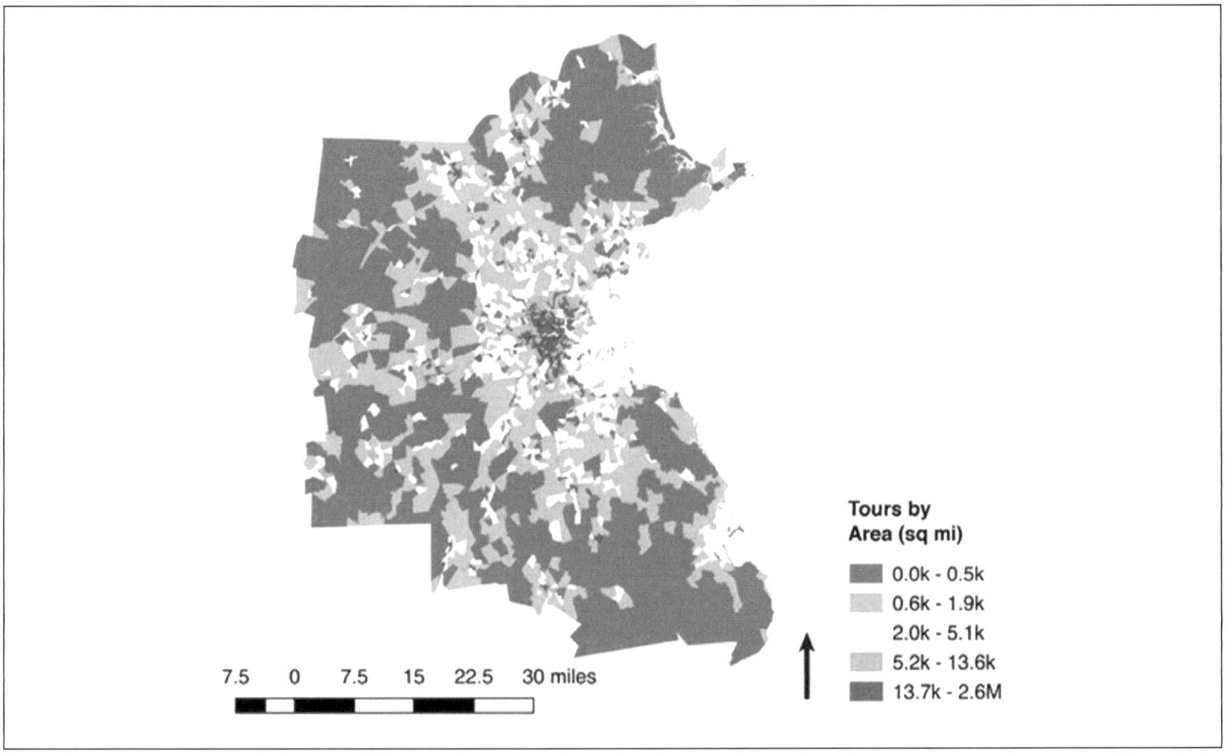

Figure 4. Tour destinations per TAZ.

app-based survey platform (*38*). Subjects will install FMS and fill out a pre-survey, daily validation surveys, and daily SP surveys for three consecutive weeks.

Research under way is focusing on calibrating the DAS models using traffic counts. In addition, a Within-day framework is developed to represent how agents change their Pre-day DAS as the simulation advances due to network changes, delays, and unforeseen events. This framework currently includes route-choice and mode-choice models, which account for the potentially incorrect behavioral assumption that individuals choose their activity mode and destination before the start and end time. These models are being extended to account for re-scheduling activities and re-planning the rest of the day. Currently, the Within-day framework is also useful to model the effect of real-time information (e.g., travel advisors) and incidents.

In this framework, escort activities are modeled separately without accounting for any household interactions. These activities are supposed to capture pick-ups, drop-offs, and trips that are performed to accompany someone. Research under way is focusing on modeling social interactions particularly through joint household activities. A household DAS framework to be integrated into the Day Pattern Level is being proposed. The sequential models in this framework represent whether or not the household will have any joint activities during the day, what types of activities, the number of tours associated with each of these activities, which household members are participating in these tours, and whether or not the participating members travel together (e.g., carpooling). The output of the joint household models will then feed into the individual Day Pattern Level models.

Overall, the presented and implemented model benefits from its framework and implementation environment. The model allows for complex activity patterns and activity purpose interaction, and takes into account the time-dependence of the value of duration. The simulation software it is implemented in, SimMobility, is streamlined with Lua files, which allow for the facilitated addition of new purposes and modes.

Acknowledgments

This work is part of a greater project, Sustainable Travel Incentives with Prediction, Optimization, and Personalization (Tripod), sponsored by the U.S. Energy Department's Advanced Research Projects Agency—Energy (ARPA-E). It was awarded by ARPA-E's one of the latest programs, the Traveler Response Architecture using Novel Signaling for Network Efficiency in Transportation (TRANSNET). The Central Transportation Planning Staff has also contributed with access to data for this project. Assel Dimitriyeva contributed to the estimation of the framework. Kakali Basak, Nishant Kumar, and Balakumar Marimuthu have helped with implementation into SimMobility.

Author Contributions

The authors confirm contribution to the paper as follows: study conception and design: CLDA, MBA; data processing: MD; analysis and interpretation of results: IVDL, MD, CLDA, AA; draft manuscript preparation: IVDL. All authors reviewed the results and approved the final version of the manuscript.

References

1. Ben-Akiva, M. E., and J. L. Bowman. Activity Based Travel Demand Model Systems. In *Equilibrium and Advanced Transportation Modeling* (Marcotte, P., and S. Nguyen, eds.), Kluwer, Montreal, Canada, 1998, pp. 27–46.
2. Rasouli, S., and H. Timmermans. Activity-based Models of Travel Demand: Promises, Progress and Prospects. *International Journal of Urban Sciences*, Vol. 18, No. 1, 2014, pp. 31–60.
3. Hägerstraand, T. What About People in Regional Science? *Papers in Regional Science*, Vol. 24, No. 1, 1970, pp. 7–24.
4. Miller, H. J. Modelling Accessibility Using Space-time Prism Concepts within Geographical Information Systems. *International Journal of Geographical Information System*, Vol. 5, No. 3, 1991, pp. 287–301.
5. Adler, T., and M. Ben-Akiva. A Theoretical and Empirical Model of Trip Chaining Behavior. *Transportation Research B*, Vol. 13, 1979, pp. 243–257.
6. Recker, W. W., M. G. McNally, and G. S. Root. A Model of Complex Travel Behaviour, Part 1, Theoretical Development. *Transportation Research A*, Vol. 20, 1986a, pp. 307–318.
7. Recker, W. W., M. G. McNally, and G. S. Root. A Model of Complex Travel Behaviour, Part 2, An Operational Model. *Transportation Research A*, Vol. 20, 1986b, pp. 319–330.
8. Ben-Akiva, M. E., J. L. Bowman, and D. Gopinath. Travel Demand Model System for the Information Area. *Transportation*, Vol. 25, 1996, pp. 241–266.
9. Bowman, J. L. Activity Based Travel Demand Model System with Daily Activity Schedules. Master of Science thesis in transportation. Massachusetts Institute of Technology, Cambridge, MA, 1995.
10. Bowman, J. L. The Day Activity Schedule Approach to Travel Demand Analysis. PhD thesis. Massachusetts Institute of Technology, Cambridge, MA, 1998.
11. Bowman, J. L., and M. Ben-Akiva. Activity-based Disaggregate Travel Demand Model System with Activity Schedules. *Transportation Research A*, Vol. 35, 2000, pp. 1–28.
12. Bowman, J. L., M. Bradley, Y. Shiftan, T. K. Lawton, and M. E. Ben-Akiva. Demonstration of An Activity-based Model System for Portland. *Proc., 8th World Conference on Transport Research*, Antwerp, Belgium, 1998.
13. Bradley, M., J. L. Bowman, and B. Griesenbeck. SACSIM: An Applied Activity-based Model System with Fine Level Spatial and Temporal Resolution. *Journal of Choice Modelling*, Vol. 3, No. 1, 2010, pp. 5–31.
14. Kitamura, R., and S. Fujii. Two Computational Process Models of Activity-travel Choice. In *Theoretical Foundations of Travel Choice Modelling* (Gärling, T., T. Laitila, and K. Westin eds.), Elsevier, Oxford, Kidlington, England, 1998, pp. 251–279.
15. Bhat, C. R., J. Y. Guo, S. Srinivasan, and A. Sivakumar. A Comprehensive Micro-simulator for Daily Activity-travel Patterns. *Proc., Conference on Progress in Activity-based Models*, EIRASS, Maastricht, the Netherlands, 2004.
16. Habib, K. A Comprehensive Utility based System of activity-Travel scheduling Options Modelling (CUSTOM) for Worker's Daily Activity Scheduling Processes. *Transportamerica A: Transport Science*, Vol. 14, No. 4, 2018, pp. 292–315.
17. Davidson, W., P. Vovsha, J. Freedman, and R. Donnelly. CT-RAMP Family of Activity-Based Models. *Proc., Australasian Transport Research Forum 2010*. Canberra, Australia, 2010.
18. Miller, E., J. Vaughan, D. King, and M. Austin. Implementation of a "Next Generation" Activity-Based Travel Demand Model: The Toronto Case. *Conference of the Transportation Association of Canada*. Charlottetown, PEI, 2015.
19. Gärling, T., K. Brännäs, J. Garvill, R. G. Golledge, S. Gopal, E. Holm, and E. Lindberg. Household Activity Scheduling. *Selected Proceedings of the Fifth World Conference on Transport Research*, Vol. 4. Western Periodicals, Ventura, CA, 1989, pp. 235–248.
20. Pendyala, R. M., R. Kitamura, C. Chen, and E. I. Pas. An Activity-based Microsimulation Analysis of Transportation Control Measures. *Transport Policy*, Vol. 4, 1997, pp. 183–192.
21. Pendyala, R. M., R. Kitamura, and D. V. G. P. Reddy. Application of An Activity Based Travel Demand Model Incorporating a Rule-based Algorithm. *Environment and Planning B*, Vol. 25, 1998, pp. 753–772.
22. Arentze, T. A., and H. J. P. Timmermans. *Albatross, a Learning-based Transportation Oriented Simulation System.* EIRASS, Eindhoven University of Technology, Eindhoven, the Netherlands, 2000.
23. Arentze, T. A., and H. J. P. Timmermans. A Learning-based Transportation Oriented Simulation System. *Transportation Research B*, Vol. 38, 2004, pp. 613–633.
24. Arentze, T. A., and H. J. P. Timmermans. *Albatross V2, A Learning-based Transportation Oriented Simulation System.* EIRASS, Eindhoven University of Technology, Eindhoven, the Netherlands, 2005.
25. Balmer, M., K. Meister, M. Rieser, K. Nagel, and K. W. Axhausen. Agent-based Simulation of Travel Demand: Structure and Computational Performance of MATSim-T. *Paper Presented at the 2nd TRB Conference on Innovations in Travel Modeling*, Portland, OR, 2008.
26. Galli, E., S. Eidenbenz, S. Mniszewski, C. Teuscher, and L. Cuellar. ActivitySim: Large-scale Agent-Based Activity Generation for Infrastructure Simulation. *Agent-Directed Simulation (ADS '09)*. San Diego, CA, 2008.
27. Auld, J., and A. Mohammadian. Activity Planning Processes in the Agent-based Dynamic Activity Planning and

Travel Scheduling (ADAPTS) Model. *Transportation Research Part A*, Vol. 46, 2012, pp. 1386–1403.
28. Adnan, M., F. Pereira, C. Lima de Azevedo, K. Basak, M. Lovric, S. Feliu, Y. Zhu, J. Ferreira, and M. Ben-Akiva. SimMobility: A Multi-scale Integrated Agent-based Simulation Platform. Presented at 95th Annual Meeting of the Transportation Research Board, Washington, D.C., 2015.
29. Siyu, L. Activity-Based Travel Demand Model: Application and Innovation. Doctoral dissertation. National University of Singapore, Singapore, 2015.
30. Ben-Akiva, M., and T. Watanatada. A Continuous Spatial Choice Logit Model. In *Structural Analysis of Discrete Data with Econometric Applications*. The MIT Press, Cambridge, MA, 1981, pp. 320–342.
31. Ben-Akiva, M., and M. Abou-Zeid. Methodological Issues in Modelling Time-of-travel Preferences. *Transportmetrica A: Transport Science*, Vol. 9, No. 9, 2013, pp. 846–859.
32. Massachusetts Department of Transportation. *Massachusetts Travel Survey 2010–2011*. Massachusetts Department of Transportation, Boston, MA, 2012.
33. Bierlaire, M., and M. Fetiarison. Estimation of Discrete Choice Models: Extending BIOGEME. *Proc., 9th Swiss Transport Research Conference (STRC)*, Monte Verità, Switzerland, 2009.
34. Lu, Y., M. Adnan, K. Basak, F. C. Pereira, C. Carrion, V. H. Saber, H. Loganathan, and M. Ben-Akiva. Simmobility Mid-term Simulator: A State of the Art Integrated Agent Based Demand and Supply Model. Presented at 94th Annual Meeting of the Transportation Research Board, Washington, D.C., 2015.
35. Ierusalimschy, R., and L. H. Figueiredo W. Celes. Lua 5.1 the evolution of Lua. *Proc., ACM HOPL III*. 2-1–2-26, 2007.
36. Fournier, N., and E. Cristofa. *An Integration of Population Synthesis Methods for Agent-based Microsimulation*. University of Massachusetts. Amherst, MA, 2017.
37. Viegas, de, and I. Lima. *Greater Boston Area GitHub Repository*, 2017. https://github.com/isabelviegas/simmobility GBApreday
38. Zhao, F., A. Ghorpade, F. C. Pereira, C. Zegras, and M. Ben-Akiva. Quantifying Mobility: Pervasive Technologies for Transport Modeling. In *Adjunct Proceedings of the 2015 ACM International Joint Conference on Pervasive and Ubiquitous Computing and Proceedings of the 2015 ACM International Symposium on Wearable Computers*. ACM, New York City, NY, 2015, pp. 1039–1044

The Standing Committee on Transportation Demand Forecasting (ADB40) peer-reviewed this paper (18-06357).

Article

TRR
JOURNAL OF THE TRANSPORTATION RESEARCH BOARD

Transportation Research Record
2018, Vol. 2672(49) 158–170
© National Academy of Sciences:
Transportation Research Board 2018
Article reuse guidelines:
sagepub.com/journals-permissions
DOI: 10.1177/0361198118799168
journals.sagepub.com/home/trr
SAGE

A Spatial Latent Profile Analysis to Classify Land Uses for Population Synthesis Methods in Travel Demand Forecasting

Elizabeth C. McBride[1], Adam W. Davis[1], and Konstadinos G. Goulias[1]

Abstract

In this paper, a new land use classification method is explored for its utility in explaining travel behavior and as a new dimension in population synthesis for travel demand forecasting. This method is based on latent profile analysis applied to 17 business establishment indicators for each of the more than 20,000 block groups in California. The method reproduces the four types of land use environments (urban, suburban, exurban, and rural) identified in a previous paper, and improves our ability to create a finer-grain geographic classification based on land use. It also offers similar indications about the difference between urban dwellers (that make more trips but travel shorter distances) and rural residents (that make fewer trips but with more vehicle miles traveled).

Population synthesis is the generation of a synthetic population with the goal of matching a real population of interest as closely as possible. It is the first step in activity-based microsimulation models, in which individuals traveling across a network are modeled for an area of interest. Residential self-selection often occurs. For this reason, land use is a key informant that can be added to population synthesis when the aim is to transfer behaviors from a sample to the entire population, simply improve recreation of the resident population, or both.

The goal of this paper is to show the value in including spatial characteristics, and specifically land use, into population synthesis methods. In addition, the aim is to demonstrate improvements made upon a method described in our previous paper written on the subject (*1*).

In the specific population synthesis context, the joint distributions of variables used as controls are often unknown. For example, in one Census block, the number of households making US$60,000 to US$70,000 a year and the number of households with children may be known, but the number of households making that amount of money who also have children may be unknown. Population synthesis provides a way to estimate these joint distributions. Incorporating land use into the synthesis allows for joint distribution variations based on regional characteristics. It allows for differing correlations between traits depending on where people live. For instance, suburbs might be expected to have a higher proportion of families with children than a downtown area. The inclusion of land use can identify such patterns.

This paper uses a latent profile analysis (LPA) method to create land use classifications that are then used in a population synthesis of the entire state of California. In summary, a new method is presented that derives types of land use surrounding residential areas, then incorporates these types in: a) exploring their influence on travel behavior, and b) integrating the types of land use in population synthesis.

Literature Review

A family of population synthesis methods emerged after 1996 based on Beckman et al., using the iterative proportional fitting (IPF) algorithm (*2*). This method uses categorical variables for which there are known values for the area of interest to populate. It selects people from a survey that is used as the "seed," then uses them to

[1]Department of Geography and GeoTrans Lab, University of California, Santa Barbara, CA

Corresponding Author:
Address correspondence to Elizabeth C. McBride: emcbride@gucsb.edu

populate the area. The method has been adopted and modified by a number of researchers over the years.

The idea of multiple control levels then became important. Originally, population synthesis was run using only household-level control variables to determine the selection and distribution of households into the synthetic population. Presently, incorporating control variables at both the household- and person-level is increasingly common. Although IPF is still performed without multilevel controls (*3*), the use and implementation of multilevel controls is an area of interest for population synthesis researchers (*4–7*).

Methods with multilevel controls that do not involve IPF have also been developed. Casati et al.'s hierarchical Markov chain Monte Carlo (hMCMC) method uses a hierarchical system of classification of individuals within the synthetic households, and uses continuous characteristics (rather than categorical, like IPF) (*8*). Ma and Srinivasan developed a method they called fitness-based synthesis that uses a fitness function to match the control variable characteristics (*9*).

The research outlined above does not take into account that the place a person lives affects how they travel, and very little work on this topic exists to date. Ballas et al. use an iterative approach that is nearly identical to IPF (*10*). Their approach to improving population synthesis was to have "local" households from the sample be more likely to be used to populate the areas they were from. The best method they found was to allow only local sample households into the population that was used as the synthesis seed. This is similar to the PopGen population synthesizer software used in the research discussed in this paper. However, this method is modified to use households from areas with similar land use classification instead of proximity.

Data

In this paper, three data sources are used: the 2012 National Establishment Time Series (NETS), the 2010 United States Census, and the 2012 California Household Travel Survey (CHTS). NETS is an annual database of business establishments in the United States (*11*, *12*), which includes extensive information about organizations, captured as annual "snapshots" based on Dun & Bradstreet archival national establishment data. The number of personnel employed by businesses in California was extracted from the NETS to be used as the basis for the land use classification performed. Block-group-level census data was collected and used to form the marginal data in the population synthesis. The United States census block group is a collection of census blocks that contains between 600 and 3,000 people (*13*). They tend to be smaller in cities (where density is higher). There are 23,212 census block groups in California. Block group counts of people and households and univariate distributions of their characteristics are used as marginal distributions by population synthesis to ensure realistic populations are created. Data from the 2010 American Community Survey was also used for household income information employed as a marginal distribution variable. The CHTS collected information about households and people in the state of California. It covered all 58 counties, and collected household- and person-level demographic information about respondents. It also included a one-day travel diary from every person (*14*). The CHTS was used as the seed for the population synthesis. The people and households in the CHTS were used to populate the state of California during the synthesis process. The total number of respondents was 36,925 households and 94,901 individuals.

Methods

Previous Methods

The first attempt at incorporating land use classification in population synthesis was documented in detail in our previous paper on the subject (*1*). In summary, the population synthesis method and software PopGen are based on IPF, which uses a seed matrix that is repeatedly revised, with each iteration producing a matrix that has closer marginal (row/column) totals to the block-group set of totals, which are used as targets to be reproduced by the synthetic population. The seed matrix provides a multidimensional matrix of household and person characteristics in survey data. In this way, a multidimensional IPF is used to develop a synthetic population. At the end of the iterative process, a population in each geographical subdivision (the block group) matches the control totals and reproduces populations that have an internal consistency of households (e.g., a young couple with children will have the right mix of ages). For this reason, the seed matrix with all household and person characteristics is generated from a sample of households. McBride et al. (*1*) uses population synthesis software PopGen 1.1 (*5*). A kernel density surface of employee density per square kilometer is created. Then, this surface is divided based on public use microdata areas (PUMAs). The employee density category of CHTS respondents' home PUMAs determines their availability for use in synthesizing block groups of the same density category. The synthesized population described in that paper will be referred to as the *Coarse Land Use* population hereafter.

The second synthetic population created using land use utilizes the same overall population synthesis technique, but with a finer land use classification and a newer software version (PopGen 2.0) (*6*, *15–17*). The method used for classification is the same as in the *Coarse Land*

Table 1. LPA Fit Indices

Number of classes	Log likelihood	BIC	ABIC	p-value of BLRT	p-value of VLMRT	Entropy	BF	cmP
1	−856163.9	1712669.6	1712395.8	na	na	na	<.001	<.001
2	−740942.3	1482407.4	1481988.7	<.001	<.001	.99	<.001	<.001
3	−688987.4	1378678.5	1378114.9	<.001	<.001	.96	<.001	<.001
4	−658690.7	1318266.0	1317557.4	<.001	.001	.95	<.001	<.001
5	−637558.4	1276182.2	1275328.7	<.001	.421	.95	<.001	1.000

Note: BIC = Bayesian information criterion; ABIC = sample-size adjusted BIC; BLRT = bootstrapped likelihood ratio test; VLMRT = Vuong-Lo-Mendell-Rubin adjusted likelihood ratio test; BF = Bayes factor; cmP = correct model probability; na = not applicable.

Use population, but at the block-group-level instead of the PUMA level. To exemplify the difference in coarseness, there are 265 PUMAs in California, versus 23,212 block groups (*18*). This population will be referred to as the *Finer Land Use* population from now on. In the new method presented here, the 23,212 block groups are used, adding land use classification for each block group.

Latent Profile Analysis

The method used to classify the data in this paper is LPA, also known as latent class cluster analysis (*19*). LPA uses a set of variables called indicators to determine the ideal number of classes to cluster observations into. LPA provides a method to determine whether there might be a better way to divide the block groups into categories with more statistical certainty for their belonging together than the previous methods provided. It uncovers the most statistically significant way to divide the data up.

Employee densities in 17 individual categories of employment were used as indicator variables. These densities were built in ArcMap 10.2.2 (*20*) using business data from NETS. Businesses from the NETS were added to a block-group-shapefile as points based on the provided XY coordinates. A kernel smoothing process, using 2 km kernels, was employed to create a density surface for each of the employee density categories. An ArcGIS function called "Zonal Statistics as Table" (*21*) was used to calculate the mean employee density per square kilometer within each block group.

Because of the large number of block groups with few people in them, there are many block groups with low values for all employee density items, leading to extremely positively skewed data. Moreover, there are a small number of block groups with extremely high employee densities in city centers where the density is much higher than anywhere else. This combination creates extremely high variance for the variables. In order to mitigate this, a log transform was applied to the data. The log transform compressed the values of the 17 observed variables in each block group and eliminated extreme differences in values of the variables in use in LPA.

The LPA was conducted using Mplus 7.4 (*22*). The order of operations for performing an LPA are as follows. A one-class model is fit, followed by a two-class, and so on, until a model is run that is not well-identified (*19*, *23*, *24*). With every run, a set of fit statistics are recorded and used to determine model identification. These are presented in Table 1. To decide the number of latent classes, the bootstrapped likelihood ratio test (BLRT) and the Vuong-Lo-Mendell-Rubin adjusted likelihood ratio test (VLMRT) should be used. Significant *p*-values indicate statistically significant improvement in model fit compared to the model with the previous number of classes. Once a model runs with a non-significant *p*-value for either the BLRT or the VLMRT, the preceding model is chosen, as long as the other fit statistics support good model fit (*19*, *23*, *24*). As Table 1 shows, the VLMRT reached a non-significant *p*-value of 0.421 with the five-class model. Based on fit criteria, class sizes, and interpretability, the four-class model was chosen. An entropy value approaching 1 indicates clear delineation of the classes. Therefore, the entropy value of 0.95 for the four-class model means the indicators discriminate well between the classes (*25*). Based on properties described in further detail below, the four classes will be referred to from now on by the names *Rural, Exurban, Suburban,* and *Urban.*

The map in Figure 1 shows the geographic distribution of the block groups in California. The maps of the older methods are included for comparison. All three land use classifications manage to capture the main urban centers (San Francisco, Sacramento, Los Angeles, and San Diego) relatively well. It should be noted that location is not considered by the LPA, so the grouping is solely based on the most statistically significant way to group the data based on the indicator variables used. Despite this, a clear, identifiable spatial pattern in the grouping is present in the resultant map. Figure 1*c* shows that the group called *Exurban* does not extend as far

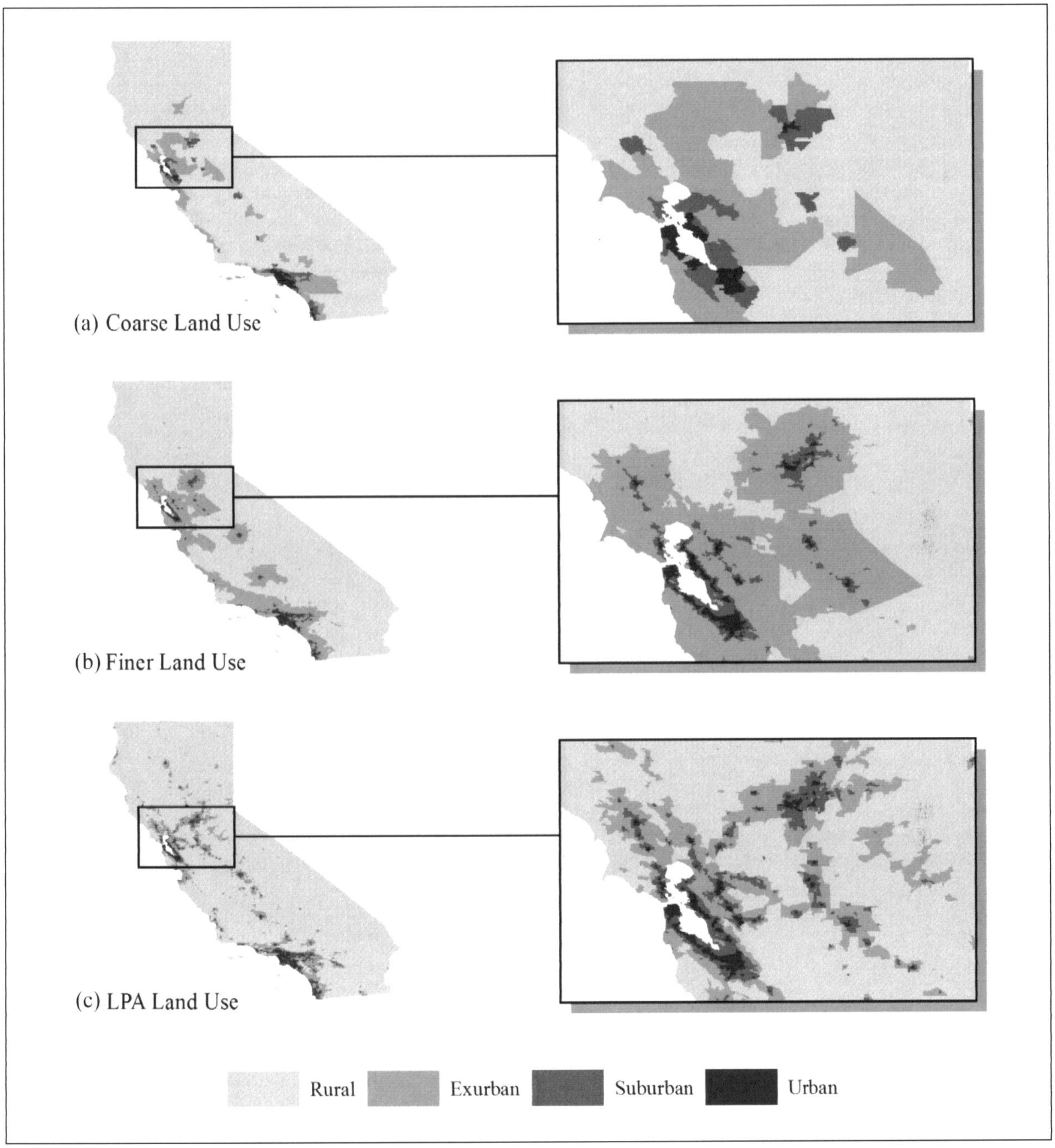

Figure 1. Comparison of land use classification schemes with three different methods of categorizing land use. Number of block groups in each category: (*a*) *Rural*: 3,222, *Exurban*: 4,743, *Suburban*: 6,965, *Urban*: 8,268; (*b*) *Rural*: 2,565, *Exurban*: 4,172, *Suburban*: 7,656, *Urban*: 8,803; (*c*) *Rural*: 1,076, *Exurban*: 2,582, *Suburban*: 10,670, *Urban*: 8,868.

beyond the city centers as it does in the *Finer Land Use* classification (Figure 1*b*). The rural area starts much closer to the urban centers. Based on empirical knowledge of the California population distribution, the LPA classification is a much more accurate depiction of the State's urban–rural landscape. The *Coarse Land Use* group is the oldest classification, which faced limitations imposed by the much coarser classification scheme (*1*).

Figure 2*a* is the item probability plot. It shows the mean value for each indicator variable in the four classes. The LPA did not result in major differences across the block groups for each of the types of

(a)

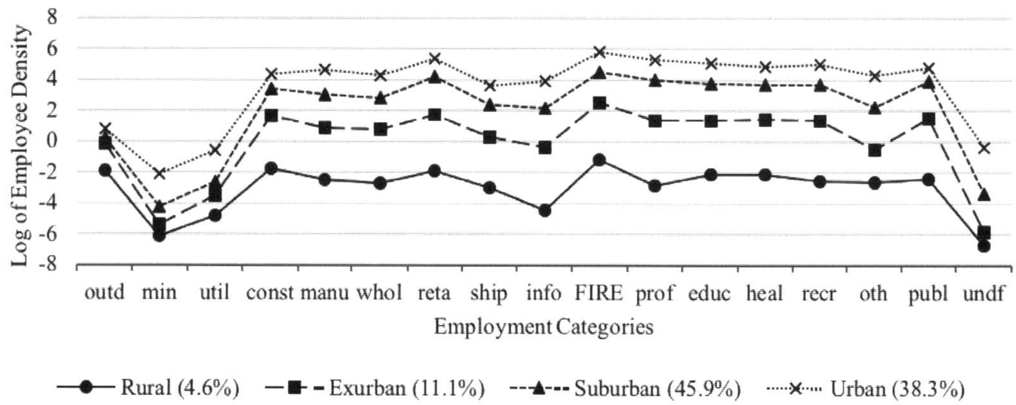

(b)

Item name	Item description
Outd	Agriculture, forestry, fishing, and hunting
Min	Mining
Util	Utilities
Const	Construction
Manu	Manufacturing
Whol	Wholesale trade
Reta	Retail trade
Ship	Transportation and warehousing
Info	Information
FIRE	Finance, insurance, real estate, and rental and leasing
Prof	Professional, scientific, management, administrative, and waste management services
Educ	Educational services
Heal	Health care
Recr	Arts, entertainment, recreation, accommodation, and food services
Oth	Other services (except public administration)
Publ	Public administration and armed forces
Undf	Undefined

NOTE: This index of variable name descriptions provides detail about what is included in each employment category.

Figure 2. Items used in LPA and their values for each class: (a) item probability plot; and (b) index of variables.

employment within each cluster. As is made clear by both figures 1 and 2, the most important factor in determining the land use classification scheme seems to be proximity to an urban center in the LPA classification. If one compares the map in Figure 1, *b* and *c*, it does appear to better pick out areas that would not be considered near urban centers. Exploring the role of the indicator variables in producing land use classes of block groups reveals that the pattern from this classification is a proxy for urban centrality. One of our objectives is to capture a more accurate spatial distribution of different land uses. The recording of employees by business establishment, headquarters, or store front is often used instead of the actual place of work. This creates a measurement error, which is discussed in the conclusions.

Despite these issues, the *LPA Land Use* classification is still substantially better than previous methods at classifying the block groups of California based on land use. The previous method (*1*) used an overall employee density instead of dividing them up by their type, and took proximity to an urban center into account in its classification. It also required that the number of people in the sample for each group be relatively even so that the synthesizer would have a good amount of people to draw from (i.e., in population synthesis census univariate distributions were used as control totals for each block and

Table 2. Censored Regression Model Results for Vehicle Miles Traveled per Person in the CHTS

Independent variables		Marginal effects	Standard error	*t* value	Pr(>|t|)
Age of householder (85 and older is the excluded category)	Age 15–24	14.00	2.42	5.78	0.000
	Age 25–34	11.42	2.08	5.49	0.000
	Age 35–44	10.62	1.97	5.41	0.000
	Age 45–54	12.46	1.85	6.74	0.000
	Age 55–59	12.56	1.80	6.97	0.000
	Age 60–64	11.93	1.76	6.78	0.000
	Age 65–74	10.03	1.72	5.83	0.000
	Age 75–84	5.98	1.79	3.34	0.001
Children present in household	Children present	–2.21	0.64	–3.44	0.001
Household income (income between US$ 0 and US$9,999 is the excluded category)	$10,000–$24,999	6.04	1.00	6.04	0.000
	$25,000–$34,999	11.34	1.04	10.86	0.000
	$35,000–$49,999	13.67	1.00	13.70	0.000
	$50,000–$74,999	18.28	0.96	19.13	0.000
	$75,000–$99,999	19.91	0.97	20.55	0.000
	$100,000–$149,999	22.01	0.96	22.91	0.000
	$150,000–$199,999	21.50	1.07	20.07	0.000
	$200,000+	22.57	1.09	20.64	0.000
Number of females in the household		–0.16	0.24	–0.70	0.486
Mean age of household		–0.11	0.02	–4.38	0.000
Measurement of land use around the residence (Urban is the excluded category)	Suburban	4.29	0.40	10.69	0.000
	Exurban	7.91	0.50	15.76	0.000
	Rural	10.16	0.65	15.60	0.000

Note: Pr(>|t|) = *p*-value for *t*-test.

drew households and individuals from a survey). This means that the number of block groups in each category was much more similar than what was found with LPA. LPA does not take any of this into consideration. All it considers is the values given to it and how those values cluster together in a statistically significant way. The number of block groups that went into each category was quite uneven. The number of block groups in the *Suburban* category was the highest (10,670). Meanwhile, in the *Rural* category there were 1,076 block groups.

Rural areas are oversampled in the CHTS, so the issue of having enough seed observations in the sample to synthesize an area properly did not become an issue, and probably would not become an issue as long as the sample is not disproportionately small compared with the number of block groups and number of people in those block groups that it is trying to synthesize.

In addition, because the groups for the LPA classification scheme were ultimately based on overall employment density, the names given to the older groups can still be applied here (*Rural*, *Urban*, etc.).

Censored Regression Models. One key question for travel behavior is whether the land use categories derived here are significant factors in explaining travel behavior. Censored regression models were run to determine whether the *LPA Land Use* categories were significant contributors to some of the travel traits of interest. This type of regression is needed because the vehicle miles traveled (VMT) and the number of trips have large masses of observations at zero. This is due to households that may stay at home all day and make no trips. The models presented here use the CHTS data. Both included the same set of traits as in McBride et al. (*1*) to test for significance: age of householder, presence of children, household income, number of females in the household, mean age of household, and the land use classes. These traits were also used because they match with what was used in the population synthesis as the marginal characteristics. The first model (Table 2) tested VMT per person as the dependent variable. The second model (Table 3) tested the number of trips per person as the dependent variable.

For the VMT per person model, everything was significant except for the number of females in the household (Table 2). This means that the miles traveled by a person are affected by the age of the householder, the presence of children in a household, the household income level, the mean age of the household, and the *LPA Land Use* classification. Categorical variables include age of householder, presence of children, household income, and land use categories. Continuous variables comprise number of females in the household (count), mean age of household.

The younger the householder is, the more miles traveled per person in a household. The reference group is the 85+ year-old householder group. A household whose householder is between 15 and 24 years old will travel

Table 3. Censored Regression Model Results for Number of Trips per Person in the CHTS

Independent variables		Marginal effects	Standard error	*t* value	Pr(>|*t*|)
Age of householder (85 and older is the excluded category)	Age 15–24	1.0	0.2	5.6	0.000
	Age 25–34	1.2	0.2	7.8	0.000
	Age 35–44	1.3	0.1	8.8	0.000
	Age 45–54	1.4	0.1	9.7	0.000
	Age 55–59	1.3	0.1	9.6	0.000
	Age 60–64	1.3	0.1	9.5	0.000
	Age 65–74	1.1	0.1	8.3	0.000
	Age 75–84	0.6	0.1	4.6	0.000
Children present in household	Children present	0.0	0.0	–0.8	0.427
Household income (income between US$ 0 and US$9,999 is the excluded category)	$10,000–$24,999	0.0	0.1	0.2	0.825
	$25,000–$34,999	0.0	0.1	0.2	0.815
	$35,000–$49,999	0.2	0.1	2.7	0.006
	$50,000–$74,999	0.4	0.1	5.5	0.000
	$75,000–$99,999	0.5	0.1	6.9	0.000
	$100,000–$149,999	0.6	0.1	8.8	0.000
	$150,000–$199,999	0.7	0.1	8.3	0.000
	$200,000+	0.7	0.1	8.4	0.000
Number of females in the household		–0.2	0.0	–8.7	0.000
Mean age of household		0.0	0.0	–6.8	0.000
Measurement of land use around the residence (Urban is the excluded category)	Suburban	–0.5	0.0	–15.9	0.000
	Exurban	–0.7	0.0	–19.0	0.000
	Rural	–1.1	0.1	–21.6	0.000

Note: Pr(>|*t*|) = *p*-value for *t*-test.

about 14 mi per person more than one whose householder is 85+ years old. If the householder is between 25 and 34, the household will travel 11.4 mi more than the 85+ year-old householder group. This pattern of decreasing miles traveled with increasing householder age continues. These results match with our expectations about how age should affect the amount of traveling done.

If children are present in the household, the number of miles traveled per person goes down. Households with children present travel 2.2 mi less than those without children. This is likely because the presence of children in a household tends to necessitate family members staying closer to home, because of limitations of having children such as dropping them off and picking them up from school, appointments, and so forth. Although having all of these travel obligations might increase the amount of traveling to an extent, it is probably offset by the limitations of needing to stay closer to home to deal with childcare duties.

Higher household income corresponds with higher miles traveled per person. The coefficients for the income categories are higher than any other group of dependent variables. They are one of the strongest indicators of higher VMT. The wealthier respondents may be making longer-distance trips because they have the flexibility to do so. As the mean age of household goes up, the miles traveled goes down.

Suburban households travel 4.29 mi more per person than urban households. Exurban households travel 7.91 mi more per person, and rural households travel 10.16 mi more per person than urban households do. Urban areas are the highest density, therefore distances between locations of interest tend to be shorter. This all means that people in urban areas will have much less of a distance to travel than people in rural areas.

In the previous paper, the same regression models were run, but with the *Coarse Land Use* classification (*1*). *Suburban* and *Exurban* categories had nearly the same values for the difference in miles traveled per person as compared to *Urban*. However, for the *Rural* group, the old-method rural households traveled 8.92 mi more, whereas in this one they travel 10.16 mi more. Finer classification exacerbates the differences in VMT between the two land use classification methods.

In relation to number of trips per person, the model (Table 3) uses the same independent variables as the previous model. Everything except for presence of children and the low-income levels were significant. Households with middle-aged householders make a higher number of trips than the younger or older householders. These households are in a life-cycle stage that requires more traveling for family members and working.

The two lowest income categories were not significant, meaning they were not significantly different than the US$ 0 to US$9,999 category. Until households get to US$35,000 or above, there is no significant difference in the number of trips they make in a day compared with the lowest income category. They similarly function to the lowest income category in relation to the number of trips they make in a day. The pattern of lower income meaning fewer trips parallels the VMT model. With less money at hand, there is less flexibility to participate in activities outside of home, like day trips, vacations, or recreational activities. In this model, a 1-mi trip is worth the same as a 50+ mi trip. This might be why travel's relationship with lower income emerges more in the VMT model.

More females in a household corresponds with fewer trips per day. Every additional female in a household corresponds with a decrease of 0.16 in the number of trips per person in a household. Past travel behavior research has shown this pattern. Gender role dichotomy leads to women spending more time in the private sphere, and men spending more time in the public sphere (*26*, *27*). Every increase of 1 year to the mean age of household leads to a decrease in the number of trips of 0.01 as the VMT also showed.

Suburban households make 0.49 fewer trips per person than urban households. Exurban households make 0.74 fewer trips per person than urban households. Rural households make 1.09 fewer trips per person than urban households. This is a consistent trait of rural households that have a lower number of trips of longer distances.

When compared with the previous paper's model on number of trips (*1*), the most defined difference is in the *Exurban* category. The *Coarse Land Use* model had −0.56 trips compared with *Urban* households, and the new model has −0.74. The old-model rural group had −0.95 trips compared with the *Urban* group, whereas in this one they have −1.09. *Suburban* is similar in both.

Findings

The synthetic population was generated using PopGen 2.0, the program described earlier (*6*, *15–17*). The household-level variables used include household annual income, number of people, age of householder, and presence of children. The person-level variables used include age and sex.

After synthesis, travel behavior characteristics were transferred to the synthetic populations from their replication sources in the CHTS. These were used to study differences in travel behavior for the different populations and to see whether the use of classification and various classification schemes has an effect on travel behavior traits.

Maps of the synthetic population created without including land use at all were not included, but if readers are interested in viewing and comparing these with the upcoming maps, they can be found in the previous paper (*1*). To summarize, they show a much more random pattern of density than what is seen in any of the other populations for all of the variables. The analysis of the maps will focus on comparing the *Finer Land Use* with the *LPA Land Use*, as these are the two methods that are most similar and that the LPA method is improving upon.

The miles traveled per person in the *LPA* population is more strongly related to the "urban-ness" of a block group than in the other populations. As Figure 3 shows, San Francisco and the surrounding area are identifiable on all three maps. However, compared with the *LPA*, the *Coarse Land Use* and *Finer Land Use* show more ambiguity in the pattern of gradual increases in miles traveled farther from city centers.

It is expected that urban populations make more trips than rural populations as the regression model shows. The map's borders of the urban areas are much more defined in the *LPA Land Use* (Figure 4*c*) than they are in the *Finer Land Use*. The higher values for the number of trips are more restricted to urban areas in the *LPA Land Use*. Although the same pattern is present in the *Coarse*- and *Finer Land Use* populations, it is not as distinct, and the borders of the urban areas are less clear.

The percent of trips on foot (Figure 5) clearly delineates urban areas. Walking for a large number of trips is only possible in "walkable neighborhoods" that are characteristic of urban areas, in which the distances between points of interest are close enough together that it is possible to walk between them. It is a good test of whether the land use classification is helping the population synthesis to populate areas with households that have travel traits that are more characteristic of the areas they are populating. The maps show that higher percentages of walking trips are limited to the urban areas, a specific pattern that is most clearly shown in the *LPA Land Use* population (Figure 5*c*). The pattern is present in the *Coarse Land Use* and *Finer Land Use* populations (Figure 5, *a* and *b*), but it is not limited to urban areas as much. There is still some "random noise" of large, rural block-groups with high percentages of walking trips, even in the *Finer Land Use* population that most likely should not have those numbers.

Conclusions

The use of LPA for land use classification provides benefits that make it a valuable tool. It produces land use indicators that explain travel behavior and shows a finer-grain ability to transfer behavioral data to an entire population. The method allows for inclusion of all kinds of variables in the classification. This makes it easily transferrable to whatever region researchers might be interested in synthesizing. If employment variables are not available, but other variables that could contribute to a land use classification are, then those can be used. Compared with the previous methods, the LPA method

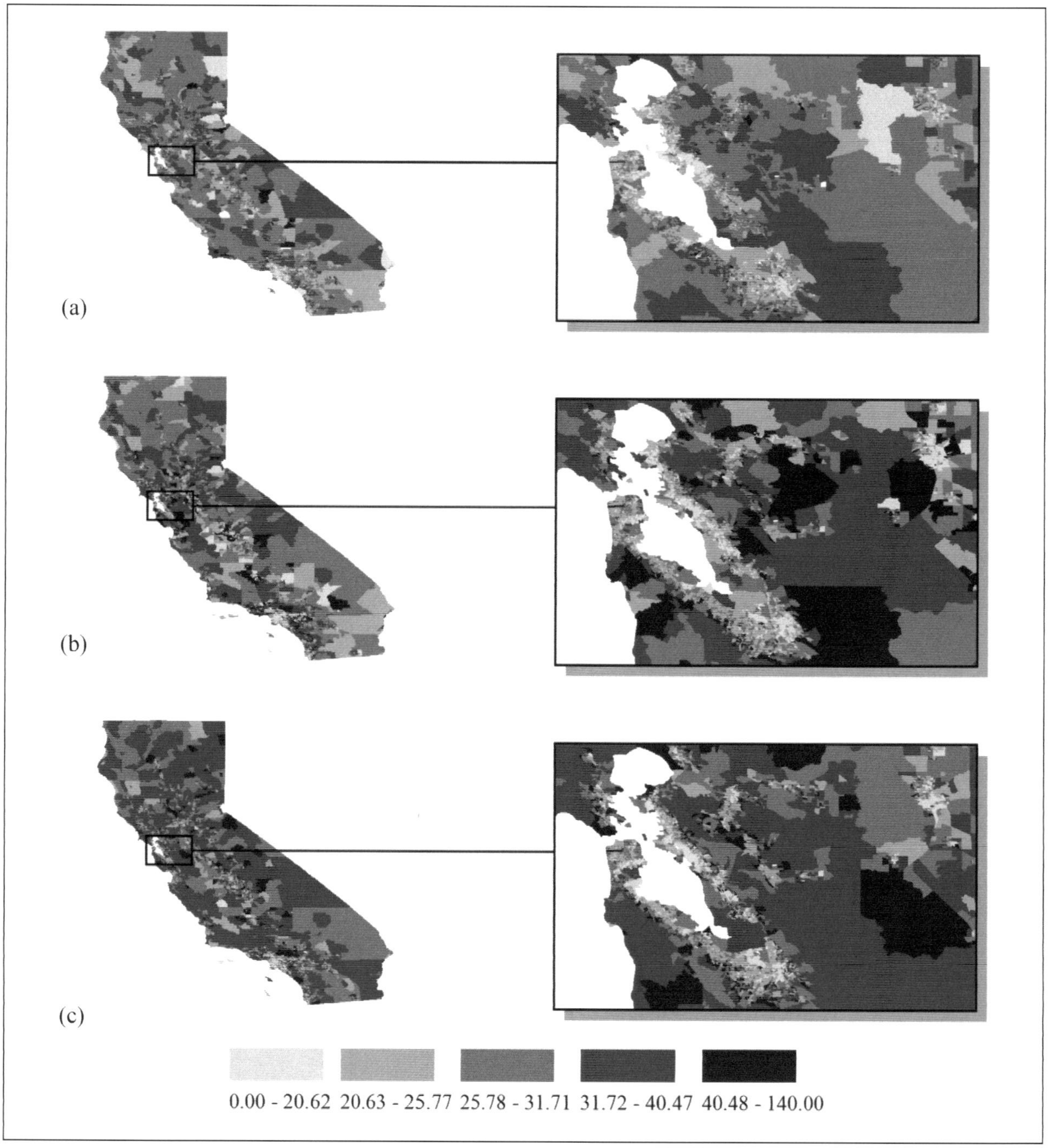

Figure 3. Miles traveled per person: (*a*) *Coarse Land Use*; (*b*) *Finer Land Use*; (*c*) *LPA Land Use*.

is also more statistically sound: it reduces the statistical uncertainty associated with the previous method, and it demonstrates the potential to explain two key behavioral traits: number of trips and VMT. Most importantly, the LPA method reproduces a spatial distribution that most closely resembles the spatial distribution that is expected for California.

In this application, we used the entire suite of 17 types of business establishments enumerating their number of employees and producing block-group-level counts that were then used as indicators for land use LPA. One possibility for the land use classification LPA is to use fewer employee categories that are more location-based, being selective about what to choose based on substantive

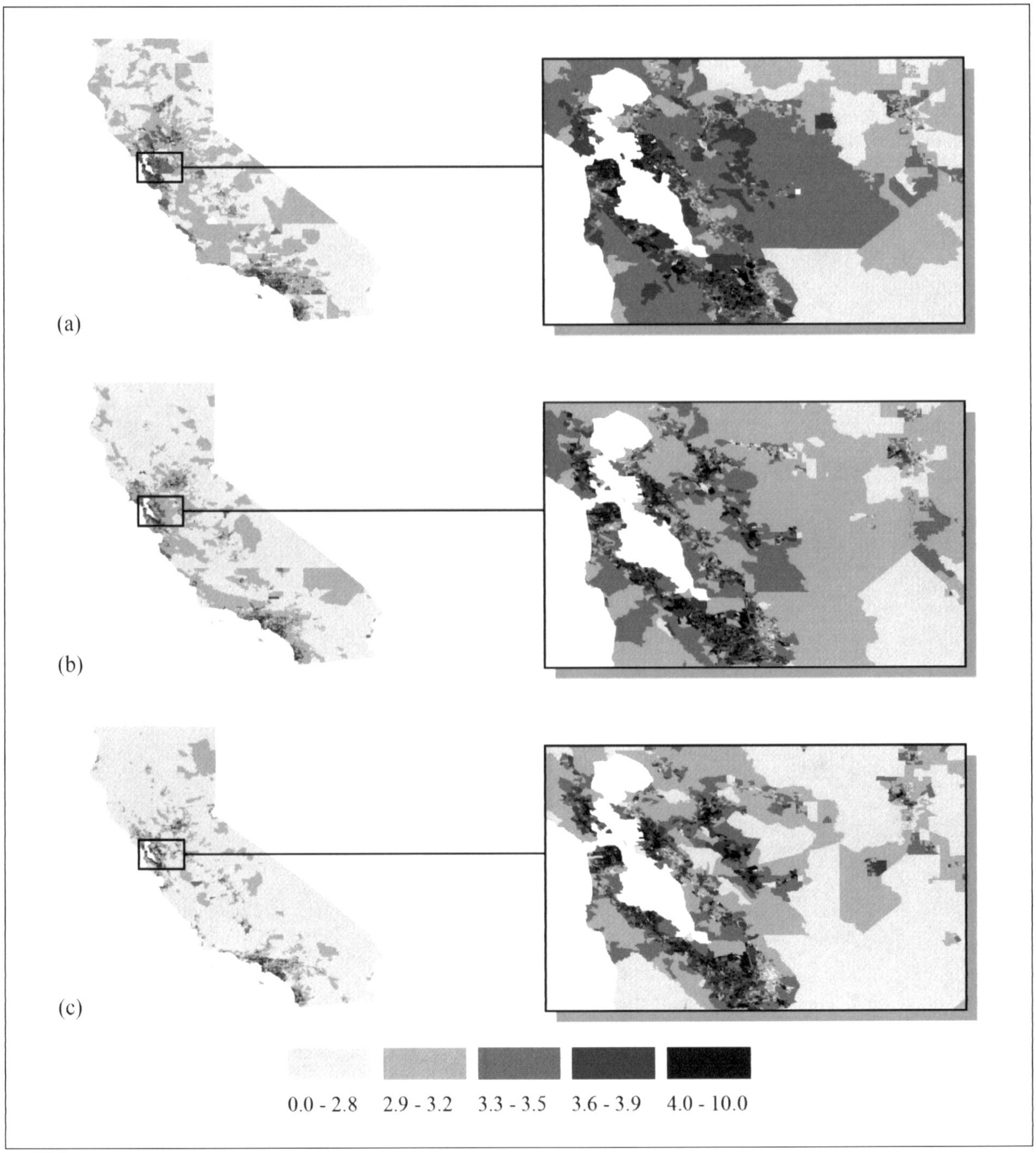

Figure 4. Number of trips per person: (*a*) *Coarse Land Use*; (*b*) *Finer Land Use*; (*c*) *LPA Land Use*.

theories about the types of employment that are more present in certain regions of California than in others. Along with including employee density categories in the classification, one could also include network density, public transportation services, distances to major employment centers, or a combination of these. This expanded repertory of data will properly place the employee densities where employees might not be present at the business "storefront." This will also enable comparison with somewhat more traditional accessibility indicators (*28*).

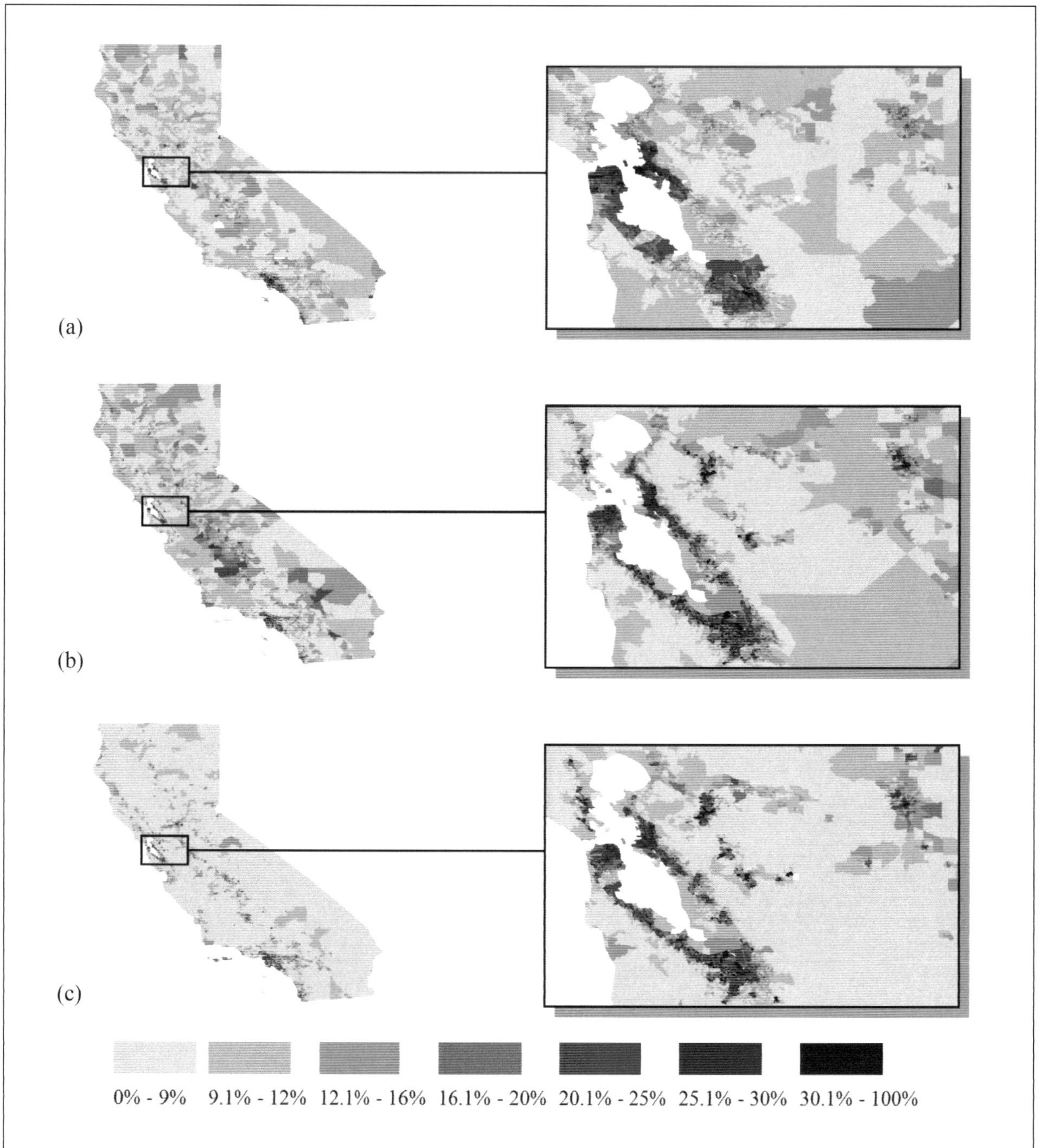

Figure 5. Percent of trips on foot: (*a*) *Coarse Land Use*; (*b*) *Finer Land Use*; (*c*) *LPA Land Use*.

In addition, taking into account the market size of each area (e.g., using something like number of employees divided by the number of residents in a block group) may also facilitate accounting for any jobs–housing imbalance. Normalizing by the population of an area means we would be comparing ratios of different employment types instead of just the number of people employed per square kilometer. Finally, NETS is a longitudinal record of business establishments that can be used further to study the evolution of land use, and, if combined with a longitudinal version of LPA, enables the study of spatial transitions in land use. All these are left as future tasks.

Author Contributions

The authors confirm contribution to the paper as follows: study conception and design: EM, AD, KG; data preparation: EM, AD, KG; analysis and interpretation of results: EM, AD, KG; draft manuscript preparation: EM, AD, KG. All authors reviewed the results and approved the final version of the manuscript.

References

1. McBride, E. C., A. W. Davis, J. H. Lee, and K. G. Goulias. Incorporating Land Use in Synthetic Population Generation Methods and Transfer of Behavioral Data. *Transportation Research Record: Journal of the Transportation Research Board*, 2017. 2668: 11–20.
2. Beckman, R. J., K. A. Baggerly, and M. D. McKay. Creating Synthetic Baseline Populations. *Transportation Research Part A: Policy and Practice*, Vol. 30, No. 6, 1996, pp. 415–429.
3. Adiga, A., A. Agashe, S. Arifuzzaman, C. L. Barrett, R. Beckman, K. Bisset, J. Chen, Y. Chungbaek, S. Eubank, S. Gupta, M. Khan, C. J. Kuhlman, E. Lofgren, B. Lewis, A. Marathe, M. V Marathe, H. S. Mortveit, E. Nordberg, C. Rivers, P. Stretz, S. Swarup, A. Wilson, and D. Xie. *Generating a Synthetic Population of the United States*. Technical Report NDSSL 15-009. Network Dynamics and Simulation Science Laboratory, Virginia Tech, Blacksburg, VA, 2015.
4. Auld, J., and A. Mohammadian. Efficient Methodology for Generating Synthetic Populations with Multiple Control Levels. *Transportation Research Record: Journal of the Transportation Research Board*, 2010. 2175: 138–147.
5. Pendyala, R. M., K. C. Konduri, and K. P. Christian. *PopGen 1.1 User's Manual*. Lulu.com Publishers, Morrisville, NC, 2011.
6. Konduri, K. C., D. You, V. M. Garikapati, and R. M. Pendyala. Enhanced Synthetic Population Generator That Accommodates Control Variables at Multiple Geographic Resolutions. *Transportation Research Record: Journal of the Transportation Research Board*, 2016. 2563: 40–50.
7. Zhu, Y., and J. Ferreira, Jr. Synthetic Population Generation at Disaggregated Spatial Scales for Land Use and Transportation Microsimulation. *Transportation Research Record: Journal of the Transportation Research Board*, 2014. 2429: 168–177.
8. Casati, D., K. Müller, P. J. Fourie, A. Erath, and K. W. Axhausen. Synthetic Population Generation by Combining a Hierarchical, Simulation-Based Approach with Reweighting by Generalized Raking. *Transportation Research Record: Journal of the Transportation Research Board*, 2015. 2493: 107–116.
9. Ma, L., and S. Srinivasan. Synthetic Population Generation with Multilevel Controls: A Fitness-Based Synthesis Approach and Validations. *Computer-Aided Civil and Infrastructure Engineering*, Vol. 30, No. 2, 2015, pp. 135–150.
10. Ballas, D., G. Clarke, D. Dorling, H. Eyre, B. Thomas, and D. Rossiter. SimBritain: A Spatial Microsimulation Approach to Population Dynamics. *Population, Space and Place*, Vol. 11, No. 1, 2005, pp. 13–34.
11. Feldman, M. *National Establishment Time-Series*. http://maryannfeldman.web.unc.edu/data-sources/longitudinal-databases/national-establishment-time-series-nets/. Accessed October 1, 2018.
12. Walls & Associates. *National Establishment Time - Series (NETS) Database: 2012 Database Description*. Denver, CO, 2012.
13. United States Census Bureau. *Geographic Terms and Concepts - Block Groups*. United States Census Bureau. https://www.census.gov/geo/reference/gtc/gtc_bg.html.
14. California Department of Transportation. *2010–2012 California Household Travel Survey Final Report Appendix*. California Department of Transportation, Sacramento, CA, 2013.
15. Ye, X., K. C. Konduri, R. M. Pendyala, B. Sana, and P. Waddell. A Methodology to Match Distributions of Both Household and Person Attributes in the Generation of Synthetic Populations. Presented at 88th Annual Meeting of the Transportation Research Board, Washington, D.C., 2009.
16. Bar-Gera, H., K. C. Konduri, B. Sana, X. Ye, and R. M. Pendyala. *Technical Report 09-1354: Estimating Survey Weights with Multiple Constraints using Entropy Optimization Methods*. Transportation Research Board of the National Academies, Washington, D.C., 2008.
17. Mobility Analytics Research Group. *PopGen: Synthetic Population Generator* [Online]. Mobility Analytics Research Group. http://www.mobilityanalytics.org/popgen.html. Accessed October 1, 2018.
18. Goulias, K. G., A. W. Davis, E. C. McBride, K. Janowicz, and R. Zhu. *Long Distance Travel in CHTS and Social Media Augmentation*. Final Report, Contract Number: 65A0529, Task Order 047. University of California, Santa Barbara. Department of Geography, GeoTrans and STKO Laboratories, California Department of Transportation, Sacramento, CA, 2017.
19. Masyn, K. E. Latent Class Analysis and Finite Mixture Modeling. In T. D. Little (ed.), *The Oxford Handbook of Quantitative Methods: Statistical Analysis, Vol. 2*. Oxford University Press, New York, NY, 2013, pp. 551–611.
20. Environmental Systems Research Institute. *ArcGIS Desktop: Release 10.2*. Environmental Systems Research Institute, Redlands, CA, 2014.
21. Environmental Systems Research Institute. *Zonal Statistics as Table—Help and ArcGIS for Desktop*. Environmental Systems Research Institute, Redlands, CA, 2016. http://desktop.arcgis.com/en/arcmap/10.3/tools/spatial-analyst-toolbox/zonal-statistics-as-table.htm. Accessed October 1, 2018.
22. Muthén, L., and B. Muthén. *Mplus User's Guide*, 7th ed. Muthén & Muthén, Los Angeles, CA, 1998.
23. Nylund, K. L., T. Asparouhov, and B. O. Muthen. Deciding on the Number of Classes in Latent Class Analysis and Growth Mixture Modeling: A Monte Carlo Simulation Study. *Structural Equation Modeling: A Multidisciplinary Journal*, Vol. 14, No. 4, 2007, pp. 535–569.

24. Asparouhov, T., and B. O. Muthén. Using Mplus TECH11 and TECH14 to Test the Number of Latent Classes. *Mplus Web Notes*, No. 14, 2012, pp. 1–17.
25. Celeux, G., and G. Soromenho. An Entropy Criterion for Assessing the Number of Clusters in a Mixture Model. *Journal of Classification*, Vol. 13, No. 2, 1996, pp. 195–212.
26. Turner, T., and D. Niemeier. Travel to Work and Household Responsibility: New Evidence. *Transportation*, Vol. 24, No. 4, 1997, pp. 397–419.
27. Kwan, M.-P. Space-Time and Integral Measures of Individual Accessibility: A Comparative Analysis using a Point-Based Framework. *Geographical Analysis*, Vol. 30, 1998, pp. 191–216.
28. Chen, Y., S. Ravulaparthy, K. Deutsch, P. Dalal, S. Y. Yoon, T. Lei, K. G. Goulias, R. M. Pendyala, C. R. Bhat, and H.-H. Hu. Development of Indicators of Opportunity-Based Accessibility. *Transportation Research Record: Journal of the Transportation Research Board*, 2011. 2255: 58–68.

The Standing Committee on Transportation Demand Forecasting (ADB40) peer-reviewed this paper (18-06631).

Sage Publications USA
sagepub.com

ISBN 978-1-5108-8255-3

CURRAN ASSOCIATES INC.
proceedings
.com

11th Edoardo Amaldi Conference on Gravitational Waves (AMALDI 11)

Journal of Physics: Conference Series
Volume 716

Gwangju, South Korea
21 – 26 June 2015

IOP
Institute of Physics